LIBRARY
APR 10 1992
Atomic Energy of Canada Limited

D1524634

Etiology of Human Disease at the DNA Level

Cover photo: A digital image of a human metaphase chromosome spread hybridized with biotinylated DNA specific for chromosome 8 and detected with fluorescein-labeled avidin. The resulting grey scale image was pseudocolored with a blue to red gradient representing low to high fluorescent intensity. Courtesy of Martin Ferguson and David Ward, Yale University.

Etiology of Human Disease at the DNA Level

Editors

Jan Lindsten, M.D., Ph.D.

Department of Clinical Genetics
Karolinska Hospital
Stockholm, Sweden

Ulf Pettersson, M.D., Ph.D.

Department of Medical Genetics
University of Uppsala
Uppsala, Sweden

NOBEL SYMPOSIUM 80

RAVEN PRESS NEW YORK

Raven Press, Ltd., 1185 Avenue of the Americas, New York, New York 10036

Made in the United States of America

Library of Congress Cataloging-in-Publication Data

Etiology of human disease at the DNA level / edited by Jan Lindsten, Ulf Pettersson.
p. cm.
Consists of papers presented at the Nobel Symposium on the "Etiology of Human Disease at the DNA Level," in Björkborn, Karlskoga, Sweden, June 11 and 14, 1990, and sponsored by the Nobel Foundation and the Alfred Nobel's Björkborn Foundation.
Includes bibliographical references and index.
1. Medical genetics—Congresses. I. Lindsten, Jan E. II. Pettersson, Ulf. III. Nobelstiftelsen. IV. Alfred Nobel's Björkborn Foundation. V. Nobel Symposium on the "Etiology of Human Disease at the DNA Level" (1990 : Björkborn, Karlskoga, Sweden)
[DNLM: 1. Chromosome Mapping—congresses. 2. Gene Therapy—congresses. 3. Genetic Markers—congresses. 4. Hereditary Diseases—diagnosis—congresses. 5. Hereditary Diseases—etiology—congresses. 6. Hereditary Diseases—therapy—congresses. QZ 50 E847 1990]
RB155.E828 1991
616'.042—dc20
DNLM/DLC
for Library of Congress 91-8911
ISBN 0-88167-761-2 CIP

9 8 7 6 5 4 3 2 1

Contents

Contributors

Antonio Baldini, M.D. *Department of Human Genetics, Yale University School of Medicine, 333 Cedar Street, New Haven, Connecticut 06510*

S. Gwyn Ballard, Ph.D. *Department of Human Genetics, Yale University School of Medicine, 333 Cedar Street, New Haven, Connecticut 06510*

Alan H. Beggs, Ph.D. *Division of Genetics, The Children's Hospital, 300 Longwood Ave., Boston, Massachusetts 02115*

Graeme I. Bell, Ph.D. *Howard Hughes Medical Institute Research Laboratory, Department of Biochemistry and Molecular Biology, The University of Chicago, 5841 S. Maryland Ave., N237A, Box 391, Chicago, Illinois 60637*

John W. Belmont, M.D., Ph.D. *Institute for Molecular Genetics and Howard Hughes Medical Institute, Baylor College of Medicine, Houston, Texas 77030*

Sir Walter F. Bodmer, M.R.C.P. *Imperial Cancer Research Fund Labs, P.O. Box 123, Lincoln Inn Fields, London WC2A 3PX, U.K.*

David Botstein, Ph.D. *Department of Genetics, Stanford University School of Medicine, Stanford, California 94305-5120*

Frederick M. Boyce, M.D., Ph.D. *Division of Genetics, The Children's Hospital, 300 Longwood Ave., Boston, Massachusetts 02115*

Ann Boyle *Department of Molecular Biophysics and Biochemistry, Yale University School of Medicine, 333 Cedar St., New Haven, Connecticut 06510*

Michael S. Brown, M.D. *Department of Molecular Genetics, University of Texas Southwestern Medical Center at Dallas, 5323 Harry Hines Blvd., Dallas, Texas 75235*

C. Brunn *Johns Hopkins Hospital, William Welch Medical Library, 1830 E. Monument St., 3rd Floor, Baltimore, Maryland 21205*

Mario R. Capecchi, Ph.D. *Howard Hughes Medical Institute, Department of Human Genetics and Biology, University of Utah, 337 South Biology Building, Salt Lake City, Utah 84112*

C. Thomas Caskey, M.D. *Institute for Molecular Genetics and Howard Hughes Medical Institute, Baylor College of Medicine, Houston, Texas 77030*

Denis Cournoyer, M.D. *Department of Medicine, McGill University, Montreal, Canada H3G 1Y6; Division of Hematology, Montreal General Hospital, 1650 Avenue Cedar, Montreal, Quebec, Canada H2L 1H3*

Nancy J. Cox, Ph.D. *Howard Hughes Medical Institute Research Laboratory, Department of Biochemistry and Molecular Biology, The University of Chicago, 5841 S. Maryland Ave., N237A, Box 391, Chicago, Illinois 60637*

Earl W. Davie, Ph.D. *Department of Biochemistry, University of Washington, J405 Health Sciences Building, SJ-70, Seattle, Washington 98195*

Stefan S. Fajans, M.D. *Department of Internal Medicine, Division of Endocrinology and Metabolism, University of Michigan Medical Center, Ann Arbor, Michigan 48109*

Chris A. Feener *Division of Genetics, The Children's Hospital, 300 Longwood Ave., Boston, Massachusetts 02115*

***Frederick A. Fletcher, M.Sc.** *Institute for Molecular Genetics, Baylor College of Medicine, Houston, Texas 77030*

Joseph L. Goldstein, M.D. *Department of Molecular Genetics, University of Texas Southwestern Medical Center at Dallas, 5323 Harry Hines Blvd., Dallas, Texas 75235*

Fiona R. Green *Arterial Disease Research Unit, The Charing Cross Sunley Research Centre, 1 Lurgan Ave., Hammersmith, London SW6 6JX, U.K.*

James F. Gusella, M.D. *Molecular Neurogenics Laboratory, Massachusetts General Hospital; Department of Genetics, Harvard University, Boston, Massachusetts 02114*

Stephen Humphries, Ph.D., M.R.C.Path. *Arterial Disease Research Unit, The Charing Cross Sunley Research Centre, 1 Lurgan Ave., Hammersmith, SW6 6JX, U.K.*

Rudolf Jaenisch, M.D. *Whitehead Institute for Biomedical Research, Massachusetts Institute of Technology, Nine Cambridge Center, Cambridge, Massachusetts 02142*

Cecily H. Kelleher, M.D. *Department of Health Promotion, University College Galway, Galway, Ireland*

**Present address:* Department of Experimental Hematology, Immunex R&D Corporation, 51 University Street, Seattle, Washington 98101.

Louis M. Kunkel, Ph.D. *Division of Genetics, The Children's Hospital, 300 Longwood Ave., Boston, Massachusetts 02115*

***Peter Lichter, Ph.D.** *Department of Human Genetics, Yale University School of Medicine, 333 Cedar St., New Haven, Connecticut 06510*

R. Lucier, Ph.D. *Johns Hopkins Hospital, William Welch Medical Library, 1830 E. Monument St., 3rd Floor, Baltimore, Maryland 21205*

Grant R. MacGregor, Ph.D. *Howard Hughes Medical Institute, Baylor College of Medicine, Houston, Texas 77030*

Hugh O. McDevitt, M.D. *Department of Microbiology and Immunology, Sherman Fairchild Science Building, Stanford University School of Medicine, Stanford, California 94305-5402*

Tom W. Meade, D.M., F.R.C.P., F.F.C.M. *MRC Epidemiology and Medical Care Unit, Northwick Park Hospital, Harrow HA1 3UJ, U.K.*

Joan Menninger *Department of Human Genetics, Yale University School of Medicine, 333 Cedar St., New Haven, Connecticut 06510*

Marilyn Monk, Ph.D. *MRC Mammalian Development Unit, University College London, Wolfson House, 4 Stephenson Way, London NW1 2HE, U.K.*

Kateri A. Moore, D.V.M. *Institute for Molecular Genetics, Baylor College of Medicine, Houston, Texas 77030*

Richard C. Mulligan, Ph.D. *Whitehead Institute for Biomedical Research, Cambridge, Massachusetts, and Department of Biology, Massachusetts Institute of Technology, Nine Cambridge Center, Cambridge, Massachusetts 02142*

Marsha Newman *Howard Hughes Medical Institute and Departments of Biochemistry and Molecular Biology, The University of Chicago, 5841 S. Maryland Ave., Chicago, Illinois 60637*

Richard J. O'Reilly, M.D. *Department of Pediatrics, Memorial Sloan-Kettering Cancer Center, 1275 York Ave., New York, New York 10021*

P. L. Pearson, Ph.D. *Johns Hopkins Hospital, 1830 E. Monument St., 3rd Floor, Baltimore, Maryland 21205*

****Maurizio Scarpa, M.D.** *Department of Immunology, University College and Middlesex School of Medicine, London, U.K. W1*

**Present address:* Deutsches Krebsforschungszentrum, In Neuenheimer Feld 280, D-6900 Heidelberg, Germany

***Present address:* Department of Pediatrics University of Padova, Via Gustiniani 3, 35100 Padova, Italy.

Mitsuko Seino, R.N. *Howard Hughes Medical Institute Research Laboratory, Department of Biochemistry and Molecular Biology, The University of Chicago, 5841 S. Maryland Ave., N237A, Box 391, Chicago, Illinois 60637*

Susumu Seino, M.D., D.M.Sc. *Howard Hughes Medical Institute Research Laboratory, Department of Biochemistry and Molecular Biology, The University of Chicago, 5841 S. Maryland Ave., N237A, Box 391, Chicago, Illinois 60637*

Oliver Smithies, Ph.D. *Department of Pathology, School of Medicine, University of North Carolina at Chapel Hill, 7525 Brinkhous-Bullitt Building, Chapel Hill, North Carolina 27599-7525*

Judith R. Snyder *Division of Genetics, The Children's Hospital, 300 Longwood Ave., Boston, Massachusetts 02115*

Angela E. Thomas, M.B., M.R.C.P., M.R.C.Path. *The Charing Cross Sunley Research Centre, 1 Lurgan Avenue, Hammersmith, SW6 6JX, U.K.*

David C. Ward, Ph.D. *Departments of Human Genetics and Molecular Biophysics and Biochemistry, Yale University School of Medicine, 333 Cedar St., New Haven, Connecticut 06510*

Robert Williamson, Ph.D., F.R.C.Path. *Department of Biochemistry and Molecular Genetics, St. Mary's Hospital Medical School, Paddington, London W2 1PG, U.K.*

Debra J. Wolgemuth, Ph.D. *Department of Genetics and Development, College of Physicians and Surgeons of Columbia University, New York, New York 10032*

Song-hua Wu, M.D. *Howard Hughes Medical Institute Research Laboratory, Department of Biochemistry and Molecular Biology, The University of Chicago, 5841 S. Maryland Ave., N237A, Box 291, Chicago, Illinois 60637*

Preface

In addition to awarding the Nobel prizes, the Nobel Institution is involved in various scientific activities. Most important of these are the Nobel symposia, which are sponsored by the Nobel Foundation and its Symposium Committee.

The Nobel symposia are traditionally arranged according to certain rules, the most important one being that the topics should be of great current scientific interest. Furthermore, the meetings are small and closed; only 40–50 internationally leading scientists are invited.

The present Nobel symposium on the etiology of human disease at the DNA level is number 80 in this series of symposia. It was held at Alfred Nobel's Björkborn, which is located in the middle of Sweden, between June 11 and 14, 1990. The meeting covered an extremely interesting aspect of biomedical research. The field of human molecular genetics is advancing at an enormous pace. We were fortunate to attract the leading scientists in the field to the symposium; they discussed the molecular basis of human disease from all possible angles. The chapters in this book represent the forefront of human genetics; the book should be of wide interest to students and scientists in the field of human genetics.

Jan Lindsten
Ulf Pettersson

Stockholm and Uppsala, Sweden
October 1990

Acknowledgments

We would like to thank all the participants for their stimulating contributions, which are presented in this book. We would also like to thank the Nobel Foundation and Alfred Nobel's Björkborn Foundation for sponsoring the symposium and Raven Press for publishing these proceedings. Special thanks goes to Gertie Ågren, Anita Lundmark, and Ann-Mari Dumanski for their excellent help with practical arrangements.

Etiology of Human Disease at the DNA Level

Mapping of Human Disease Traits

Etiology of Human Disease at the DNA Level,
edited by Jan Lindsten and Ulf Pettersson.

1

Linkage Genetics in Humans: Origins and Prospects

David Botstein

Department of Genetics, Stanford University School of Medicine, Stanford, California 94305

The rediscovery of Mendel in the first years of this century was immediately followed by the recognition that some human traits, indeed some human diseases, are inherited according to Mendel's laws. It soon became unmistakably clear that the mechanisms of heredity in humans are entirely typical of those in all higher eukaryotes. Among the earliest post-Mendelian discoveries was the principle of genetic linkage and the idea of a linkage map (Sturtevant, 1913). Yet linkage mapping in humans was not practiced on a large scale until after 1980, even though the applicability of the principle and, indeed, a number of essential statistical methods for detecting linkage had been in place since the 1930s (see Ott, 1985, for a review). It is the purpose of this short chapter to sketch briefly how linkage mapping was finally applied to inherited human diseases, to place the ideas behind human genetic linkage mapping in their proper historical context, and to give some inkling of the future and limits to linkage mapping in humans.

LINKAGE MAPPING WITH DNA POLYMORPHISMS

The limiting factor that made linkage mapping difficult between the 1930s and the 1980s, even for diseases obviously inherited in a simple Mendelian way, was the supply of adequately polymorphic genes that could serve as markers. Figure 1 illustrates that Mendelian inheritance and, indeed genetic linkage, can be observed readily, given only loci (marked A, B and C) such that the four alleles at each locus in the parents (A1, A2 in the father and A3, A4 in the mother; B1, B2 in the father and B3, B4 in the mother; and so on) can all be distinguished. As shown in Fig. 1, when all the alleles can be distinguished, it is easy to see segregation of the parental alleles (each child gets either A1 or A2 from the father and A3 or A4 from the mother, etc.).

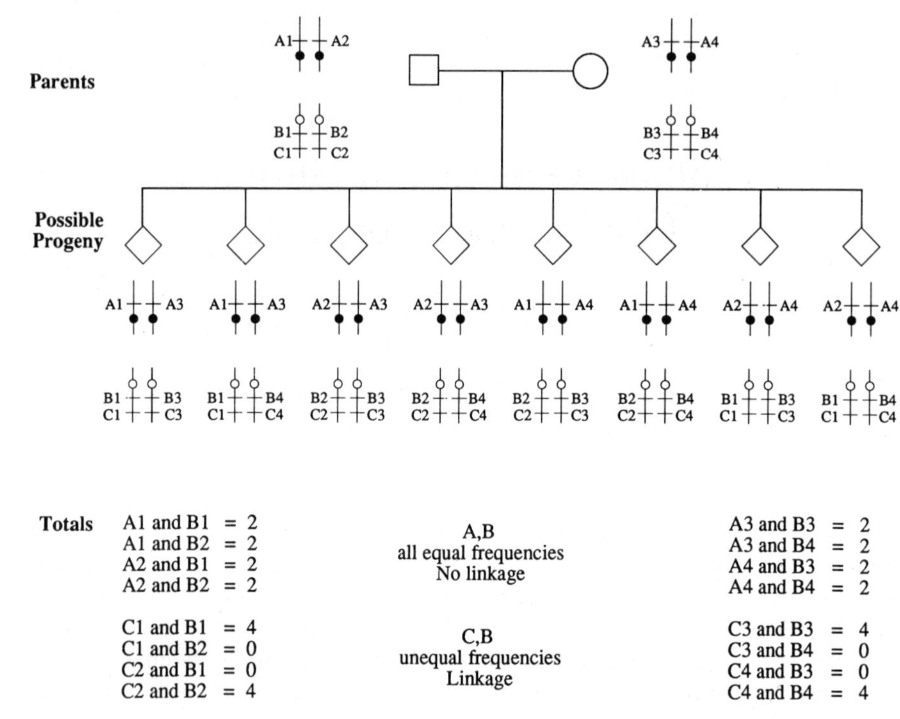

FIG. 1. Mendelian inheritance, and genetic linkage, can be observed given loci (A, B, C) such that the four alleles at each locus in the parents can be observed (A1, A2 in the father and A3, A4 in the mother, etc.). See text for details.

Independence of inheritance of alleles residing on different chromosomes is equally easily seen (e.g., A1 from the father is accompanied by B2 as frequently as B1; the combinations A3,B3; B3,B4; A4,B3; A4,B4 are equally frequently inherited from the mother; etc.). Linkage is easily observed as well, as the inheritance of C1 is highly correlated with the inheritance of B1; C2 is correlated with B2 (both from the father); on the mother's side we see correlation of C3 with B3 and C4 with B4 in Fig. 1. Note also that we see, from these correlations, not only that loci B and C are linked, but also the so-called linkage phase (i.e., the fact that C3 and B3 lie on the same chromosome in the mother).

Recombinant DNA technology provided a source of polymorphic markers in the form of restriction fragment length polymorphisms (RFLPs). In 1980 my colleagues and I (Botstein et al., 1980) noted that if there is enough variation among the DNA sequences of humans, differences in pattern of digestion by sequence-specific endonucleases (the *restriction enzymes*) would be found from individual to individual. These differences in restriction fragment length could be used as codominant genetic markers, just as we used the hypothetical alleles A1, etc. in Fig. 1. The method of detection proposed in 1980 (and still the most common) is to use single-copy DNA probes derived from human genomic clones as hybridization probes in gel-transfer experiments by the method of Southern (1985). Genomic DNA is extracted from white blood cells, cut with restriction enzymes, separated by size by gel electrophoresis, "blotted" onto filter paper, hybridized with labeled (radioactive or fluorescent) DNA probe, and analyzed in comparison with other samples from the same family. It should be emphasized that the DNA probe used to elicit the RFLP has a dual nature: it is a *genetic marker* that can be placed on a *genetic map* by *linkage* via polymorphism it reveals; it is also a *physical marker* that can be placed on a *physical map* because it is a single-copy DNA sequence. RFLPs thus bind together the genetic and physical maps of the human genome.

Of course, real RFLP markers are never completely polymorphic, and thus not every mating is "informative," in the sense of allowing the distinction between parental alleles. For a given locus, the inheritance of the marker could be followed in only a fraction of the families under study; in the remainder the marker is homozygous in key individuals, and thus yields no information. For this reason, we emphasized the importance of the "informativeness" of markers for mapping, proposing a measure of usefulness we called the *polymorphism information content* (PIC). The problem of informativeness of markers made us also propose that a map consisting mainly of polymorphic markers be constructed, so that even if a marker near a particular region turned out to be uninformative, the next marker on the map could be used in its stead, albeit with some loss of resolution. We proposed in 1980 that standard likelihood measures ("LOD scores") would serve admirably for disease mapping as well as for the construction of RFLP linkage maps.

DISEASE MAPPING WITH RFLP MARKERS

In 1983, the first autosomal disease gene was mapped using RFLPs: Gusella et al. (1983) found that the gene causing Huntington's disease is linked to a RFLP marker located near the end of the short arm of chromosome IV. Since then, a large number of other disease genes, including notably the gene for the recessive disease cystic fibrosis, have been mapped (Table 1). Because the RFLP markers are useful in physical as well as genetic mapping, they are crucial in the isolation of actual DNA sequences corresponding to the disease gene; the particular recent success in isolating the gene that causes cystic fibrosis was an example of this use of RFLP markers (Rommens et al., 1989; Kerem et al., 1989).

Despite the success in mapping some of the major inherited disease genes by linkage mapping using individual markers, theoretical analysis shows that

TABLE 1. *Diseases mapped using RLFPs*

Disease	Chromosome	Reference
Duchenne's muscular dystrophy	X	Davies et al., 1983
Huntington's disease	4	Gusella et al., 1983
Retinoblastoma	13	Cavenee et al., 1983
Cystic fibrosis	7	Tsui et al., 1985; Knowlton et al., 1985; White et al., 1985
Adult polycystic kidney disease	16	Reeders et al., 1985
Familial colon cancer	5	Bodmer et al., 1987
von Recklinghausen's neurofibromatosis	17	Barker et al., 1987; Seizinger et al., 1987
Bilateral acoustic neurofibromatosis	22	Rouleau et al., 1987
Multiple endocrine neoplasia type 2A	10	Simpson et al., 1987

use of a set of mapped markers whose linkage relationships are known and that span all of the genome will be more efficient and powerful (Lander and Botstein, 1986a). In large part, this is because of the increased power of linkage tests with markers flanking a disease gene as compared with single markers lying to one side. With single markers, single crossover events can completely reverse the relationship between the marker and the disease gene, but with flanking markers, only a double-crossover (an exceedingly rare event) will suffice to switch the marker-disease gene relationship completely.

Several techniques that take advantage of a complete linkage map have been suggested that allow either disease gene mapping without family study (*homozygosity mapping*; Lander and Botstein, 1987), mapping of diseases showing heterogeneity of cause (*simultaneous search*; Lander and Botstein, 1986b) or even mapping of genes contributing to quantitative traits (Lander and Botstein, 1988). This last method, which is applicable only to model systems, has been used to map several traits specified by as many as five contributing quantitative trait loci in tomato (Paterson et al., 1988).

In 1987, the first reasonably complete genetic linkage map of the human genome was published by Donis-Keller et al. This map has been preceded by maps of individual chromosomes and chromosome arms, notably the X chromosome (Drayna and White, 1985). The Donis-Keller et al. map covers, in the sense of showing continuous linkage, about 95% of the genome. However, not all the markers in it are highly informative and routine use as suggested above is not quite yet a reality. Improvement of the map, both with respect to informativeness of markers and density of markers, is a short-term goal of the Human Genome Initiative. As the maps become better, with more informative markers spaced at even intervals, the techniques involving the entire map can be applied to diseases with complex etiologies (using simultaneous search) and very rare recessive diseases (homozygosity mapping).

INTELLECTUAL ORIGINS OF THE RFLP MAPPING STRATEGY

As indicated above, the idea of a linkage map based on frequency of recombination is very old, having been published by Sturtevant in his historic paper of 1913. In the same paper, Sturtevant put forward the principle that double-crossovers will be rare, and he used this principle to order markers. Thus there is nothing in the idea of linkage mapping per se that is new in RFLP mapping. Quite to the contrary, it is very much in the tradition of genetic mapping as understood by Sturtevant and generations of geneticists who followed him.

Physical markers that can be scored in genetic crosses, a central attribute of RFLP markers, are also not new. The first use of a physical marker in a genetic cross was by Creighton and McClintock in 1931; this paper was the first to correlate cytological crossing over and genetic recombination. Correlations between genetic and physical (usually cytogenetic) markers became a major preoccupation of geneticists of all kinds since then.

Molecular polymorphisms were also old antecedents of the RFLP strategy. The first such marker in humans was hemoglobin-S, the protein with an altered beta-chain that is the cause of sickle-cell anemia (Neel, 1949; Pauling et al., 1949). Many protein polymorphisms were subsequently discovered in humans and it was these that were the main markers in pre-RFLP linkage studies.

The first correlations between physical and genetic markers at the DNA level were carried out with bacteriophage λ (Davis and Parkinson, 1971; Parkinson and Davis, 1971). These papers describe the generation and physical mapping of deletion and substitution mutations by formation of DNA heteroduplexes followed by measurements of the molecules under the electron microscope. Some of these very deletions and substitutions led to, and even are still present in, today's recombinant DNA vectors.

The same year saw the publication of the first map of a genome with restriction enzymes (Danna and Nathans, 1971). It was quickly understood, since this was very much in the tradition of bacteriophage and animal virus genetics, that maps of the recognition sites of these DNA-sequence-specific endonucleases constituted an important new tool for physical mapping with a technical simplicity and theoretical resolving power quite beyond that of DNA electron microscopy. The method was limited, however, to small genomes in which all the fragments produced after a digest could be separated from each other on a gel.

The first use of a difference in restriction fragment recognition sites as a passive genetic marker in genetic crosses was done by Sambrook's group at Cold Spring Harbor (Grodzicker et al., 1974). Using simple restriction mapping as introduced by Danna and Nathans (1971), they located temperature-sensitive (*ts*) mutations on the adenovirus genome by crosses between strains (Ad2 and Ad5) that differed in their restriction maps. By making crosses between *ts* strains that also differed in restriction pattern, they could select temperature-independence and then score the presence or absence of particular restriction enzyme recognition sites. The design of this experiment goes back to Creighton and McClintock (1931), i.e., the first correlation between genetic and physical maps.

The method for extracting, by hybridization, restriction enzyme cleavage pattern information from complex genomes with thousands of fragments was invented by Southern (1975). His method, which involves transfer ("blotting") of DNA from the medium of size separation of fragments (usually agarose gel) to filter paper on which hybridization is performed, was one of the central elements of the revolutionary methods now lumped under the rubric *recombinant DNA technology.* It was first applied to the case of a single-copy DNA sequence in a mammalian genome by Botchan et al. in 1976. They followed the integration of SV40 into cellular DNA by gel-transfer hybridization using viral DNA as probe, and found many obviously single-copy integrants. This paper made clear the possibility of following single genes by gel-transfer, and was the basis for our expectation that polymorphism in restriction fragment length would be routinely detectable, since the different integration sites of SV40 were readily detected.

Having surveyed the origins of the elements of RFLP analysis, namely, linkage mapping with physical markers, restriction fragment length differences as molecular markers, and gel-transfer to visualize the restriction fragment length difference, we come to the first use of RFLPs as genetic markers in complex genomes. Two groups discovered and applied the technique to genes of yeast (*Saccharomyces cerevisiae*) independently of each other; both applications were published in 1977. Petes and Botstein (1977) found a polymorphism in the restriction pattern of the ribosomal DNA of yeast in a diploid strain that was heterozygous with respect to this property. By sporulating that strain and performing tetrad analysis, it was possible to show that all the 100-rDNA copies comprise a single tandem array at a single locus (subsequently mapped to chromosome XII). Olson et al. (1977) sought ways to distinguish the eight different genes in yeast that encode tyrosine tRNAs that can mutate to become ochre suppressors. Taking advantage of the different restriction fragment lengths of the different genes, and performing crosses between strains that were polymorphic at one or more of the loci of these genes, Olson et al. were able to map these genes to 8 loci on six different chromosomes.

Application to human genetics began with Kan and Dozy (1978), who used a restriction fragment length difference revealed by probing with the β-globin gene itself to carry out antenatal diagnosis of sickle-cell anemia. The principle of linkage was not directly invoked, and no linkage mapping was proposed. This was followed by the proposal of Botstein et al. (1980) to find RFLPs deliberately, construct a linkage map, and use it to find disease genes by linkage. As mentioned above, the first such mapping of a gene by RFLP analysis was Huntington's disease (Gusella et al., 1983).

Thus we see that RFLP linkage mapping grew naturally out of classical and molecular genetics. Its intellectual antecedents are the same basic papers that are the antecedents of much of genetic analysis, and much of what we call recombinant DNA technology.

In this context, it must be remarked that the commonly used concept of "reverse genetics," as used by some in the human genetics community, makes no historical sense. "Reverse" genetics is meant by human geneticists to mean finding of a gene by its effect (i.e., phenotype) followed by its mapping (by linkage) and only subsequently by molecular isolation of the gene's DNA or protein product. The paradigm case is cystic fibrosis. Yet Sturtevant mapped genes by their phenotypic effects, and the DNA corresponding to one of his genes (white eyes) was only recently isolated. Surely we do not want to say Sturtevant practiced "reverse" genetics! He was, after all, nothing less than the *inventor* of linkage mapping!

If the above is not enough reason to abandon the term "reverse genetics," consider the problem posed by Charles Weissman, who in 1978 (2 years before the publication of the RFLP mapping idea; Weissman, 1978) proposed "reversed genetics" as a term for an approach in which " . . . a mutation is first generated in a predetermined area of the genome by site-directed mutagenesis and the effect of the lesion is then studied either in vivo or in vitro" (Weissman et al., 1979). This use of the term "reversed genetics" at least makes historical sense (Sturtevant is now facing forward again). Nevertheless, in view of the confusion, and lest we never know whether we are coming or going, in forward or reverse direction, let us agree to abandon entirely the idea of "reverse" genetics.

In conclusion, the idea of mapping genes using polymorphic DNA markers has allowed the mapping of many disease genes and promises to allow the mapping of even more. The idea grew naturally out of the history of genetics and molecular biology. As the human DNA marker map becomes better, in the sense of more polymorphic markers at shorter intervals that are easy to use, we can envision finding genes that contribute to the inheritance of more complex diseases than the simple Mendelian. We can also look forward to the use of RFLP or other DNA markers in aligning the physical and genetic maps of the human genome. This may turn out to be of major importance to the Human Genome Initiatives that are underway in the

United States and around the world. The DNA markers have, in reality, made it possible to have a real genetics of humans, in the sense of Sturtevant and McClintock, in the sense of *Drosophila* and yeast—a genetics based on linkage and phenotype as well as molecules.

REFERENCES

Barker D, Wright E, Nguyen K, et al. Gene for von Recklinghausen neurofibromatosis is in the pericentromeric region of chromosome 17. *Science* 1987;236:1100–1102.

Bodmer WF, Bailey CJ, Bodmer J, et al. Localization of the gene for familial adenomatous polyposis on chromosome 5. *Nature* 1987;328:614–616.

Botchan M, Topp W, and Sambrook J. The arrangement of simian virus 40 sequences in the DNA of transformed cells. *Cell* 1976;9:269.

Botstein D, White RL, Skolnick M, and Davis RW. Construction of a genetic linkage map in man using restriction fragment length polymorphisms. *Am J Hum Genet* 1980;32:314–331.

Cavenee WR, Dryja TP, Phillips RA, et al. Expression of recessive alleles by chromosome mechanisms in retinoblastoma. *Nature* 1983;305:779.

Creighton HB and McClintock B. A correlation of cytological and genetical crossing-over in *Zea mays*. *Proc Natl Acad Sci USA* 1931;17:492–497.

Danna KJ and Nathans D. Specific cleavage of SV40 DNA by restriction endonucleases of *Hemophilus influenzae*. *Proc Natl Acad Sci USA* 1971;68:2913.

Davis RW and Parkinson JS. Deletion mutants of bacteriophage lambda III. The structure of att-p. *J Mol Biol* 1971;56:403.

Davies KE, Pearson PL, Harper PS, et al. Linkage analysis of two cloned DNA sequences flanking the Duchenne muscular dystrophy locus on the short arm of the human X chromosome. *Nucleic Acids Res* 1983;11:2303.

Donis-Keller H, Green P, Helms C, et al. (1987) A genetic linkage map of the human genome. *Cell* 51:319–337.

Drayna D and White R. The genetic linkage map of the human X chromosome. *Science* 1985;230:753–758.

Grodzicker T, Williams J, Sharp P, and Sambrook J. Physical mapping of temperature-sensitive mutations of adenoviruses. *Cold Spring Harbor Symp Quant Biol* 1974;39:439.

Gusella JF, Wexler NS, Conneally PM, et al. A polymorphic DNA marker genetically linked to Huntington's disease. *Nature* 1983;306:234.

Kan YW and Dozy AM. Antenatal diagnosis of sickle-cell anemia by DNA analysis of amniotic-fluid cells. *Lancet* 1978;2:910.

Kerem B, Rommens JM, Buchanan JA, et al. Identification of the cystic fibrosis gene: genetic analysis. *Science* 1989;245:1073–1080.

Knowlton RG, Cohen-Haguenauer O, Van Cong N, et al. A polymorphic DNA marker linked to cystic fibrosis is located on chromosome 7. *Nature* 1985;318:380.

Lander E and Botstein D. Mapping complex genetic traits in humans: new methods using a complete RFLP linkage map. *Cold Spring Harbor Symp Quant Biol* 1986a;51:49–62.

Lander ES and Botstein D. Strategies for studying heterogeneous genetic traits in humans by using a linkage map of restriction fragment length polymorphisms. *Proc. Natl. Acad. Sci. USA* 1986b;83:7353–7357.

Lander ES and Botstein D. Homozygosity mapping: a way to map human recessive traits with the DNA of inbred children. *Science* 1987;236:1567–1570.

Lander ES, and Botstein D. Mapping Mendelian factors underlying quantitative traits using RFLP linkage maps. *Genetics* 1988;121:185–194.

Morgan TH. Sex limited inheritance in *Drosophila*. *Science* 1910;32:120.

Neel JV. The inheritance of sickle cell anemia. *Science* 1949;110:56–66.

Olson MV, Montgomery DL, Hopper AK, et al. Molecular characteristics of the tyrosine tRNA genes of yeast. *Nature* 1977;267:639.

Ott J. *Analysis of Human Genetic Linkage*. Johns Hopkins University Press, Baltimore, 1985.

Parkinson JS and Davis RW. A physical map of the left arm of the lambda chromosome. *J Mol Biol* 1971;56:425.

Paterson AH, Lander ES, Hewitt JD, et al. Resolution of quantitative traits into Mendelian factors by using a complete RFLP linkage map. *Nature* 1988;335:721.

Pauling L, Itano HA, Singer SJ, and Wells IC. Sickle cell anemia, a molecular disease. *Science* 1949;110:534–548.

Petes TD and Botstein D. Simple Mendelian inheritance of the reiterated ribosomal DNA of yeast. *Proc. Natl. Acad. Sci. USA* 1977;74:5091–5095.

Reeders ST, Breuning MH, Davies KE, et al. A highly polymorphic DNA marker linked to adult polycystic kidney diseases on chromosome 16. *Nature* 1985;317:542.

Rommens JM, Iannuzzi MC, Kerem B, et al. Identification of the cystic fibrosis gene: chromosome walking and jumping. *Science* 1989;245:1059–1065.

Rouleau GA, Wertelecki W, Haines JL, et al. Genetic linkage of bilateral acoustic neurofibromatosis to a DNA marker on chromosome 22. *Nature* 1987;329:246–248.

Seizinger BR, Rouleau GA, Ozelius LJ, et al. Genetic linkage of von Recklinghausen neurofibromatosis to the nerve growth factor receptor gene. *Cell* 1987;49:589–594.

Simpson NE, Kidd KK, Goodfellow PJ, et al. Assignment of multiple endocrine neoplasia type 2A to chromosome 10 by linkage. *Nature* 1987;328:528–530.

Southern EM. Detection of specific sequences among DNA fragments separated by gel electrophoresis. *J Mol Biol* 1975;98:503–517.

Sturtevant AH. The linear arrangement of six sex-linked factors in *Drosophila*, as shown by their mode of association. *J Exp Zool* 1913;14:43–49.

Tsui LC, Buchwald M, Barker D, et al. Cystic fibrosis locus defined by a genetically linked polymorphic DNA marker. *Science* 1985;230:1054.

Weissmann C. Reversed genetics. *Trends Biochem. Sci.* May 1978:109–111.

Weissman C, Nagata S, Taniguchi T, Weber H, and Meyer F. The use of site directed mutagenesis in reversed genetics. In: *Genetic Engineering: Principles and Methods*. Volume 1. Eds. Jane K. Setlow and Alexander Hollaender. Plenum Press, New York, pp. 133–150, 1979.

White R, Woodward S, Leppert M, et al. A closely linked genetic marker for cystic fibrosis. *Nature* 1985;318:382.

Etiology of Human Disease at the DNA Level,
edited by Jan Lindsten and Ulf Pettersson.

2

Genetics of Colorectal Cancer at the Somatic and Germ Line Levels

Walter F. Bodmer

Imperial Cancer Research Fund, London WC2A 3PX, England, U.K.

Colorectal carcinomas are now the most common cancers not due to smoking in most western developed societies. There has been comparatively little improvement in the survival of colorectal cancer patients for many years and new approaches are clearly needed to improve both prevention and treatment of this disease. There are now many lines of evidence to support the basic assumption that cancer develops through a series of genetic changes progressing from the initiated cell, whose progeny eventually give rise to the malignant disease. The definition of this series of genetic changes and their functional effects underlies a fundamental understanding of the disease process. This must be an essential basis for improving the prognosis of colorectal cancer.

There are a variety of approaches that can be used to look for the genetic changes involved in tumour progression. These include the search for mutations or other changes in the expression of known oncogenes, mainly identified through the oncogenic viruses, the search for allele loss in tumours as compared to normal tissue (1) and the study of familial cancers. Knudson (2) suggested that there should be a correspondence between genetic changes in somatic cells that are steps in the progression to a cancer, and changes in the same genes in the germ line which may sometimes give rise to inherited familial cancer susceptibility. This idea implies that the identification of a gene involved in an inherited susceptibility may also identify a gene that is one of the steps in the somatic progression of sporadic forms of the same tumour. The development of somatic cell genetic analysis, especially using interspecific hybrids, and recombinant DNA technology and through this the availability of an essentially unlimited range of polymorphic markers at the DNA level for family analysis have revolutionised the possibilities for the genetic analysis of cancers (3). This paper will briefly review the present status of the genetics of colorectal cancer with a special emphasis on familial colorectal cancers including in particular the inherited susceptibility adenomatous polyposis coli (APC).

GENETICS OF ADENOMATOUS POLYPOSIS COLI

APC (sometimes also called familial adenomatous polyposis or FAP) is a dominantly inherited susceptibility to colorectal cancer which occurs with an overall frequency of 1 in 5,000 to 1 in 10,000 that appears not to vary substantially between different groups of populations. Affected individuals develop, usually during adolescence, up to several hundreds or even thousands of adenomatous polyps in the colon and rectum. These polyps, if left untreated, will inevitably give rise to one or more adenocarcinomas, supporting the hypothesis that colorectal cancers generally arise from precancerous adenomas. Screening the offspring of affected individuals by sigmoidoscopy, beginning in the early teens, has so far been used to identify individuals within a family who are at risk of developing a carcinoma. Prophylactic colectomy can then eliminate much of this risk, although there is a residual problem because of the development of tumours elsewhere. Following Knudson's ideas, the localisation and eventual identification of the APC gene should provide not only the basis for early diagnosis and counselling of APC families, but should also lead to the identification of a gene that may be important for a high proportion of sporadic colorectal carcinomas.

The APC gene has been localised to chromosome 5q21 following a case report (4) of a patient who had APC and other abnormalities, and who had a visible deletion of a band on the long arm of chromosome 5. Following this clue, DNA markers which had been mapped to chromosome 5 using somatic cell hybrids were shown to be closely linked to APC in family studies. Furthermore, in situ annealing located the APC gene to the region of chromosome band 5q21 (5–8).

The discovery of further deletions similar to that originally described by Herera (4) provides a basis for the systematic isolation of DNA probes in the vicinity of the APC gene (9). In one case, an Australian family has been described in which two multiply abnormal brothers with APC had a visible deletion in the 5q21 region and a mother with APC, who died of an inoperable carcinoma before her karyotype could be established (10). This family confirms the dominant inheritance of susceptibility by deletion of gene function. Using a combination of somatic cell hybrid analyses of the deleted 5q chromosomes, and cloning the ends of long DNA fragments defined by the rare cutting restriction enzyme *Bss*hII, new genetic markers within these deletions, and so close to the APC gene, are being identified. The markers are assigned to the deletion by their presence or absence in human-hamster somatic cell hybrids which contain either a deleted chromosome 5 or the normal chromosome 5, in general with little or no other human material. The data clearly show that the deletions, although indistinguishable cytologically, are different at the molecular level and that one of them obtained from a girl who is too young for a clear-cut diagnosis of APC, clearly encompasses the APC gene showing that the girl is definitely at risk with respect to adenomatous polyposis coli. The range of polymorphic markers now available ensures that nearly all families with APC are informative, and so should be counselled with respect to the risks of individual family members.

A variety of techniques are now being used to obtain further clones close to the APC gene with a view to isolating the gene itself. These include the analysis of clones obtained from radiation hybrids which have incorporated small fragments of chromosome 5, chromosome microdissection of the region around chromosome 5q21 and the use of alu-PCR-based techniques for identifying clones that are present on chromosome 5 but absent from deletions encompassing the APC gene (11).

Chromosome 5 Marker Loss Around the APC Gene in Sporadic Colorectal Carcinomas

The assignment of the APC gene to chromosome 5 led immediately to the demonstration that at least 20–40% of sporadic colorectal carcinomas show allele loss for markers on chromosome 5 (12,13). The pattern of loss of markers along the chromosome suggests a focus at 5q21, and so that it is indeed the APC gene that is involved in a high proportion of sporadic colorectal carcinomas, as predicted by Knudson's hypothesis (14). The overwhelming majority of adenomas from APC patients, however, retain two alleles at informative loci (12,13). The fact that polyps from APC patients mostly do not show evidence of further changes on chromosome 5 led Bodmer (5) to suggest that heterozygosity for the APC deficiency might on its own give rise to the polyposis observed in APC patients. A threshold effect based, for example, on negative control of growth factor production could be an explanation. Cells in APC individuals will on average produce half as much APC gene product as in normal individuals and so, by chance, a small proportion of APC cells may produce an amount of product that is below the threshold required to contain growth, while the probability of this happening in normal individuals may be negligible (15). On this hypothesis, a polyp, once it has arisen, can provide the opportunity for further genetic changes to take place which then lead to the progression to an overt adenocarcinoma. Such a model could explain observations that have suggested associations between c-myc expression levels and chromosome 5 allele loss in sporadic colorectal carcinomas (16) and elevated levels of p53 in normal cells from APC patients (17).

Colorectal carcinoma derived cell lines which show chromosome 5 allele loss are a source of APC mutations and provide a basis for the functional analysis of the APC gene through complementation or abnormalities in the messenger RNA derived from candidate DNA sequences.

Chromosome 17 Allele Loss and p53 Mutations

The observation that the short arm of chromosome 17 was sometimes missing in colorectal carcinomas led to the demonstration that a high proportion of colorectal carcinomas showed allele loss for markers on the short arm of chromosome 17 (18). The fact that this was in the neighbourhood of the gene for p53 (19) provided the clue for Vogelstein and his colleagues to show that the changes on chromosome 17 were due to mutations in the p53 gene (20,21). Specifically it was shown that p53 gene mutations in colorectal and other cancers occurred in conserved regions of the gene and might be present in well over 50% of colorectal cancers. These observations have now been extended to lung, breast and other adenocarcinomas, making p53 the most commonly mutated oncogene in a wide variety of human cancers (22).

The presence of mutant forms of p53 is mostly correlated with increased levels of expression which can readily be detected by suitable monoclonal antibodies, either on tissue sections or in tumour-derived cell lines. Earlier observations (23) of elevated levels of p53, thought at that time to be normal, can now be explained by the presence of mutated forms of p53 which stabilise the protein and, through this, lead to high levels of expression.

A study of 10 colorectal-carcinoma-derived cell lines showed that 8 had mutations as detected either by direct sequence analysis or by chemical cleavage at mismatched base pairs. Three mutations were at the same conserved base pair position (amino acid 273 Arg to His). The mutant cell lines all lacked a normal sequence and so were either homozygous or hemizygous for the p53 mutation. This is consistent with the observation of 70–80% of 17p allele losses in colorectal carcinomas. Premalignant lesions from sporadic adenomas as well as polyps from APC cases did not react with the monoclonal antibodies to p53, indicating that at least early premalignant lesions do not have p53 mutations of the type that lead to overexpression of the protein.

The extreme specificity of the p53 mutations suggests that they have been specifically selected for, most probably through their effects on counteracting the activity of the normal p53 product. The high frequency of chromosome 17 allele loss and the homo- or hemizygosity of all the mutant cell lines indicates that, following the initial selection of a p53 mutation, there must still be a strong selective disadvantage associated with the presence of the remaining wild-type p53 gene leading to selection to eliminate this. The fact that deletions of p53 are not selected for suggests that, at least in the case of colorectal cancer, reducing the wild-type level of p53 product by a factor of two, in contrast to the likely situation for the APC gene, is not enough to give a significant advantage of the outgrowing tumour. The "dominant negative" p53 mutations most probably function by sequestering the normal p53 gene product to an extent that results in an effective level of available normal product which is well below 50% of the normal level. The situation may be different in other tumours such as osteosarcomas (24) and chronic myeloid leukaemia in blast crisis (25), in which p53 deletions have commonly been found. In these cases the baseline level of p53 may be sufficiently low that deletion of function rather than a dominant negative mutation is frequently the first step affecting p53 expression.

Other Genetic Changes in Colorectal Carcinomas

Allele losses for other chromosomes have also been consistently observed in colorectal carcinomas. These include mainly chromosomes 8, 18, and 22, (perhaps only in Japan) as well as lesser losses for chromosomes 5, 6, 12, and 15 (26,27). The most exciting recent development is the discovery of the gene DCC described by Vogelstein and colleagues (28) on chromosome 18 which explains the 18q allele loss. This extraordinarily large gene appears to be a member of the cell adhesion molecule family. This fits in with data suggesting an important role for extracellular matrix receptors in the control of differentiation, and, therefore, a low level of expression of such receptors in many colorectal carcinomas, together with other results that suggest an interaction between extracellular matrix receptors and members of the CEA family, another branch of the family of adhesion molecules within the immunoglobulin supergene family Pignatelli (29,30).

Another somewhat different type of genetic change in colorectal carcinomas is the comparatively frequent loss of individual HLA alleles as described by Smith et al. (31). This suggests a relatively high frequency of T-cell attack on tumours, leading to selection for changes in HLA expression in order to escape the effects of such an immune response. Following the discovery by Townsend and others that cytotoxic T cells respond to proteolytically processed proteins which may be intracellular molecules, it is clear that T cells could respond to dominant oncogene mutations such as in p53 and K-*ras*, another common mutation in positions 12 or 13 in colorectal carcinomas (32,33). Such immune response can now be looked for systematically and, for the first time, suggests the possibility of rational immunomodulation of the immune responses for the treatment of tumours and perhaps even vaccination against dominant oncogene changes (34).

NON-POLYPOSIS INHERITED COLORECTAL CANCER

There are several reasonably well-defined dominantly inherited susceptibilities to colorectal cancer in addition to APC. These include the Peutz-

Jeghers syndrome and the cancer family or Lynch II syndrome. The latter is intriguing in that these families express multiple types of tumours including especially colorectal, breast and ovarian adenocarcinomas (35). There has been a suggestion, based on linkage with the Kidd blood group, that the cancer family syndrome gene may be on chromosome 18, an intriguing finding in relation to the 18q allele loss identified with the DCC gene. However, so far, there has been no confirmed genetic linkage for any of these dominantly inherited syndromes. The fundamental nature of the p53 and K-*ras* gene products makes it unlikely that mutations in these could be inherited through the germ line to give rise to specific cancer susceptibilities. The DCC gene is, however, a more plausible candidate, especially since this and the APC gene are so far found to be altered only in colorectal carcinomas. Clearly any of the genes involved in these syndromes could lead to the discovery of still more recessive genetic changes in sporadic colorectal carcinomas.

Although these other inherited colorectal carcinoma syndromes may be a heterogeneous collection, this heterogeneity should be sorted out by genetic linkage studies guided by careful clinical description. It is possible, however, that the clinical picture may exaggerate the genetic heterogeneity. Thus it is now clear that there is no fundamental genetic distinction between polyposis and Gardner's syndrome, originally described as polyposis with other features, based on the fact that there is no heterogeneity between polyposis and Gardner's syndrome with respect to the 5q21 linkage. Thus it appears that the same gene, although perhaps with different mutations, can be associated with a variety of phenotypic manifestations.

There may be a further important class of low penetrance, relatively common genes influencing the incidence of colorectal cancer (36). Adenomas occurring in small number may, in this case, reflect a more penetrant expression of such genes. This situation falls into the category of multifactorial inheritance for which conventional formal linkage analysis is unlikely to be possible. This parallels studies of associations between HLA and disease as, for example, in the case of rheumatoid arthritis or Hodgkin's disease, in which sib pair analysis has been used to indicate significant effects of an HLA-linked gene (37). When the penetrance of a disease susceptibility gene is quite low, perhaps only a few percent, unaffected individuals, who are a mixture of those without the susceptibility genotype and those with it but who have not contracted the associated disease, are essentially unusable for linkage analysis. It is to circumvent this problem that multiple case family studies, which are an extension of sib pair linkage analysis, must be used. They simply ask the question whether the expected distribution of a genetic marker in a pedigree amongst any set of affected individuals departs significantly from what is expected from Mendelian segregation. Expected distributions can be determined empirically by Monte Carlo simulation. The potential power of this approach is that it makes no assumptions as to the nature of the inheritance of the trait involved because only information on the affected individuals is taken into account. It will only work, however, if there are enough polymorphic markers to span the genome at intervals that are sufficiently close to be likely to give a significant distortion as an indication of linkage between the marker and a low penetrance susceptibility. As was first clearly pointed out by Solomon and Bodmer (38), DNA polymorphisms for restriction enzyme sites (now detected in other ways), through providing an effectively unlimited range of genetic markers, have solved this problem and should revolutionize our ability to study the genetic determination of complex traits, such as low penetrance inherited susceptibility to colorectal cancer.

CONCLUSIONS

Human gene mapping using somatic cell hybrids, DNA probes, in situ hybridisation and other techniques has provided the basis for the enormous advances in our understanding of the genetics of colorectal and other cancers as briefly reviewed in this paper. Now we can use molecular genetic approaches to find the relevant functional genes and their products identified by translocations, by allele losses appearing in tumours or by genetic linkage studies on familial cancers. This is the route by which the retinoblastoma, DCC, and Wilms's tumour gene products have been found, and by which it has been shown that p53 is the commonest oncogene in human carcinomas. While the search for the functional gene itself is still long and cumbersome, the technology of long-range DNA analysis by pulsed field gel electrophoresis, overlapping cosmid clones and yeast artificial chromosomes (YACs) is being developed very rapidly. YAC cloning in particular should help to cover regions of a few million base pairs comparatively rapidly, and so fill in the gap between the molecular level and what can be seen down a microscope or measured as a detectable recombination fraction of 1 to a few percent in human families. Of course, if we knew where all the functional genes were along the chromosomes then, having found a marker close to our particular disease, such as APC, we could simply look up the functional genes in the neighbourhood and very quickly identify plausible candidates. Those could be identified, for example, by their pattern of tissue expression, and then tested for their relevance to APC by looking for mutations in affected individuals or in colorectal-carcinoma-derived cell lines. Once we have the complete human genome analysed, it should be possible to identify all the genetic changes involved in any particular colorectal carcinoma, and then base approaches to prevention and treatment on the fundamental knowledge of the changes that have occurred during tumour progression.

REFERENCES

1. Cavanee WK, Dryja TP, Phillips RA, et al. Expression of recessive alleles by chromosomal mechanisms in retinoblastoma. *Nature* 1983;305:779–784.
2. Knudson AG. Mutation and cancer: Statistical study of retinoblastoma. *Proc Natl Acad Sci USA* 1971;68:820–823.
3. Bodmer WF. Somatic cell genetics and cancer. In L.M. Franks (ed.): *Cancer Surveys,* Vol. 7 No. 2. Oxford University Press, Oxford, England, 1988;239–250.
4. Herrera L, Kakatis S, Gibas L, et al. Gardner syndrome in a man with an interstitial deletion of 5q. *Am J Med Genet* 1986;25:473–476.
5. Bodmer WF, Bailey CJ, Bodmer J, et al. Localization of the gene for familial adenomatous polyposis on chromosome 5. *Nature* 1987;328:614–619.
6. Leppert M, Dobbs M, Scambler P, et al. The gene for familial polyposis coli maps to the long arm of chromosome 5. *Science* 1987;238:1411–1413.
7. Meera Khan P, Tops CMJ, v.d. Broek M, et al. Close linkage of a highly polymorphic marker (D5S37) to familial adenomatous polyposis (FAP) and confirmation of FAP localisation on chromosome 5q21 q22. *Hum. Genet.* 1988;79:183–185.
8. Nakamura Y, Lathrop M, Leppert M, et al. Localization of the genetic defect in familial adenomatous polyposis within a small region of chromosome-5. *Am. J. Hum. Genet.* 1988;43:638–644.
9. Varesco L, Thomas H, Fennel S, et al. CpG island clones from a deletion encompassing the gene for adenomatous polyposis coli. *Proc Natl Acad Sci USA* 1989;86;10118–10122.
10. Hockey KA, Mulcahy MT, Montgomery P, and Levitt S. Deletion of chromosome 5q and familial adenomatous polyposis. *J Med Genet* 1989;26:61–68.
11. Cotter FE, Hampton GM, Nasipuri S, Bodmer WF, and Young BD. Rapid Isolation of human chromosome-specific DNA probes from a somatic cell hybrid. *Genomics* 1990;7:001–007.
12. Solomon E, Voss R, Hall V, et al. Chromosome 5 allele loss in human colorectal carcinomas. *Nature* 1987;328:616–619.
13. Kerr IB, Murday VA, Hiorns LR, Bussey HJR, and Bodmer WF. Prevalence of Ki-ras mutation and chromosome 5 allele loss in colorectal carcinomas arising in cases of familial adenomatous polyposis. *Cold Spring Harbor Cancer Cells* 1989;7:241–244.
14. Ashton-Rickardt PG, Dunlop MG, Nakamura Y, et al. High frequency of apc loss in sporadic colorectal carcinoma due to breaks clustered in 5q21 22. *Oncogene* 1989;4:1169–1174.
15. Bodmer WF. Genetic analysis of tumour suppression. In G Bock, J Marsh (eds): *Ciba Foundation Symposium 142* 1989;93–98.
16. Erisman MD, Scott JK, and Astrin SM. Evidence that the familial adenomatous polyposis gene is involved in a subset of colon cancers with a complementable defect in c-myc regulation. *Proc Natl Acad Sci USA* 1989;86:4264–4268.
17. Kopelovich L. and DeLeo AB. Elevated levels of p53 antigen in cultured skin fibroblasts from patients with hereditary adenocarcinoma of the colon and rectum and its relevance to oncogenic mechanisms. *J Natl Cancer Inst* 1986;77:1241–1246.
18. Fearon ER, Hamilton SR, and Vogelstein B. Clonal analysis of human colorectal tumours. *Science* 1987;238:193–197.
19. Lane DP. Oncogenes of the DNA tumour viruses: Their interactions with host proteins. In D. Glover (ed.): *Frontiers in Molecular Biology: Oncogenes.* I.R.L. Press, Oxford, 1989.
20. Baker SJ, Fearon ER, Nigro JM, et al. *Science* 1989;244:217–221.
21. Nigro JM, Baker SJ, Preisinger AC, et al. *Nature* 1989;246:491–494.
22. Rodrigues N, Rowan A, Smith ME, et al. p53 mutations in colorectal cancer. *Proc Natl Acad Sci USA* (in press).
23. Crawford LV, Pim DC, Gurney EG, Goodfellow P, and Taylor-Papadimitriou J. Detection of a common feature in several human tumour cell lines—a 53,000-Dalton protein. *Proc Natl Acad Sci USA* 1981;78:41–45.
24. Masuda H, Miller C, Koeffler HP, Battifora H, and Kline MJ. Rearrangement of the p53 gene in human osteogenic sarcomas. *Proc Natl Acad Sci USA* 1987;84:7716–7719.
25. Ahuja H, Bar-Eli M, Advani SH, Benchimol S, and Cline MJ. Alterations in the p53

gene and the clonal evolution of the blast crisis of chronic myelocytic leukemia. *Proc Natl Acad Sci USA* 1989;86:6783–6787.
26. Vogelstein B, Fearon ER, Hamilton SR. et al. Genetic alterations during colorectal tumor development. *N Engl J Med*, 1988;319:525–532.
27. Okamoto M, Ikeuchi T, Iwama T, et al. Loss of constitutional heterozygosity in colon carcinoma from patients with familial polyposis coli. *Nature* 1988;331:273–277.
28. Vogelstein B, Fearon ER, Cho KR, et al. Identification of a chromosome 18q gene that is altered in colorectal cancers. *Science* 1990;247:49–56.
29. Pignatelli M, Smith MEF, and Bodmer WF. Low expression of collagen receptors in moderate and poorly differentiated colorectal adenocarcinomas. *Br J Cancer* 1990; 636–638.
30. Pignatelli M, and Bodmer WF. Integrin cell adhesion molecules and colorectal cancer. *J Pathol* (in press).
31. Smith MEF, Marsh SGE, Bodmer JG, Gelsthorpe K, and Bodmer WF. Loss of HLA-A,B,C allele products and lymphocyte function-associated antigen 3 in colorectal neoplasia. *Proc Natl Acad Sci USA* 1989;86:5557–5561.
32. Bos JL, Fearon ER, Hamilton SR, et al. Prevalence of ras gene mutations in human colorectal cancers. *Nature* 1987;327:293–297.
33. Forrester K, Almoguera C, Grizzle WE, Han K, and Perucho M. Detection of high incidence of K-ras oncogenes during human and carcinogenesis. *Nature* 1987;327:298–303.
34. Bodmer WF. Immune attack against cancer—a new look. UICC Cancer Magazine 1990; (in press)
35. McKusick V. *Mendelian Inheritance in Man.* Catalogs of Autosomal Dominant, Autosomal Recessive, and X-Linked Phenotypes. The Johns Hopkins University Press, Baltimore.
36. Bishop T, Cannon-Albright LA, Skolnick MH, et al. Common inheritance of susceptibility to colonic adenomatous polyps and associated colorectal cancers. *N Engl J Med* 1988;319:533–537.
37. Bodmer WF. The human genome sequence and the analysis of multifactorial traits. Ciba Found. Symp. 1987;130:215–228.
38. Solomon E, Bodmer WF. Evolution of sickle variant gene. *Lancet* 1979 Apr 28 1(8122):923.

Etiology of Human Disease at the DNA Level,
edited by Jan Lindsten and Ulf Pettersson.

3

Databases To Serve the Genome Program and the Medical Genetics Community

P. L. Pearson, R. Lucier, and C. Brunn

William Welch Medical Library, Johns Hopkins University,
Baltimore, Maryland 21205

Ever since mapping and sequencing the entire human genome was first proposed in the mid-1980s as being a technically difficult but feasible project (1), it has been argued generally that data management is one of the most, if not *the* most important component of the genome program. Several factors have contributed to this conclusion, including the necessity to organize the huge amount of information that will derive from the genome program and to make the information available for further analysis. There was a naive belief in the earlier discussions on planning the genome program that DNA sequence analysis was the only type of activity to be considered. However, several subsequent appraisals of the problem clearly considered the construction of physical and genetic maps to be important forerunners to sequencing the genome (2,3), for two reasons. First, when the order and location of parts of the genome were defined first, it permitted a significantly more efficient sequencing of the genome in an ordered and economical fashion. Second, sequencing technology was not yet advanced far enough to permit large-scale sequence analysis in a cost-effective and timely fashion. Accordingly the decision was taken to concentrate much of the effort for the first 5 years of the program on mapping the human genome.

Although the human genome program is regarded by some to have officially started in 1989, the organization underlying the gene mapping activities goes back some 16 years earlier (4). At that time, a group of human geneticists came together at a workshop held in New Haven in 1973 to compile the sum total of all map information on human genes. Subsequently, the mapping community gathered at workshops organized at 1- to 2-year intervals to make new updates of the human gene mapping information. The workshops have become an important vehicle for organizing human gene mapping information. One of the main organizers of the workshops, Dr. Frank Ruddle, recognized the necessity of maintaining a database for storing human gene map information and its associated literature. Accordingly, he initiated a database at Yale in the mid-1970s using local resources. For some years the database was supported by NIH funding at a modest level. However, in 1985 the Howard Hughes Medical Institute (HHMI) took over the funding and gave the possibility of extending the scope of the database. For example, one of the major extensions provided at that time was that of making the information accessible online. The database, known as the Human Gene Mapping Library (HGML), became used as the prime repository for all human gene mapping information and as a source of information for organizing the gene mapping workshops. Important features included the use of international standards for nomenclature, map location based on chromosome banding, a description of the genetic markers and the type of variation or polymorphism that could be detected using such markers. HGML was used as the basis for coordinating information for the HGM8, HGM9, HGM9.5, and HGM10 meetings. A specific interface for multiple data entry and editing of HGML data by chromosome committee chairs was implemented for the HGM9.5 and HGM10 meetings (5).

Although the HGML was much more useful for catering to the needs of the gene mapping community than any previous database, it suffered from certain drawbacks, namely, the database operated under a hierarchical management system called SPIRES that only ran on IBM mainframe computers and was not transportable onto mini- and microcomputers. Further, the data structures could not be modified easily to encompass new types of mapping data and respond to the changing needs of the scientific community in a timely fashion. In 1989 the Howard Hughes Medical Institute evaluated the needs of the community for human gene mapping information. It concluded that HGML no longer embodied the functionality needed for a database serving as the primary repository for the enormous amount of mapping information that would emerge from the human genome program over the coming years.

After considering different possibilities, HHMI approached Johns Hopkins University Medical School (JHU) to see whether it would be prepared to take over the responsibility of organizing a new gene mapping database. One of the main reasons for choosing JHU as a site for the new database was the possibility of combining the mapping information with the genetic disease information already existing within another information system at JHU, namely the Online Mendelian Inheritance in Man (OMIM). This genetic disease catalogue, started by Victor McKusick in the mid-1960s, has been consistently updated by McKusick for the last 25 years and is arguably the most valuable compendium of human genetic disease information currently available (6). In 1985 a collaboration between the National Library of Medicine and the Welch Medical Library of JHU resulted in the McKusick catalogue being put into an online accessible database. In 1987 the HHMI took over the support of OMIM and was concerned that the new database to be established at JHU embody both the gene mapping information in

HGML and disease information in OMIM. The new database was called the Genome Data Base (GDB) and work commenced on designing and building the new data structures in mid-1989.

The following is an account of the overall database requirements for the human genome program. Special consideration is given to the gene mapping databases and to the role fulfilled by GDB. Included are some of the design features of GDB, how it will be used to fulfil the needs of the gene mapping and medical genetics communities, what the relationships are to other databases needed for genome analysis and in what ways the database will be developed to keep abreast of changing scientific requirements.

DATABASES: PUBLIC AND PRIVATE

In broad terms, two generic types of database will initially be of major importance for the genome program. This distinction is based on whether the databases are concerned with sequence information, i.e., defining what things are, as opposed to databases for mapping information, i.e., defining where things are. It is anticipated that other types of database will become more important in future including those concerned with the secondary and tertiary structure of proteins. However, at this juncture development of the mapping and primary sequence databases must receive the highest priority if they are going to meet the immediate demands of the human genome program. Another distinction that can be made is between "public" and "private" databases (Fig. 1). Public databases attempt to encompass consensus

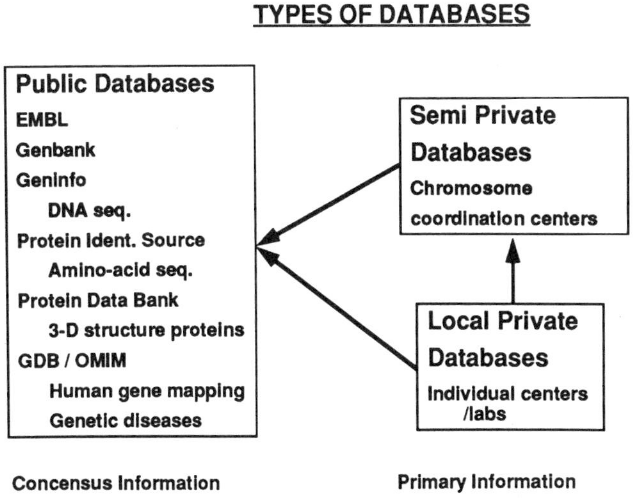

TYPES OF DATABASES

Public Databases
EMBL
Genbank
GenInfo
DNA seq.
Protein Ident. Source
Amino-acid seq.
Protein Data Bank
3-D structure proteins
GDB / OMIM
Human gene mapping
Genetic diseases

Semi Private Databases
Chromosome coordination centers

Local Private Databases
Individual centers /labs

Concensus Information

Primary Information

FIG. 1. Different types of databases. GDB, Genome Data Base; OMIM, Online Mendelian Inheritance in Man.

information and make the information available to a wide range of users, whereas private or semiprivate databases contain raw or less processed data. In addition, the information is usually accessed by a much smaller user group.

Currently there are three public primary sequence databases, two of which, EMBL and Genbank, exchange data and synchronize their contents. Further, they permit direct data submissions. The third, Geninfo, run by the National Library of Medicine (NLM), is currently set up to capture primary sequence information from the literature and does not permit direct data entry. Up to now the human gene mapping information has been made available through the HGML at New Haven. However, from September 1990 onwards, the HGML is ceasing to operate and the GDB at Johns Hopkins will provide that service to the community. The mapping information will be distributed with the genetic disease information present in OMIM. In addition the GDB will provide a tool for gathering the gene mapping information both during and prior to the gene mapping workshops by providing direct data entry and editing facilities for each of the individual chromosome committees.

While public databases will cater to aspects involving the whole genome, private or semi-private databases will be organized around individual chromosomes and serve the immediate data requirements of those laboratories involved in physical and genetic mapping activities on the respective chromosome. One of the major problems arising from an independent and parallel development of so many databases is a lack of common data standards and means of map representation. If the chromosome-specific databases are to play the pivotal role envisaged for them in permitting coordination of the mapping information on each chromosome, agreement must be reached quickly on issues of nomenclature, syntax, and data representation. The Joint DOE/NIH Informatics Task Force (7), called into being to make recommendations on the use of informatics for the genome program, should draw up suitable recommendations on this matter without delay.

It has become clear that the enormous amount of information emerging from the genome program will prevent the majority of the information from being published in conventional journals. Indeed, some journals are already refusing to publish large quantities of sequence data or even articles describing conclusions from sequence information unless the sequence has first been deposited in an appropriate database. It may be anticipated that the nature of publication will change such that deposition of data in a public database will constitute an acceptable mode of publication in future.

DATA FLOW INTO AND WITHIN THE GENOME DATA BASE

The GDB operates under a commercially available relational database management system, namely Sybase. It provides the possibility of remote

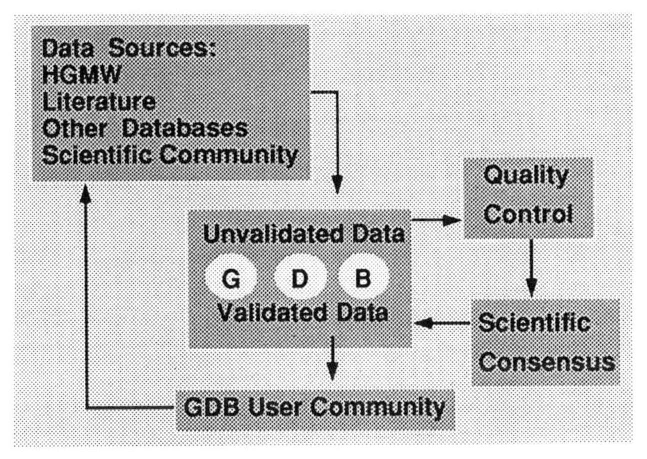

FIG. 2. Dynamic treatment of data by the Genome Data Base (GDB). New entries are put through an editorial loop and validated. HGMW, Human Gene Mapping Workshops.

data entry and access via Internet and Telenet communication networks. Data is entered from various sources, including the literature, direct laboratory entry, and persons responsible for collating the map information on individual chromosomes such as the chromosome committee chairpersons involved in the Human Gene Mapping Workshops (HGMW). Extensive data entry, editing, and perusal interfaces have been developed and a description of these will be presented elsewhere. GDB also has its own editorial staff responsible for maintaining various sorts of data including nomenclature,

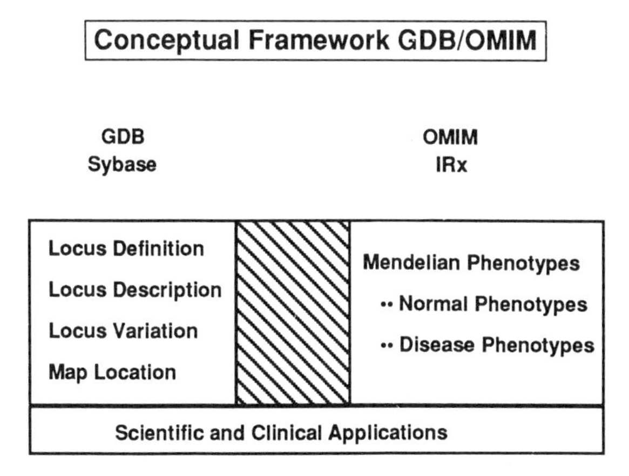

FIG. 3. Conceptual framework of the Genome Data Base (GDB) and Online Mendelian Inheritance in Man (OMIM).

physical and genetic map information, DNA polymorphisms, and DNA probes. As shown in Fig. 2, GDB provides the ability to treat data dynamically by cycling new, invalidated entries through an editorial loop, at which stage the data are flagged as being validated.

Relationship Between GDB and OMIM

As stated above, GDB is a compendium of gene mapping and disease phenotype information. The map information is stored within the Sybase structures of GDB and the disease information in the IRx structures of OMIM. A link has been established between the two types of information system such that the user can move transparently from the map to the disease information and back again (Fig. 3). In the future, medical genetics applications will be developed permitting diagnosis of genetic diseases based on map and DNA probe information stored in the database.

Types of Genetic Markers

Currently, GDB recognizes three types of genetic markers or loci, namely, genes, fragile sites and anonymous DNA segments. Each has a unique symbol to identify it and various members of the GDB editorial staff are responsible for ensuring a standardized use marker nomenclature. The convention used for defining the symbols for DNA segments is given in Fig. 4. Notable is the use of a number defining the chromosomal origin of a DNA segment, followed by a single letter indicating the complexity of DNA, and finally an

DNA SEGMENT DEFINITION

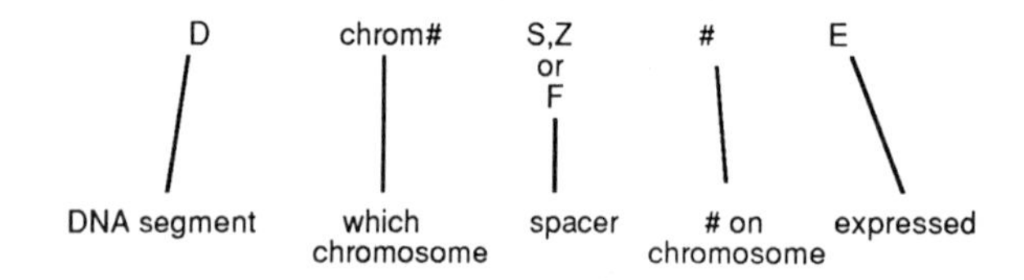

ATTRIBUTES

- which chromosome
- arbitary number on chromosome
- map location
- complexity DNA; simple,repetitive,family
- probes
  - clones
  - PCR primers, STSs
  - ASOs

FIG. 4. DNA segment definition in GDB.

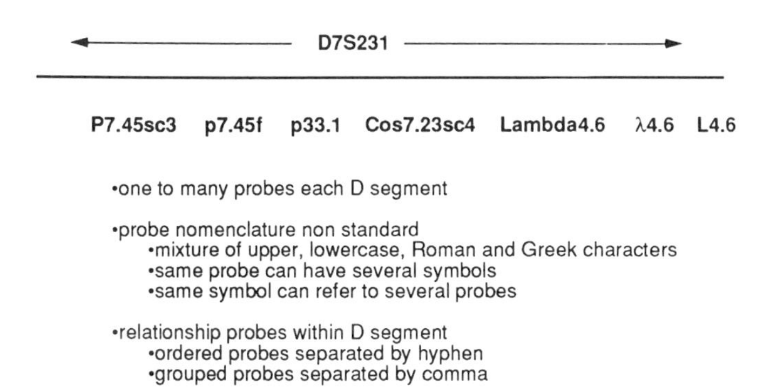

FIG. 5. Relationship between DNA segments and DNA probes in GDB.

arbitrary number assigned in chronological order. The relationship between DNA segments and DNA probes is depicted in Fig. 5. It is clear that each DNA segment can contain one to many probes. In the future several new types of genetic markers or loci require definition, including chromosome break points, meiotic crossovers, restriction sites and even partial maps comprised of groups of markers. A possible convention for defining chromosome break points is given in Fig. 6 using a system analogous to that used for DNA segments.

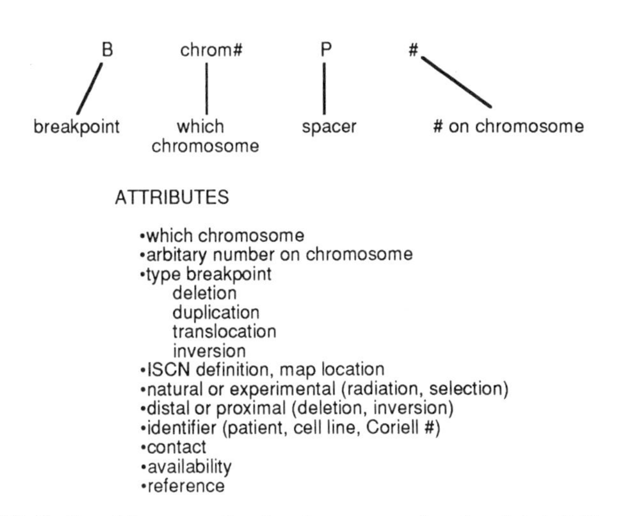

FIG. 6. Possible convention for chromosome break point definition.

Map Information

The current version of GDB contains map information based on the cytogenetic location of markers along chromosomes. Although the convention for doing this is well defined, being based on a chromosome banding nomenclature first established 19 years ago (8), the system has outlived its

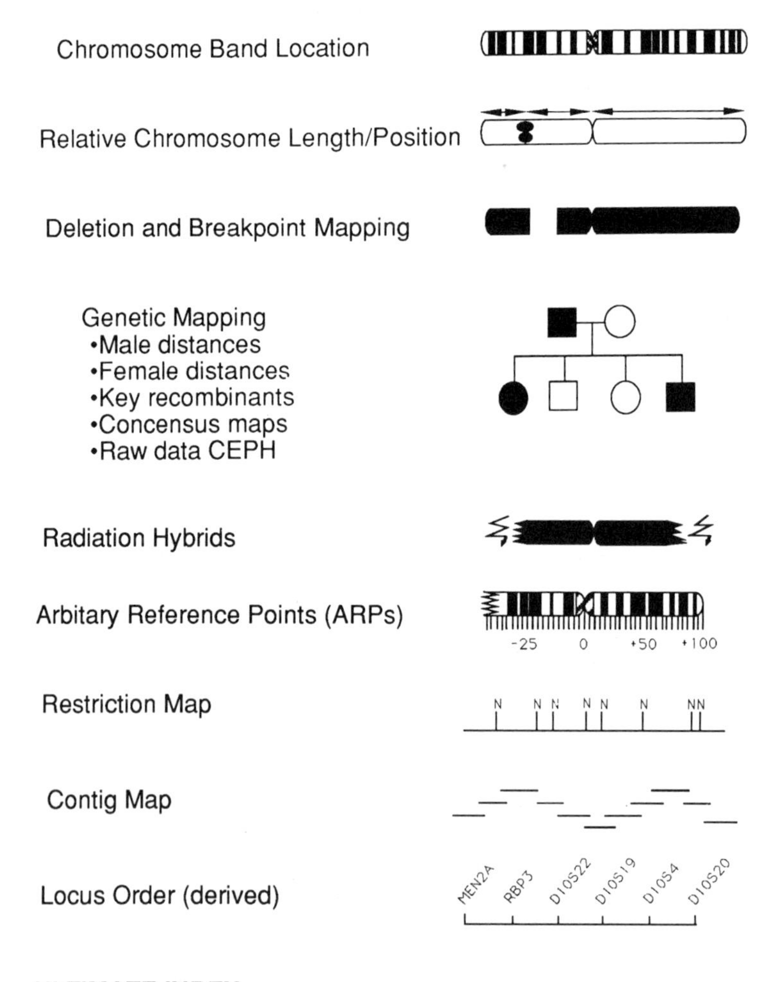

FIG. 7. Mapping methods and indices to be incorporated into GDB.

usefulness as a stand-alone system and will have to be supplemented by other methods of mapping information. Figure 7 provides an overview of the various types of mapping information that will be incorporated into GDB.

One of the major problems will be one of representing the information for all the various types of mapping strategy in a standard fashion. Since all maps provide order information we intend to represent all types of map information within GDB primarily as order information. Distance and probability information will be appended as secondary attributes to the order information. A proposal for doing this is given in Fig. 8. The convention is quite powerful because it permits redefinition of an ordered partial map as a locus within a higher order map. Such flexibility will permit integration of various types of map information and also assist in map construction.

Development Time Scale

The first version of GDB was planned for release in September 1990. Figures 9 and 10 give an overview of the content of GDB at the time of release 1 and the intended content at the time of release 2, respectively. Plans for creating the data structures needed for the new information for release 2 are already well advanced.

Geographical Distribution and Use of GDB

The database will be used for capturing the mapping information collected at the HGM workshops held in Oxford in September, 1990 (HGM10.5) and

LOCUS ORDER INFORMATION

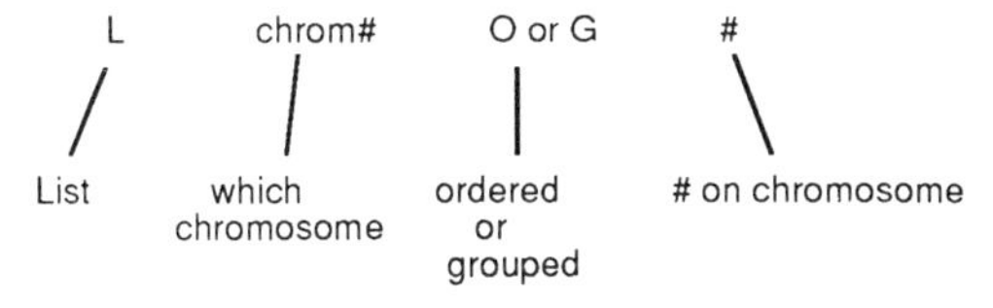

ATTRIBUTES

- •list loci
- •which chromosome
- •ordered or grouped
- •# on chromosome
- •parent of list
- •# loci in list
- •symbol each locus
- •position each locus in list
- •polarity of list

- •distances between adjacent loci
- •distances between end loci
- •probability on local order
- •probability on total order
- •probability on local distance
- •probability on total distance
- **•type of map: meiotic, physical etc.,**
- •reference

FIG. 8. Locus order information to be included in GDB.

GDB RELEASE 1 SEPT. 90

CONTENT

- Chromosome band location
- Genes
- Anonymous DNA segments
- Fragile sites
- DNA polymorphisms
- DNA probes, oligos, PCR primers etc.,
- OMIM information, genetic disease description
- References

TOOLS AVAILABLE

- Editing interfaces
  DNA, nomenclature, chromosome committees
- General user interface
- Communication -remote access
  Internet, telenet, Remote site distribution (ICRF London)
- Remote site maintenance
  JHU-ICRF synchronisation

FIG. 9. New information release of the Genome Data Base, September, 1990, JHU, Johns Hopkins University; OMIM, Online Mendelian Inheritance in Man.

GDB RELEASE 2 AUGUST 91

ADDITIONAL CONTENT

- Linkage, genetic map information
- Order information
  (various types physical map information)
- Expanded locus definition
  (breakpoints, meiotic X-overs, restriction sites, partial maps etc.
- Comparative mapping information on mouse
  (collaboration with Jackson Lab)

ADDITIONAL TOOLS

- Interface for direct submission of HGM11 abstracts
- Direct electronic entry and extraction
  (database to database)
- Enhanced general user interface for expanded content
- Enhanced editorial interfaces
- Communications- additional remote sites
  (DKFZ- Heidelberg)
- Flat file distribution
- Tools to compare human gene maps and diseases with those of mouse

FIG. 10. New information release of the Genome Data Base, planned for August, 1991.

in London in August 1991 (HGM11). Data perusal, entry and editing facilities will be available in both Europe and North America following the HGM10.5 workshop by access through Telenet or Internet communication networks. However, by making use of the Sybase facility for dividing the database into a "front end" or user terminal and into a "back end" or database proper, it will in principle be possible to maintain a single database fed by front ends distributed around the world.

CONCLUSIONS

This brief overview gives an impression of the ways in which GDB has been designed to meet the needs of the gene mapping community. Hopefully the data structures will provide the flexibility for new types of mapping information to be incorporated readily as the genome program develops. However, it is also clear that new database technologies will have to be implemented as the mass and complexity of data analysis increases. Conservative estimates would limit the use of existing relational technology to the next 5 years.

ACKNOWLEDGMENTS

The authors wish to thank the Howard Hughes Medical Institute for their generous financial support of the GDB project.

REFERENCES

1. Health and Environmental Research Advisory Committee (1987) *The Human Genome Initiative*.
2. National Research Council, Committee on Mapping and Sequencing the Human Genome. *Mapping and Sequencing the Human Genome*. National Academy Press, Washington, DC, 1988.
3. U.S. Congress, Office of Technology Assessment. *Mapping Our Genes, The Genome Projects: How Big, How Fast*, 1988.
4. 1973 New Haven Conference First International Workshop on Human Gene Mapping. *Cytogenet. Cell Genet.* 1974;13:1–216.
5. Mador ML, Cavanaugh ML, Chan HS, et al. The HGM10 information management system. *HGM10 Cytogenet. Cell Genet.* 1989; 51:3–7.
6. McKusick VA. *Mendelian Inheritance in Man,* 9th edition. The Johns Hopkins University Press, Baltimore and London, 1990.
7. U.S. Department of Health and Human Services and U.S. Department of Energy. *Understanding Our Genetic Heritance, The Human Genome Project: The First Five Years FY 1991–1995*, 1990.
8. 1971 Paris Conference. Standardization in human cytogenetics. *Cytogenet Cell Genet* 1972;11:313–362.

Etiology of Human Disease at the DNA Level,
edited by Jan Lindsten and Ulf Pettersson.

4

Role of Homeotic and Other Development-Regulating Genes in the Molecular Genetics of Human Malformations

*Robert Williamson and **Debra J. Wolgemuth

*Department of Biochemistry and Molecular Genetics, St. Mary's Hospital Medical School, Imperial College London, Paddington, London W2 1PG, England, U.K.; **Department of Genetics and Development, College of Physicians and Surgeons of Columbia University, New York, New York 10032*

Dysmorphic syndromes have not yet been studied extensively using molecular genetic techniques. Most congenital malformations are described as multifactorial, with environmental and genetic components and a moderate recurrence risk; many are familial, although few are due to a single locus defect. There are, however, rare large families in which a syndrome that is usually multifactorial or sporadic segregates in a straightforward Mendelian manner. Such families may be studied using reverse genetics techniques to identify the gene, product and pathway involved, and this in turn can be used to illuminate the more common cases which are not due to inherited single locus events. Together with the use of transgenic animals to model these events, molecular genetics may permit the interpretation of not only the genetic determinants of developmental pathways, but also the way in which the environment interacts with these to cause common phenocopies of rare Mendelian malformations.

There are approximately 600 inherited diseases for which the affected protein is known; for most of these, the gene coding for the protein has now been cloned and the mutation(s) causing the defect has been defined. For the even more numerous Mendelian conditions in which the protein is not known, such as cystic fibrosis, Duchenne muscular dystrophy and neurofibromatosis, reverse genetics is being used to identify new genes which can then be analysed for the causative mutational events.

In this paper, we will discuss another group of diseases, morphological defects which occur in a recognisable pattern: dysmorphic syndromes or malformations. They are often disorders of form affecting one or more organs or tissues, resulting from abnormal embryonic development, and usually involve a group of associated pattern defects. Dysmorphic syndromes on the one hand merge into the class are caused by internal or external factors, such as an interruption of blood supply to a developing tissue or to physical pressure (disruptions and deformations), and on the other into defined single gene defects.

Molecular genetics is essentially reductionist, and it may be useful to look at dysmorphic syndromes from this perspective, to attempt to dissect them into component simple parts, and to use these data to model causation. Implicit in the concept of a syndrome is a constellation of single signs, and in most cases only a proportion are present in a single case. This is true for DiGeorge, Treacher-Collins, and Down's syndromes, and many others, and has led to the concept of a contiguous gene defect, in which a deletion, duplication or dysfunction of a chromosome *region* including several genes leads to the syndrome, with each affected gene being involved in causing one or another sign (Emanuel, 1988). The precise extent of the genetic event (a deletion, for instance) determines which signs present in a given patient. This concept has been best elaborated for the Wilms's tumour/aniridia/genito-urinary defects/mental retardation (WAGR) syndrome on chromosome 11; it is perhaps most interesting to apply this approach to Down's syndrome, which includes many clinical features each of which may also occur separately, such as increased incidence of congenital heart defect, Hirschsprung's disease and leukaemia.

It is also worth noting that conditions which are clearly dysmorphic syndromes, such as the characteristic facies and bone features of beta-thalassaemia major, are not usually described as such when their single gene aetiology is understood. Finally, it is also possible to see an overlapping relationship between the rare multifactorial dysmorphic syndrome and environmental conditions, as the correlation between the presentation of phocomelia (in which some of the long bones of the arm are missing) and the thalidomide syndrome shows, presumably because interference in many different ways, perhaps at different points, with a common pathway can cause similar end results. As our knowledge of the underlying genetic and developmental pathways of morphogenesis increases, it can be anticipated that the generic term "dysmorphic syndrome" will be used less, and will be replaced by accurate descriptions of single and multiple genetic and environmental effects.

RARE AND COMMON MALFORMATIONS

Dysmorphologies are important as a group; approximately 1 child in 50 is affected by a major condition in this category (with serious cosmetic defect and/or requiring clinical treatment), although estimates vary widely because of a lack of consensus as to what should be included (Emery and Rimoin,

1983; Connor and Ferguson-Smith, 1987). For instance, in East Africa, postaxial polydactyly is very common, affecting 3% of neonates, but it can be minor, with only a flesh tab, and is usually treated easily at birth by removing the extra digit. While this is clearly a malformation, it also is clearly not a major clinical problem. The lower estimates of incidence often include only those conditions which require continuing clinical care.

Many dysmorphic syndromes are described using the misleading term "multifactorial," because genetic and environmental aetiologies may be involved, separately or in combination. However, this obscures the fact that, while a condition may be multifactorial, each case separately may not be. For instance, X-linked cleft palate (Moore et al., 1987) can be caused by a single gene defect, in which case it is virtually independent of environmental factors. It can be purely environmental, as when caused by excessive alcohol intake by the mother at a critical stage of pregnancy, in which case the genetic background is probably of little consequence. It can be caused by a combination of genetic and environmental factors, which is probably true for the majority of cases. Finally, some cases may be caused by random (stochastic) processes during cell growth and migration (Kurnit et al., 1987). While it may be correct to describe a clinical condition as complex (or multifactorial), each affected individual will have his or her own aetiology, which may be completely genetic or completely environmental, is likely to be more simple, and will in any case be completely defined. Complexity does not imply insolubility.

For any condition that is defined by the inheritance of a single gene in a mendelian fashion (even in one family), the classical rules of heredity apply, and analysis is simple. Multifactorial disorders have more complex inheritance patterns. The recurrence risk can be assessed in a number of ways, but for humans (in whom controlled breeding experiments are both unethical and impractical) the determination of "heritability" based upon the proportion of close relatives sharing a particular clinical condition is fraught with problems. This has been discussed extensively for complex disorders such as schizophrenia, for which the extent of genetic predisposition is still unclear, although the familial nature of the disease is not in doubt (Kendler, 1988). The recent article on "The Recessive Gene for Attending Medical School" (showing that students of medicine often come from medical families) was only half a joke (McGuffin and Huckle, 1990). The power of molecular genetics lies in its ability to simplify this analysis by addressing the question of inheritance in a completely reductionist fashion—does a phenotype in a family or set of families cosegregate with a DNA allele or not? This in turn allows molecular approaches to provide cogent answers which cannot be obtained by looking at phenotype/phenotype or phenotype/environment relationships.

Of course, if there are several genes involved in causation of a phenotype, these will have different "weights" in a population in which the condition is polygenic or multifactorial than in family studies, in which one, or at most a small number, of genes (in comparison with the number in the population) are involved. However, even if many genes can determine a phenotype, often it is found that a small subset determines much of the variation in the population; this has been shown both in principle (Ott, 1990), and in practice for hypercholesterolaemia (Sing and Boerwinkle, 1987). By looking at individual families, it is possible to identify these genes one by one; each is (at least) important in the family in which the mutation is found, and may well be one of the genes which is important in the population. Even if this is not the case, it may still illuminate a pathway (as for LDL receptor defects, mutations which are not major determinants of hypercholesterolaemia in the population but which allowed the dissection of the role of the metabolism of LDL.)

THE RARE PARADIGMAL FAMILY

It is possible to consider rare disorders which segregate in a mendelian fashion as paradigms of more common conditions which are multifactorial. For instance, cleft lip and palate are very common; the recurrence risk is small (perhaps 2–3%), and the condition is more common in girls than in boys. Investigators in Iceland identified a very large family, with over 100 members, in which secondary cleft palate segregates as an X-linked disorder; boys are affected, and female carriers of the gene are tongue-tied. Penetrance is greater than 80%, the presentation is fairly uniform within gender, and the data are compatible with a single gene mutation. We have studied the cosegregation of gene probes and the phenotype in this family, and have shown that the mutation is located on the X chromosome at Xq21.3-Xq22 (Moore et al., 1987; Ivens et al., 1988).

This family is clearly atypical—cleft palate is rarely monogenic, it usually affects girls, and it is heterogeneous in presentation. How can such a family be described as paradigmal? The existence of this large family provides an entry into study of the developmental genetics of the system of palatal fusion through the elucidation of one key step where clefting is known to result if a mutation occurs, even though this particular step may not be one that is commonly implicated in the disease process.

CHROMOSOME DELETIONS GIVE CLUES TO GENE LOCATION

In some dysmorphic syndromes, chromosomal deletions of variable extent occur, as in the case of DiGeorge syndrome, a rare disorder characterised by absent thymus and parathyroids, heart defects and craniofacial dysmorphologies, most of which are associated with a putative failure of cell migration from the neural crest to the first branchial arch. The syndrome is associated in approximately 10% of those affected with a visible and variable deletion of part of chromosome 22, including two patients with an interstitial deletion at 22q11 (Greenberg et al., 1988). Therefore, it seems likely that there is a major locus for a gene or genes involved in causing DiGeorge syndrome at 22q11.

Twenty-seven DNA markers have been mapped to the region of 22q11, and have identified several sequences which are located within these visible deletions in the two patients (Carey et al., 1990). The same probes have been analysed in cases of DiGeorge syndrome without visible cytogenetic abnormalities, and it was shown that there are submicroscopic deletions which only affect a small proportion of the probes. A smallest region of overlap is being established, within which the gene(s) that are mutated or deleted so as to cause DiGeorge syndrome will be identified by pulse field mapping and HTF island identification.

It has also been proposed that DiGeorge syndrome is an example of a contiguous gene syndrome (Emanuel, 1988, although it should be noted that the syndrome can be caused by a translocation, which presumably affects only a single gene, as well as by a deletion). To test the hypothesis, the chromosome 22 probes mapping to the locus can in turn be used as candidate markers in families which do not have DiGeorge syndrome, but in which isolated thymic dysfunction, or congenital heart abnormalities, are found, in an attempt to determine whether these single features are due to the mutation of one gene only from a linear array which accounts for the syndrome as a whole.

SINGLE GENE MUTATIONS AND CHROMOSOMAL TRANSLOCATIONS

A similar approach was adopted in the study of Greig cephalopolysyndactyly, an autosomal dominant condition with polysyndactyly of the hands and feet and a characteristic facial appearance (Brueton et al., 1988). There were two families in which the dysmorphology segregates with an apparently balanced translocation at 7p13. The gene coding for the epidermal growth factor receptor is located at 7p13, and restriction fragment length polymorphisms (RFLPs) for this gene were used to study cosegregation with Greig syndrome in seven families with no chromosomal abnormality. Cosegregation was found with no recombination. In this case, there is no deletion; it is likely that the translocation has disrupted a single gene, causing the syndrome, and that the noncytogenic cases involve other mutations, perhaps single base changes, in the same gene.

The pattern of malformation seen in Greig syndrome in humans is very similar to that in a mouse mutant, extra toes. The gene that causes extra toes has been mapped to mouse chromosome 13, in the middle of a region which shows homology to chromosome 7p13 in humans (Winter and Huson, 1988). Since the experimental approaches to mouse genetics allow options for determining gene position which are not possible in human studies, this homology should permit more rapid isolation of the mutated gene. The full potential of this approach will become apparent when it is possible to create transgenic mice which have different features and severities of Greig syndrome by introducing different mutations into the gene. This in turn should lead to a clear understanding of the way in which this gene, and its corresponding protein, function in normal development.

HOMEOTIC GENES AND DYSMORPHOLOGY

Recent breakthroughs in the molecular genetic analysis of pattern formation during development in less complex animals such as *Drosophila* have provided another approach to identifying the genetic basis of congenital abnormalities in higher organisms, including humans. Dramatic transformations of body segment identity, known as homeotic mutations, were observed for many years at the classical genetics level in *Drosophila* (Lewis, 1978). Molecular cloning of the genes involved, notably of the *Antennapedia* gene (McGinnis et al., 1984; Scott and Weiner, 1984) revealed the presence of a domain of sequence similarity, known as the homeobox, among several different *Drosophila* genes involved in segmentation, segment identity, and positional determination in the egg. The homeobox is a 183-bp domain which has been identified by sequence similarity in the genome of organisms from *C. elegans* to man. Several features of the structural properties of this motif strongly suggest that it encodes DNA-binding proteins that function as transcription factors (Gehring, 1987). This raises the possibility that homeobox-containing genes may function as major regulatory genes during development, dictating the pattern of expression of other genes responsible for the structural manifestations of segmentation and pattern formation.

The high level of conservation of this domain at the nucleotide level has made it possible to use the homeo domain to "fish out" homologous genes from more complex genomes, and has provided the opportunity to study the expression of the genes thus identified during mammalian development and cellular differentiation. The pattern of expression of homeobox-containing genes during mammalian development has been most extensively studied in mice, although comparative analyses with other vertebrate systems has received increasing attention, thus permitting evolutionary aspects of function to be considered. At least 30 murine homeobox genes have been identified, most of which are members of the Antennapedia class of homeobox proteins, known as Hox genes (HOX in humans). These genes are organised in four clusters located on mouse chromosomes 6 (Hox-1), 11 (Hox-2), 15 (Hox-3) and 2 (Hox-4, formerly referred to as Hox-5). The clustering of these genes and their linear order within the cluster is also conserved from *Drosophila* to man (Akam, 1989).

The murine homeobox-containing genes exhibit highly ordered and temporally and spatially restricted patterns of expression during embryonic development (reviewed in Holland and Hogan, 1988), further supporting the notion that they play roles in regulating embryogenesis. For example, the mouse gene Hox-1.4, which has been extensively studied by our laboratory, is expressed during midgestation development, most abundantly in the spinal cord, as far anterior as the posterior myelencephalon. Other notable sites of embryonic expression include the mesenchymal layer of the developing gut, and like several other Hox genes, the somites which will give rise to the vertebral structures. Interestingly, none of the Hox-1 genes identified to date has been shown to be expressed more anterior than the region of the hindbrain. Little is known about adult tissue-specific expression of the Hox genes, except for Hox-1.4, which has been shown to exhibit a cell lineage specificity of expression in the meiotic and postmeiotic male germ line (Wolgemuth et al., 1986, 1987).

Several of the Hox genes have been shown to map in the vicinity of previously identified mutations of mouse development. For example, the Hox-1 cluster maps close to *hypodactyly* (chromosome 6); *hammer-toe* and the Hox-7.1 gene are close to one another (chromosome 5); and Hox-4 (formerly Hox-5) maps near to *rachiterata* on chromosome 2. However, none has yet been shown to be allelic, in contrast to the serendipitous apparent identity shown by the Pax-1 gene and *undulated* (Balling et al., 1988). Careful characterisation of the structural organisation as well as the pattern of expression of these mouse genes has nonetheless proved to be extremely important, providing the basis from which to design experiments to alter their structure and/or pattern of expression utilising the transgenic mouse system.

MANIPULATION OF HOMEOBOX GENES IN TRANSGENIC MICE AND THEIR USEFULNESS IN STUDIES OF HUMAN DYSMORPHOLOGIES

The first correlation of altered expression of a homeobox-containing gene and aberrant development in mammals was observed in experiments in which the Hox-1.4 gene was overexpressed in transgenic mice (Wolgemuth et al., 1989). Multiple copies of the Hox-1.4 gene were stably integrated into the genomes of several lines of transgenic mice. In some of these lines, mice were born that developed congenital megacolon, a condition most commonly manifested by the distal colon being defective in innervation by the enteric nervous system. In humans, the best characterised form of congenital megacolon is known as Hirschsprung's disease.

Although a definitive cause-effect mechanism has yet to be elucidated, there was a striking correlation of elevated levels of expression of the transgenic Hox-1.4 genes in the developing gut of the midgestation embryos and the development of the megacolon phenotype. It remains to be determined if the transgenes are also expressed in the gut in a temporally inappropriate way (since the gut is a site of normal, albeit low level, expression of Hox-1.4) or if elevated levels of Hox-1.4 might affect the expression of other Hox genes also known to be expressed in the developing gut. The issue of auto- and crossregulation of development-regulating genes is of growing experimental focus and consideration. It will also be of interest to correlate the temporal and spatial pattern and levels of altered Hox-1.4 expression with the corresponding phenotype of abnormality in the distal colon, which results in megacolon among the various transgenic lines.

Whatever the exact correlation of altered expression of Hox-1.4 and development of the congenital abnormality of megacolon turns out to be, the observation of the phenotype in six independent lines of transgenic mice argues for an association of altered Hox-1.4 expression and the congenital megacolon phenotype. The phenotype is variable in its severity, although consistent within a given line. This is curiously analogous to the variable severity of congenital megacolon among affected humans.

Recently, the expression of another mouse Hox gene, Hox-1.1, has been altered in transgenic mice, this time under the control of a relatively weak but putatively ubiquitously expressed promoter for the chicken β-actin gene (Balling et al., 1989). Initial analysis of the resulting phenotypes from what apparently constituted ectopic expression of Hox-1.1 during midgestation development (i.e., in head and brain regions of the embryo) suggested that craniofacial abnormalities were induced. Specifically, mice were born with multiple craniofacial abnormalities such as cleft palate, open eyes at birth, and nonfused pinnae. The similarity of these abnormalities with those observed after retinoic acid exposure during gestation is intriguing, and again draws our attention to the possible multiple aetiology of a given dysmorphic syndrome.

Further, more detailed analysis of the transgenic mice ectopically expressing the Hox-1.1 gene has suggested possible bona fide homeotic transformations within the vertebral column (Kessel et al., 1990). Although the interpretation of the nature of the pathological defects and their relationships to more primitive structures evolutionarily is not totally clear, there is the suggestion that altered expression of a homeobox gene during mouse embryogenesis can and will result reproducibly in a particular congenital abnormality. In the case of ectopic Hox-1.1 expression, an argument is made for transformation of more posterior structures into an additional anterior structure, the proatlas. This phenotype, which was observed in several but by no means all of the putatively ectopically expressing embryos, was thought to result from a dominant, ectopic expression of Hox-1.1 during embryonic days 7.5 or 8.0.

Skeletal malformations have attracted the attention of mouse and human geneticists for many years, given the relative ease of identifying the abnormalities in mice and the consequences of the abnormalities in humans. Thus the association of such a phenotype with the deliberate altered expression of a specific gene, particularly one containing a homeobox, has focussed additional consideration of correlation with existing mouse or human mutations. Interestingly, and possibly surprisingly given the apparent correlation of Pax-1 and *undulated,* none have been demonstrated between any skeletal abnormality and Hox-1.1 to date. Nonetheless, the power of being able to manipulate the expression of specific homeobox genes in the experimentally amenable murine system argues strongly for its thorough exploration, hand in hand with concomitant molecular genetic analysis of human dysmorphologies for delineating the function of these mammalian developmental regulating genes in normal and abnormal morphogenesis.

A HOMEOTIC GENE MAPS CLOSE TO THE WOLF-HIRSCHHORN SYNDROME LOCUS

The human gene for HOX7 maps to chromosome 4p16.1, and is deleted in some (but not all) cases of Wolf-Hirschhorn disease (Ivens et al., 1990). The mouse homolog Hox7 is expressed in the neural tube, the cephalic neural crest, and the developing heart and limb buds. Wolf-Hirschhorn syndrome is characterised by mid-line fusion defects (failure of neural tube closure), mental retardation, and heart and limb anomalies. It may be chance that the pattern of expression in mouse mirrors so closely the sites of the defects in humans, but clearly Wolf-Hirschhorn syndrome has the potential to be the first human disease caused by a mutation in a homeobox-containing gene (Slack, 1985).

TREACHER-COLLINS SYNDROME—ANOTHER NEURAL CREST DISORDER?

Another disorder which may involve the migration of neural crest cells during embryogenesis is Treacher-Collins syndrome. This is an autosomal dominant defect of craniofacial development which is often associated with bilateral conductive deafness, hypoplasia of the facial bones, antimongoloid slant and cleft palate. The incidence is approximately 1 in 50,000; about half the cases arise as new mutations. All of the affected tissues are derived from the first and second branchial arches, implicating neural crest ontogeny, but the underlying defect and chromosomal location are unknown.

In the case of Treacher-Collins syndrome, an attempt to use a chromosomal translocation to define candidate regions for the mutation was less successful. Dr. D. Callen, of Adelaide, found a family in which the disease appeared to be cosegregating with a balanced translocation between chromosomes 6 and 16. Because the translocation was balanced, it was not possible to predict whether chromosome 6 or 16 was more likely to be involved. Therefore, probes from the relevant regions of both chromosomes 6 and 16 were followed through several large Treacher-Collins families (once again, without karyotypic abnormalities), to look for cosegregation (Dixon and Williamson, in press). In this case, the chromosomal clue was wrong; neither region showed any probe which segregated with the syndrome. In principle, this could be due to genetic heterogeneity—perhaps the family with the translocation is unique in being affected at a locus which could, rarely, cause the disease, although it is caused more often by a mutation at another site. However, our exclusion was "confirmed" biologically when a further affected child was born to the family, in this case without the translocation.

IMPRINTING AND PRADER-WILLI AND ANGELMAN'S SYNDROMES

It is axiomatic in classical Mendelian genetics that DNA in an individual is expressed irrespective of its parental origin; most genes function (or dysfunction) equally whether they come from father or mother. However, this is not always the case (reviewed by Hall, 1990). Consider the dysmorphic Prader-Willi syndrome (short stature, small hands and feet, obesity, mental retardation, and a rather characteristic facial appearance) caused by the deletion of sequences from the distal portion of chromosome 15q. This syndrome only occurs if the deletion is of the paternal chromosome 15. If the same region, as determined both cytogenetically and by DNA analysis, is deleted from the maternal chromosome 15, a different dysmorphic syndrome results—Angelman's syndrome, with more severe mental retardation, repetitive symmetrical ataxic movements, happy disposition and a specific (but different) facial appearance.

This is a clear illustration that gene expression depends not only upon simple structural considerations such as gene sequence and number; there is a hierarchy of controls which also involve a genetic memory of chromosomal origin from the previous generation. This, in turn, can lead to quite specific dysmorphic syndromes, depending upon whether a gene is paternally or maternally derived, even though the same gene is probably involved. Imprinting may be important in Down's syndrome, of which more below, in which most cases have an extra copy of a maternal chromosome 21; it is not clear whether there are clinical differences which can be correlated with its parental origin, nor, indeed, whether imprinting still can be demonstrated in Down's and other aneuploidies.

DOWN'S SYNDROME

Down's syndrome is perhaps the most fascinating of the dysmorphic syndromes in many ways. It is common (1 in 600 births, more common than cystic fibrosis and muscular dystrophy combined); the incidence varies little between different countries, and (unlike midline defects) is not related to socioeconomic factors. Trisomies of all other chromosomes (except chromosome 1) are also common early in embryonic life, but all, apart from those affecting chromosomes 13, 18 and 21, die in utero, and trisomies 13 (Patau) and 18 (Edwards) rarely survive for more than a few weeks after birth. Down's syndrome is unique among the common severe genetic disorders in that there are no abnormal genes present, only extra copies of a small set of normal genes.

Down's syndrome is remarkably homogeneous; those with the condition are much more similar to each other than to their parents or their national group, and it is extremely easy to identify a person with trisomy 21 by their appearance. The Down's syndrome facies (tongue, eyes, head, and neck shape) is pathognomic; the movements of a person afflicted with Down's syndrome (who appears "bottom-heavy" and has a shuffling, penguin-like walk) are also characteristic, as evidenced by the fact that it is usually possible to identify someone with Down's syndrome from behind.

Clinically, Down's syndrome is associated with a high incidence of cardiac malformations (about half of Down's syndrome cases are affected, in particular with a specific type of septal defect), Hirschsprung's disease, biliary atresia and other developmental abnormalities of the gastrointestinal tract, neuropathological changes similar to those occurring in early onset Alzheimer's disease, and an increased risk of leukaemia, as well as the characteristic facies, hypotonia and fingerprints. The mental retardation found in Down's syndrome is also characteristic—each genetic or environmental cause of mental retardation gives a different and often distinct clinical and personal phenotype, although this is seldom stated.

An apparent exception to the homogeneous nature of Down's syndrome occurs when cases are due to partial trisomies arising from translocations, which account for 2–5% of the total number of cases (McCormick et al., 1989). Most of these are Robertsonian translocations, which tend to be less severe than full trisomies; approximately 30 cases have been reported which have translocations involving only a portion of chromosome 21q. These may be very mild, but to date there have been no studies as to the association of specific features of Down's syndrome with trisomy of a particular portion of the "Down's obligate region." This approach would treat Down's syndrome as a contiguous gene defect, analogous to WAGR syndrome on chromosome 11. Another interesting approach to Down's syndrome would be the study of the parental origin of the extra chromosome; while the majority of cases of Down's syndrome involve an extra copy of the maternal chromosome 21, approximately 20% have an extra paternal chromosome. The maternal or paternal chromosomes may occur as a result of nondisjunction at either first or second meiotic division. Each class of Down's syndrome (those that are paternal or maternal, those with identical or nonidentical chromosomes 21) should be studied separately now that the tools are available.

CONCLUSIONS

The study of single gene defects using reverse molecular genetics has been one of the most exciting aspects of clinical research during the 1980s. It seems certain that the coming decade will provide equally important insights, but with the focus now on developmental genetics and neurogenetics, and that these studies will involve the elucidation of the failures of function, genetic and environmental, that cause dysmorphic syndromes.

ACKNOWLEDGMENTS

This work was supported by grants from the Medical Research Council and the Wellcome Trust, and many of the ideas were developed while R.W. was a sabbatical Ludwig Schaefer Visiting Professor at Columbia University, New York. R.W. thanks his colleagues Louise Brueton, Alisoun Carey, Mike Dixon, Ellie Dow, Jane Hewitt, Al Ivens, Gudrun Moore, and Pete Scambler for helpful discussions, and Anna Kessling and Liz Thompson for critical review of the manuscript, the responsibility for which, however, remains with the authors alone. We have given minimal references, often from our own work; it goes without saying that others have made as great or greater contributions, but this is not a definitive review so much as a speculative note.

REFERENCES

Akam M. Hox and HOM: homologous gene clusters in insects and vertebrates. *Cell* 1989;57:347–349.

Balling R, Deutsch U, Gruss P. *undulated,* a mutation affecting the development of the mouse skeleton, has a point mutation in the paired box of *Pax-1*. *Cell* 1988;55:531–535.

Balling R, Mutter G, Gruss P, Kessel M. Craniofacial abnormalities induced by ectopic expression of the homeobox gene Hox-1.1 in transgenic mice. *Cell* 1989;58:337–347.

Brueton L, Huson SM, Winter RM, Williamson R. Chromosomal localisation of a developmental gene in man: direct DNA analysis demonstrates that Greig cephalopolysyndactyly maps to 7p13. *Am J Med Genet* 1988;31:799–804.

Carey AH, Roach S, Williamson R, et al. Localisation of 27 DNA markers to the region of human chromosome 22q11-pter deleted in patients with the DiGeorge syndrome and duplicated in the der22 syndrome. *Genomics* 1990;7:299–306.

Connor JM, Ferguson-Smith MA. *Essential Medical Genetics*. 2nd ed. Oxford: Blackwell, 1987, p 121.

Dixon M, Williamson R. *Am J Hum Genetics* in press.

Emanuel BS. Molecular cytogenetics: toward dissection of the contiguous gene syndromes. *Am J Hum Genet* 1988;43:575–578.

Emery AEH, Rimoin DL. *Principles and Practice of Medical Genetics*. 1st ed. Edinburgh: Churchill Livingstone, 1983, pp 2–3.

Gehring WJ. Homeo boxes in the study of development. *Science* 1987; 236:1245–1252.

Greenberg F, Elder FFB, Haffner P, Northrup H, Ledbetter D. Cytogenetic findings in a prospective series of patients with DiGeorge anomaly. *Hum Genet* 1988;43:605–611.

Hall JG. Genomic imprinting: review and relevance to human diseases. *Am J Hum Genet* 1990;46:857–873.

Holland PWH, Hogan BLM. Expression of homeobox genes during mouse development: a review. *Genes Dev* 1988;2:773–782.

Ivens A, Moore G, Williamson R, et al. X-linked cleft palate: the gene is localised between polymorphic DNA markers DXYS12 and DXY17. *Hum Genet* 1988;78:356–358.

Ivens A, Flavin N, Williamson R, et al. The human homeobox gene HOX7 maps to chromosome 4p16.1 and may be implicated in Wolf-Hirschhorn syndrome. *Hum Genet* 1990;84:473–476.

Kendler KS. Familial aggregation of schizophrenia and schizophrenia spectrum disorders: evaluation of conflicting results. *Arch Gen Psychiatry* 1988;45:377–383.

Kessel M, Balling R, Gruss P. Variations of cervical vertebrae after expression of a Hox-1.1 transgene in mice. *Cell* 1990;61:301–308.

Kurnit DM, Layton WM, Matthysse S. Genetics, chance and morphogenesis. *Am J Hum Genet* 1987;41:979–995.

Lewis EB. A gene complex controlling segmentation in *Drosophila*. *Nature* 1978;276:565–570.

McCormick MK, Schinzel A, Petersen MB, et al. Molecular genetic approach to the characterisation of the "Down syndrome region" of chromosome 21. *Genomics* 1989;5:325–331.

McGinnis W, Levine, ML, Hafen, E, et al. A Conserved DNA Sequence in Homeotic Genes of the *Drosophila Antennapedia* and *Bithorax* Complexes. *Nature* 308:428–433.

McGuffin P, Huckle P. Simulation of Mendelism revisited: the recessive gene for attending medical school. *Am J Hum Genet* 1990;46:994–999.

Moore GE, Ivens AC, Chambers J, et al. Linkage of an X chromosome cleft palate gene. *Nature* 1987;326:91–93.

Ott J. Cutting a Gordian knot in the linkage analysis of complex human traits. *Am J Hum Genet* 1990;46:219–221.

Scott MP, Weiner AJ. Structural relationships among genes that control development: sequence homology between the *Antennapedia, Ultrabithorax* and *fushi tarazu* loci of *Drosophila*. *Proc Natl Acad Sci USA* 1984;81:4115–4119.

Sing CF, Boerwinkle E. Genetic architecture of inter-individual variability in apolipoprotein, lipoprotein and lipid phenotypes. In Bock G, Collins GM (eds): *Molecular Approaches to Human Polygenic Disease*. CIBA Foundation Symposium 130. Chichester: John Wiley & Sons, 1987, pp 99–127.

Slack JM. Homeotic transformations in man. *J Theor Biol* 1985;114:463–490.

Winter R, Huson SM. Greig cephalopolysyndactyly syndrome: a possible mouse homologue (Xt-extra toes). *Am J Med Genet* 1988; 31:793–798.

Wolgemuth DJ, Engelmyer E, Duggal RN, et al. Isolation of a mouse cDNA coding for a developmentally regulated testis specific transcript containing homeobox homology. *EMBO J* 1986;5:1229–1235.

Wolgemuth DJ, Viviano C, Gizang-Ginsberg E, et al. Differential expression of the homeobox-containing gene Hox-1.4 during mouse male germ cell differentiation and embryonic development. *Proc Natl Acad Sci USA* 1987;84:5813–5817.

Wolgemuth DJ, Behringer RR, Mostoller MP, Brinster RL, Palmiter RD. Transgenic mice overexpressing the mouse homeobox-containing gene Hox-1.4 exhibit abnormal gut development. *Nature* 1989;337:464–467.

Molecular Characterization of Genetic Defects

Etiology of Human Disease at the DNA Level,
edited by Jan Lindsten and Ulf Pettersson.

5

Searching for Dystrophin Gene Deletions in Patients with Atypical Presentations

*,**,†Louis M. Kunkel, *,**Judith R. Snyder,
*,**,†Alan H. Beggs, *,†Frederick M. Boyce, and
*,** Chris A. Feener

**Division of Genetics, The Children's Hospital, Boston, Massachusetts 02115;
**Howard Hughes Medical Institute at the Children's Hospital,
Boston, Massachusetts 02115; †Department of Pediatrics, Harvard Medical
School, Boston, Massachusetts 02115*

Duchenne's and Becker's muscular dystrophies (DMD, BMD) are severe muscle wasting disorders that are inherited in an X-linked recessive manner (1,2). The gene disrupted to yield these diseases has recently been identified (3–5) and the encoded protein product, dystrophin, described (6). Dystrophin is a large (427-kD) cytoskeletal protein with homology to the spectrins, alpha-actinin, and a recently described relative called B3 or DRP (7–9). The dystrophin gene is extremely large (~2,400 kb) (10), and is located at Xp21 in the human genome. In part due to its large size, the majority (~65%) of mutations causing DMD and BMD are intragenic deletions with another 5% being duplications (5,11). Recently, the molecular basis for the differences between DMD and BMD has been elucidated (12,13). The more severe Duchenne form is caused by deletions that result in frame-shifts of protein translation because a nonintegral number of triplet codons are lost. Thus, only severely truncated, and presumably nonfunctional, dystrophin is produced in muscle tissue of these patients. Conversely, patients with milder BMD generally have deletions that remove an integral number of codons so the reading frame is maintained allowing production of internally deleted proteins (12,13). Western blot analysis of dystrophin in muscle biopsies confirms this as muscle from DMD patients contains low or undetectable levels while BMD patients generally have dystrophin of altered size and/or reduced abundance (14).

The primary sites of dystrophin expression are skeletal, smooth, and cardiac muscle; however, significant levels of protein are also found in the brain (15,16). Although sensitive PCR assays have detected rare mRNA transcripts in other cell types (17), no other tissue is thought to express physiologically significant levels of this protein. Analysis of cDNA by both cloning and PCR assays has demonstrated that alternative splicing at the amino and carboxy termini generates several different isoforms in various tissues (18). In particular, transcription in the brain apparently starts with a different promoter, which is weakly used in muscle (18,19). This finding was particularly intriguing in light of the fact that about one-third of DMD patients exhibit some degree of cognitive impairment (1). Genomic mapping of this brain promoter places it about 100 kb 5′ to the muscle promoter (Boyce et al., in preparation). The large distances between exons in this region suggest that there should be patients with deletions of one or the other of these promoters. To help our understanding of the role of dystrophin in the brain, we would like to identify patients with deletion of the brain promoter with retention of a functional muscle promoter. We predict that these patients will not have classic symptoms of muscular dystrophy, but rather, that they would likely exhibit some degree of mental retardation (MR), which might be inherited in an X-linked recessive fashion with variable penetrance.

In a large collaborative effort to substantiate the effects of deletions on the translational reading frame of dystrophin, several hundred deletions were mapped and many of the intron-exon borders were cloned and sequenced (13). In analyzing these data, we were struck by the fact that although DMD is 9–10 times more common than BMD, the preponderance of one type of border predicted that most deletions should result in the milder BMD phenotype. Furthermore, although frame-shifting deletions that caused DMD were spread throughout the locus, it soon became apparent that there were no small inframe deletions in the proximal portion of the central rod domain of dystrophin (from exons 14 to 44) where most of the known exon borders (13/17) were all the same (Fig. 1). To explain this apparent paradox, we propose that our sample suffered from a bias of ascertainment such that inframe deletions in this region cause a phenotype different from that normally observed in DMD/BMD. As a result, they would not be expected to present in neuromuscular clinics.

Thus, our deletion analyses have revealed two interesting areas of the dystrophin gene that do not seem to be commonly deleted in patients with classical muscular dystrophy. If these "missing deletions" are due to a bias of ascertainment, then we expect that either they are not deleterious, or that they cause some phenotype other than classic DMD or BMD. We predict that loss of only the brain promoter is likely to cause X-linked mental retardation; however, predicting the effects of small deletions involving exons 14–44 is somewhat more problematic. Because DMD and BMD patients often have an associated cardiomyopathy (CM) (reviewed in 2,20), we hypothesize that these missing deletions might cause CM without the skeletal muscular weakness usually associated with DMD/BMD. In support of this are several reports of families with BMD in whom CM was the most prom-

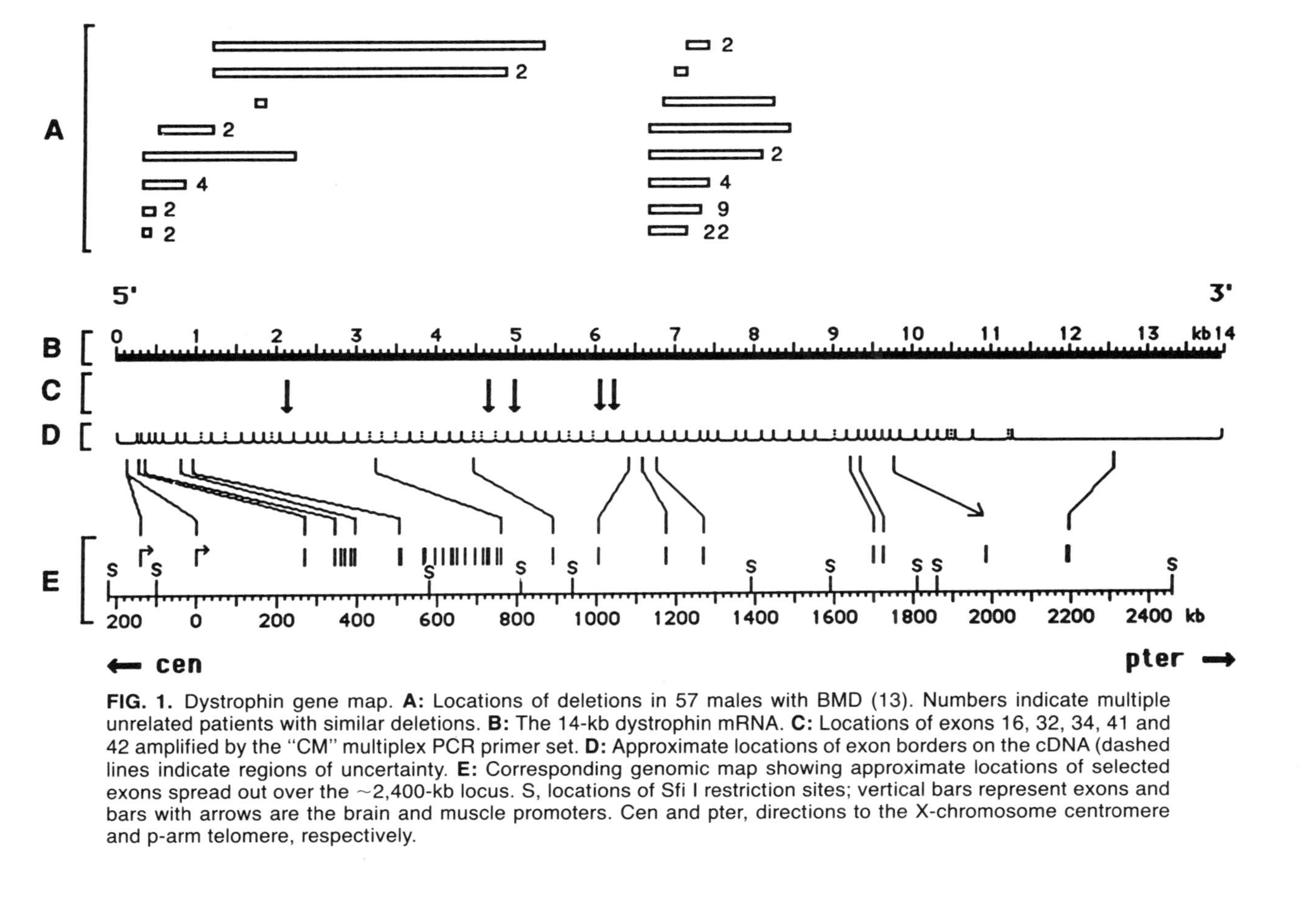

FIG. 1. Dystrophin gene map. **A:** Locations of deletions in 57 males with BMD (13). Numbers indicate multiple unrelated patients with similar deletions. **B:** The 14-kb dystrophin mRNA. **C:** Locations of exons 16, 32, 34, 41 and 42 amplified by the "CM" multiplex PCR primer set. **D:** Approximate locations of exon borders on the cDNA (dashed lines indicate regions of uncertainty. **E:** Corresponding genomic map showing approximate locations of selected exons spread out over the ~2,400-kb locus. S, locations of Sfi I restriction sites; vertical bars represent exons and bars with arrows are the brain and muscle promoters. Cen and pter, directions to the X-chromosome centromere and p-arm telomere, respectively.

inent manifestation of their disease (21–23). Furthermore, some cardiac symptoms commonly experienced by DMD patients, notably certain EKG changes and mitral valve prolapse, are seen in female carriers, suggesting that cardiac tissue may be more sensitive to perturbations in dystrophin than skeletal muscle.

Here we describe our preliminary attempts to identify patients with mutations of the brain promoter or of the proximal rod domain. To do this, we have developed two multiplex PCR assays for rapid detection of deletions involving relevant exons. By analyzing samples from male patients with either mental retardation or idiopathic cardiomyopathy, we hope to maximize our chances of detecting some of these "missing deletions."

MATERIALS AND METHODS

Patient Selection

Male patients with learning disabilities or mental retardation were ascertained through the Children's Hospital cytogenetics service from their population of patients tested for fragile X syndrome. Samples from male patients (aged 18–72) with idiopathic cardiomyopathy were referred from collaborating hospitals.

DNA Isolation and PCR Analysis

DNA was isolated from whole blood anticoagulated with EDTA either by phenol extraction (24) or by boiling. To boil blood, 0.5 mL evenly suspended blood was mixed with 0.5 mL 2× Triton–sucrose (25) solution and 0.5 mL 1× sucrose/triton solution in a 1.5-mL microfuge tube. The mixture was vortexed and incubated on ice for 1 hour with intermittent shaking. Nuclei were pelleted for 10 minutes at high speed in a table-top microfuge. After discarding the supernatant, the pellet was washed with 1 mL 1× Triton–sucrose and respun as above. Then the pellet was washed 3 times with 1 mL ddH_2O and the debris pelletted in a microfuge for 2–3 minutes. The pellet was resuspended in 150 μL ddH_2O and boiled for 5 minutes. After quickly pelletting the debris in a microcentrifuge, the DNA concentration was estimated by measuring the OD_{260}.

PCR Reaction Conditions and Analysis

PCR reactions using reagents from Perkin Elmer Cetus GeneAmp kits (Norwalk, CT) contained 0.5μ*M* of each primer, 5 units Taq polymerase, and 250 ng genomic DNA (500 ng DNA from boiled blood) per 50 μL reaction. Primers were prepared and the PCR amplification was as described by (24). PCR reaction products (15 μL) were separated on either 2% NuSieve (FMC BioProducts, Rockland, ME) + 1% agarose gels (MR study) or 1.4% agarose gels (CM study) containing 0.1 μg/mL ethidium bromide and electrophoresed at 5 V/cm.

Western Blotting

Western blots were performed as described (14) on samples of cardiac muscle that had been stored frozen under liquid nitrogen. Briefly, the muscle was solubilized in SDS buffer and separated by electrophoresis on 3.5–12.5% SDS-polyacrylamide gels. After electroblotting to nitrocellulose filters, dystrophin was visualized with anti-30-kD dystrophin sheep antisera followed by alkaline phosphatase-conjugated secondary antibodies.

RESULTS AND DISCUSSION

Studies of the Brain Promoter in Males with Mental Retardation

We predicted that some proportion of males who exhibited cognitive impairment might be candidates for alterations of the upstream dystrophin promoter, without alteration of the remainder of the gene. Our cytogenetics laboratory receives about 100 blood samples per year of young males who are suspected to be affected with X-linked mental retardation associated with the fragile X syndrome. This is a frequent cause of mental retardation in males, although many other causes of X-linked mental retardation are known to exist. DNA was prepared and PCR reactions performed using primers specific for both the upstream and downstream promoters (Table 1, Fig. 2A). As indicated in Table 2, our work in progress has failed to find patients with deletions of the upstream promoter. This result presumably does not mean that there are no cognitively impaired patients without dystrophy symptoms, but rather that we have not searched extensively enough or within the right patient populations. The ideal DNA samples would be those isolated from males who clearly exhibit an X-linked pattern of inheritance for mental retardation in the absence of fragile X, rather than isolated cases without extensive family history.

Studies of the Central Rod Domain in Patients with Idiopathic Cardiomyopathy

The second group of patients who might exhibit atypical symptoms and an abnormality of their dystrophin locus are those with deletions in the central rod domain of dystrophin. We set out to search the population for "si-

TABLE 1. *Sequence of PCR primers for dystrophin gene exons*

Pri[a]	Sequence (5′–3′)	Pri	Sequence (5′–3′)
PbF1	GAAGATCTATATTTTACAACGCAGAAATGTGG	PbR	CTTCCATGCCAGCTGTTTTTCCTGTCACTC
PbF2	GAAGATCTAGAAGAGCGAGTAGATACTGAAAGAG		
16F	TCTATGCAAATGAGCAAATACACGC	16R	GGTATCACTAACCTGTGCTGTACTC
32F	GACCAGTTATTGTTTGAAAGGCAAA	32R	TTGCCACCAGAAATACATACCACACAATG
34F	GTAACAGAAAGAAAGCAACAGTTGGAGAA	34R	CTTTCCCCAGGCAACTTCAGAATCCAAA
41F	GTTAGCTAACTGCCCTGGGCCCTGTATTG	41R	TAGAGTAGTAGTTGCAAACACATACGTGG
42F	CACACTGTCCGTGAAGAAACGATGATGG	42R	CTTCAGAGACTCCTCTTGCTTAAAGAGAT

[a]Primers are named for the exon they amplify using numbering of ref. 13, Pb is the brain-specific promoter. F is forward and R is reverse relative to the coding sequence. Either PbF1 or PbF2 was used with PbR. Primers for the muscle promoter (Pm) have been previously described (24).

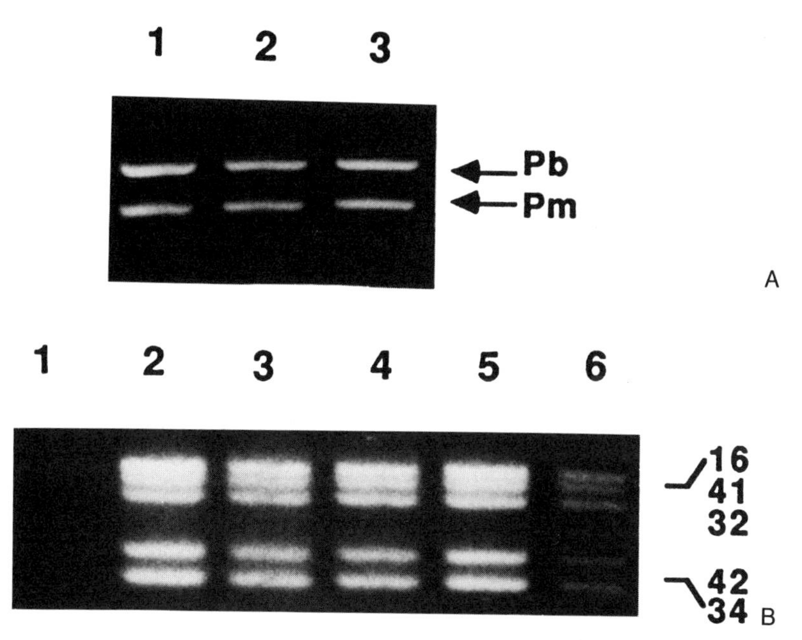

FIG. 2. Dystrophin gene analysis. **A:** PCR products from the brain and muscle promoters (Pb and Pm, respectively) in three males with mental retardation and cytogenetically normal X chromosomes. **B:** PCR products from five exons in the central rod domain (exon numbers indicated at right) in five male patients with idiopathic cardiomyopathy (lanes 2–6). Lane 1 is a no DNA control.

lent" BMD patients who might have extremely mild skeletal muscle myopathies. Instead of screening a large population of unaffected people, we decided to start by studying patients with idiopathic cardiomyopathy, as many BMD patients will develop cardiomyopathy later in life. For this study, we designed primers to amplify exons 16, 32, 34, 41 and 42 in a multiplex PCR assay (Table 1, Fig. 1C). DNA was prepared from blood samples collected from male patients who were being seen for idiopathic cardiomyopa-

TABLE 2. *DNA and protein screening results*

Patient type	Assay	Normal pattern[a]	Abnormal pattern[a]
Suspect fragile X[b]	PCR	44	0
Idiopathic cardiomyopathy (CM)	PCR	22	0
	Western blot	15	0

[a]Normal vs. abnormal refers to PCR amplification or protein analysis yielding results expected for normal.

[b]Among 57 suspected fragile X patients, some were shown by subsequent chromosome analysis to actually suffer from fragile X syndrome. These are not included in the study.

thy, and PCR amplification was performed using these primers (Fig. 2B). In addition, the battery of primers used to scan the gene for the common deletions were also used (26). The results are in Table 2. Among 22 patients screened, none were found to have deletions of their dystrophin gene.

We were also able to obtain samples of myocardium from 15 males aged 17–57 who had undergone heart transplantation for idiopathic cardiomyopathy. In this case, we examined dystrophin directly by western blot analysis (Fig. 3). In contrast to the PCR study, this approach should detect abnormalities in size regardless of the location of a deletion. As before, every patient studied had no detectable dystrophin abnormalities (Table 2).

We have described our preliminary results, testing two different hypotheses. The first supposes that there are patients with a degree of mental retardation caused by a deletion of the brain promoter without affecting the remainder of the dystrophin gene. No such deletions were found among the 57 patients screened who were referred for evaluation of fragile X. Some indeed did have fragile X syndrome, but none of the remaining 44 (Table 2) could be demonstrated to be a deletion of the brain promoter. This does not mean that such patients do not exist, only that we have not looked at sufficient numbers of patient samples.

Our second hypothesis was that there were a larger number of BMD patients than are currently being detected. We have now tested the DNA of 22 patients with cardiomyopathy whom we predict might be more likely to have deletions of their dystrophin gene. By both western blot analysis and PCR amplification of a set of exons we have yet to detect a deletion in any of these patients. Additional patients are in the process of being studied.

Although no abnormalities were detected in our preliminary study of patients with potentially unusual phenotypes for dystrophin alterations, we do

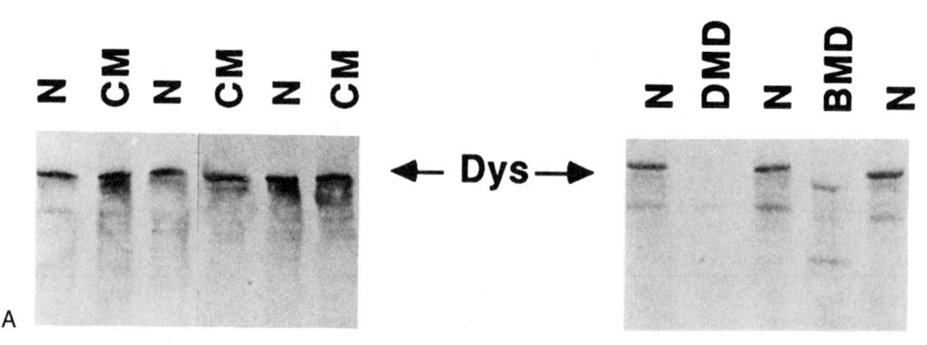

FIG. 3. Western blot analysis of dystrophin. **A:** Dystrophin in cardiac muscle from three patients who underwent transplantation for idiopathic cardiomyopathy (lanes CM) and three control hearts (lanes N). **B:** For comparison, dystrophin in skeletal muscle from patients with Duchenne's muscular dystrophy (DMD), Becker's muscular dystrophy (BMD), and normal controls (indicated above lanes).

present the sequence of those primers most likely to detect these alterations. As more patients are studied by us and others, we will be able to determine the relationship of abnormal dystrophin expression and the range of clinical symptoms observed.

ACKNOWLEDGMENTS

Our sincere thanks to the many clinical collaborators who provided patient material utilized in this ongoing project, especially to Drs. Ralph Shabetai, Angel De La Torre, Theodore Munsat and Bruce Korf. We would also like to thank the other members of our laboratory for helpful discussions and advice. We are indebted to Mark Fleming of the Howard Hughes Medical Institute Biopolymers laboratory for synthesis of oligonucleotides. We thank Nicole Picard for assistance with the manuscript and Dixon Yun for preparation of the figures. F.M.B. is the Betty Q. Banker Fellow of the Muscular Dystrophy Association. This work was supported by grants from the National Institutes of Health, Muscular Dystrophy Association, and Howard Hughes Medical Institute to L.M.K.

REFERENCES

1. Moser A. Duchenne muscular dystrophy: pathogenic aspects and genetic prevention. *Hum Genet* 1984;66:17–40.
2. Emery AEH. *Duchenne Muscular Dystrophy,* 2nd ed. New York: Oxford University Press, 1988.
3. Monaco AP, Neve RL, Colletti-Feener C, Bertelson CJ, Kurnit DM, Kunkel LM. Isolation of candidate cDNAs for portions of the Duchenne muscular dystrophy gene. *Nature* 1986;323:646–650.
4. Burghes AHM, Logan C, Hu X, Belfall B, Worton RG, Ray PN. A cDNA clone from the Duchenne/Becker muscular dystrophy gene. *Nature* 1986;324:582–585.
5. Koenig M, Hoffman EP, Bertelson CJ, Monaco AP, Feener C, Kunkel LM. Complete cloning of the Duchenne muscular dystrophy gene. *Cell* 1987;50:509–517.
6. Hoffman EP, Brown RH, Kunkel LM. Dystrophin: the protein product of the Duchenne muscular dystrophy locus. *Cell* 1987;51:919–928.
7. Koenig M, Monaco AP, Kunkel LM. The complete sequence of dystrophin predicts a rod-shaped cytoskeletal protein. *Cell* 1988;53:219–228.
8. Love DR, Hill DF, Dickson G, Spurr NK, Byth BC, Marsden RF, Walsh FS, Edward YH, Davies KE. An autosomal transcript in skeletal muscle with homology to dystrophin. *Nature* 1989;339:55–58.
9. Khurana TS, Hoffman EP, Kunkel LM. Identification of a chromosome 6 encoded dystrophin related protein. *J Biol Chem* 1990;in press.
10. Den Dunnen JT, Grootscholten PM, Bakker E, et al. Topography of the Duchenne muscular dystrophy (DMD) gene: FIGE and cDNA analysis of 194 cases reveals 115 deletion and 13 duplications. *Am J Hum Genet* 1989;45:835–847.
11. Hu X, Ray PN, Murphy EG, Thompson MW, Worton RG. Duplicational mutation at the Duchenne muscular dystrophy locus: its frequency, distribution, origin, and phenotype genotype correlation. *Am J Hum Genet* 1990;46:682–695.
12. Monaco AP, Bertelson CJ, Liechti-Gallati S, Moser H, Kunkel LM. An explanation for the phenotypic differences between patients bearing partial deletions of the DMD locus. *Genomics* 1988;2:90–95.

13. Koenig M, Beggs AH, Moyer M, et al. The molecular basis for Duchenne versus Becker muscular dystrophy: correlation of severity with type of deletion. *Am J Hum Genet* 1989;45:498–506.
14. Hoffman EP, Fischbeck KH, Brown RH. Dystrophin characterization in muscle biopsies from Duchenne and Becker muscular dystrophy patients. *N Engl J Med* 1988;318:1363–1368.
15. Chamberlain JS, Pearlman JA, Muzny DM, et al. Expression of the Duchenne muscular dystrophy gene in muscle and brain. *Science* 1988;239:1416–1418.
16. Hoffman EP, Hudecki M, Rosenberg P, Pollina C, Kunkel LM. Cell and fiber-type distribution of dystrophin. *Neuron* 1988;411–420.
17. Chelly J, Kaplan JC, Maire P, Gautron S, Kahn A. Transcription of the dystrophin gene in human muscle and non-muscle tissues. *Nature* 1988;333:858–860.
18. Feener CA, Koenig M, Kunkel LM. Alternative splicing of dystrophin mRNA generates isoforms at the carboxy terminus. *Nature* 1989;338:509–511.
19. Nudel U, Zuk D, Einat P, et al. Duchenne muscular dystrophy gene product is not identical in muscle and brain. *Nature* 1989;337:76–78.
20. Perloff JK, De Leon AC, O'Doherty D. The cardiomyopathy of progressive muscular dystrophy. *Circulation* 1966;33:625–648.
21. Kuhn E, Fiehn W, Schroder JM, Assmus H, Wagner A. Early myocardial disease and cramping myalgia in Becker-type muscular dystrophy: a kindred. *Neurology* 1979;29:1144–1199.
22. Ueda K, Okada R, Matsuo H, Harumi K, Yasuda H, Tsuyuki H, Ueda H. Myocardial involvement in benign Duchenne type of progressive muscular dystrophy. *Jpn Heart J* 1970;11:26–35.
23. Katiyar BC, Misra S, Somani PN, Chaterji AM. Congestive cardiomyopathy in a family of Becker's X-linked muscular dystrophy. *Postgrad Med J* 1977;53:12–15.
24. Beggs AH, Koenig M, Boyce FM, Kunkel LM. Detection of 98% of DMD/BMD deletions by PCR. *Hum Genet* 1990;86:45–48.
25. Kunkel LM, Smith KD, Boyer SH, et al. Analysis of human Y-chromosome-specific reiterated DNA in chromosome variants. *Proc Natl Acad Sci USA* 1977;74:1245–1249.
26. Chamberlain JS, Gibbs RA, Ranier JE, Nguyen PN, Caskey CT. Deletion screening of the Duchenne muscular dystrophy locus via multiplex DNA amplification. *Nucleic Acids Res* 1988;16:11141–11156.

Etiology of Human Disease at the DNA Level,
edited by Jan Lindsten and Ulf Pettersson.

6

The Proteins of Blood Coagulation and Their Genes

Earl W. Davie

Department of Biochemistry, University of Washington, Seattle, Washington 98195

The blood coagulation cascade involves a number of plasma proteins that participate in the intrinsic and extrinsic pathways leading to fibrin formation (Fig. 1). Most of these proteins have been recognized in patients with coagulation disorders resulting in bleeding complications. During the past 10 years, all the known proteins that participate in the coagulation cascade have been isolated and characterized and their mechanism of activation or role as cofactors have been established. Furthermore, the genes coding for these proteins have been identified and most have been sequenced. The size and chromosomal location of these genes are summarized in Table 1.

Fibrinogen is one of the largest of the coagulation proteins (M_r 340,000) and is composed of six polypeptide chains ($\alpha_2\beta_2\gamma_2$) held together by disulfide bonds (1). About 10% of the human fibrinogen contains a γ' chain that is modified on the carboxyl-terminal end (2,3). The genes coding for the three fibrinogen chains are linked and span about 45 kb of DNA (4). They are located on chromosome 4 at q2 and are arranged in the order of γ-α-β (5,6). The γ and α genes are transcribed in the same direction, while the β gene is transcribed in the opposite direction. The α gene is 5.4 kb in length and contains five exons (4). The fifth exon in this gene is very large and contains the nine tandem repeats of 13 amino acids (amino acid residues 270–374). The β gene is 8.2 kb and contains eight exons (4). This gene also contains two copies of the Alu family repeats located in intron E and the 3′ flanking region of the gene. The γ gene is 8.4 kb and contains 10 exons (7). The ninth intron in this gene codes for the γ' chain that results from the use of an alternative polyadenylation site. The amino acid sequence homology between the three chains suggests that they are derived from a common ancestral gene and this divergence occurred about 600–1,000 million years ago (8). Fibrinogen is an acute-phase protein with the three chains being under coordinate control (9). The assembly of the molecule and the formation of the disulfide bonds is poorly understood. Hopefully, the identification of inter-

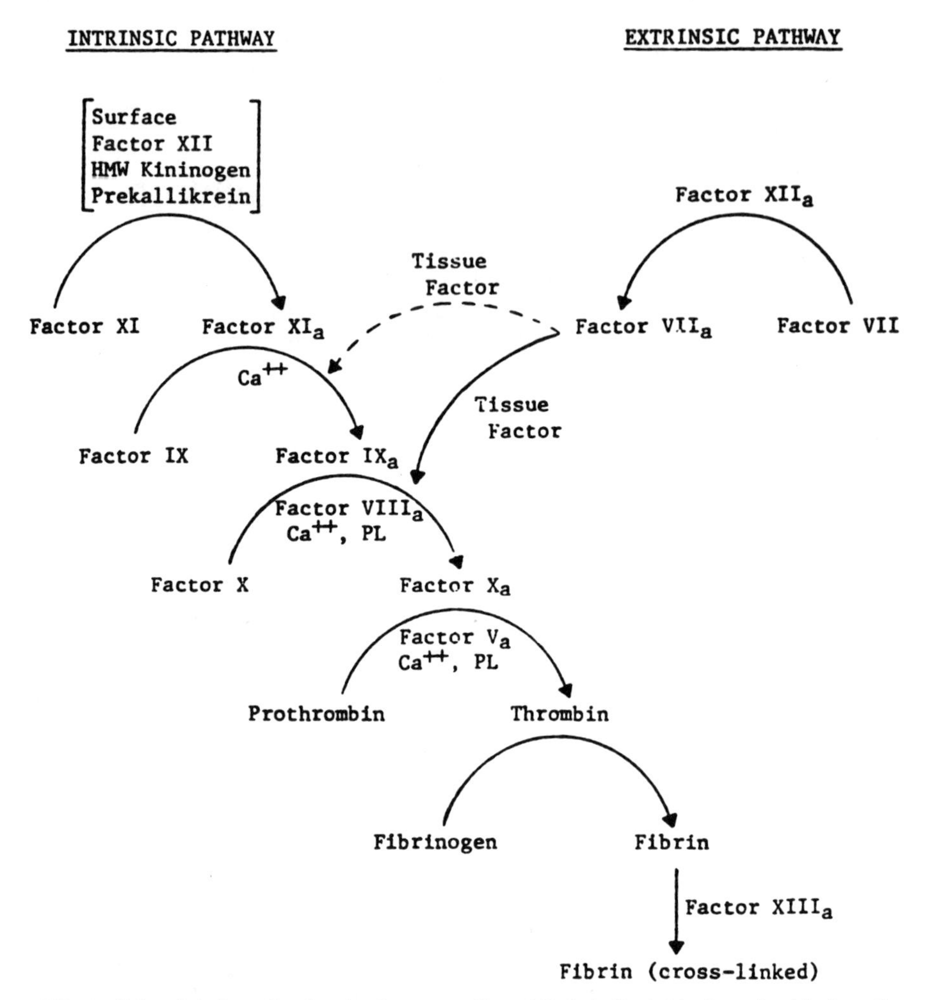

FIG. 1. Abbreviated mechanism for the generation of fibrin in the intrinsic and extrinsic pathways of the blood coagulation cascade. PL, phospholipid.

mediates in this process will be clarified now that an *in vitro* expression system has been developed (10).

The vitamin-K-dependent plasma proteins that participate in the blood coagulation cascade and its regulation include prothrombin, factors VII, IX, and X, protein C, and protein S. Each is a glycoprotein that contains 9–12 γ-carboxyglutamic acid (Gla) residues in the amino-terminal region of the protein. Furthermore, each glycoprotein, with the exception of protein S, is converted to a serine protease by minor proteolysis (11). This family of glycoproteins is synthesized in the liver as single-chain molecules with pre-

TABLE 1. *Genes for the proteins involved in blood coagulation*

Protein	Molecular weight	Chromosomal location	Size of gene (kb)
Fibrinogen	340,000	—	—
α chain	66,000	4q2	5.4
β chain	52,000	4q2	8.2
γ chain	46,500	4q2	8.4
Prothrombin	71,600	11p11–q12	21
Factor V	330,000	Only cDNA cloned	
Factor VII	50,000	13q34–qter	12.8
Factor VIII	330,000	Xq28	186
Factor IX	56,800	Xq26–27	34
Factor X	58,800	13q34–qter	27
Factor XI	120,000	4q35	23
Factor XIII	320,000	—	—
a subunit	75,000	6p24–25	>160
b subunit	80,000	1q31–32.2	28
Tissue factor	44,000	1p21–22	12.4

pro leader sequences and is secreted into the blood after considerable processing.

Prothrombin (M_r 71,600) containing 10 γ-carboxyglutamic acid residues is converted to thrombin in the presence of factor Xa, factor Va, calcium ions, and phospholipid during the coagulation cascade (Fig. 1). The protein contains two kringle domains located between the Gla and catalytic domains (12). The gene for prothrombin is located on chromosome 11 at p11-q12 (13) and is composed of 21 kb of DNA with 14 exons (14). It also has 30 copies of Alu-repetitive DNA and two copies of partial KpnI repeats. This repetitive DNA comprises nearly 40% of the gene.

Factors VII, IX, and X and protein C show a great deal of amino acid sequence homology and structural similarity (Fig. 2). Each of these proteins contains a Gla-rich region followed by two growth factor domains in the amino-terminal portion of their molecule. Like prothrombin, the carboxyl end of these vitamin-K-dependent proteins contains a serine protease or catalytic region that is homologous to pancreatic trypsin. Furthermore, each is converted to an active protease by minor proteolysis during the coagulation cascade (Fig. 1). The genes for these four proteins are also very similar in their organization (11). The coding region of each gene is interrupted by seven introns and each of these seven intervening sequences is located in the same region of the polypeptide chain (Fig. 2). Furthermore, the splice junction type at each of the introns is the same. For instance, the introns just prior, between, and just following the growth factor domains in these four proteins are all type I, i.e., the introns are always located in an amino acid codon between the first and second nucleotides. The size and DNA

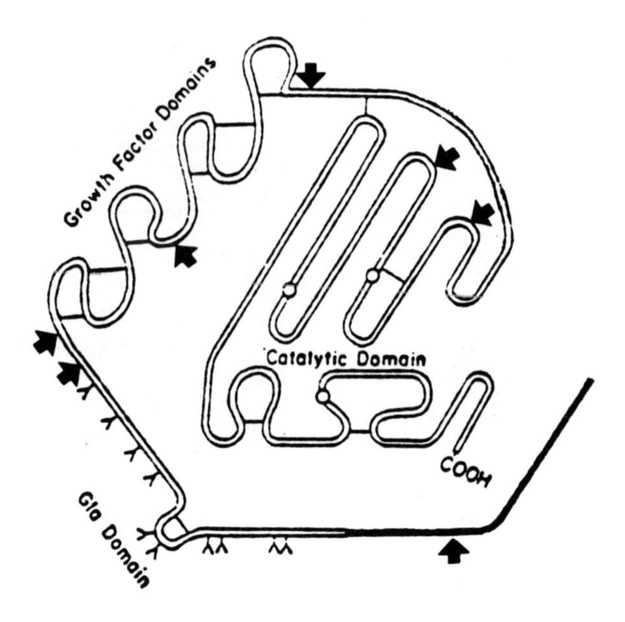

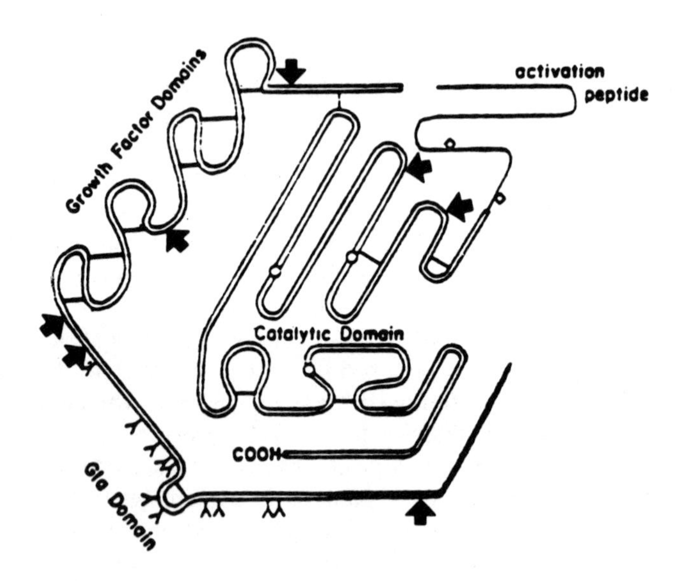

FIG. 2. Location of the introns in the genes coding for human factor VII, factor IX, factor X, and protein C. The seven introns are shown with arrows. Y refers to γ-carboxyglutamic acid present in the Gla domains. A solid line refers to the prepro leader sequence that is removed during protein biosynthesis. An open line represents a polypeptide chain for the mature protein, while a thin line corresponds to the activation peptide. Open circles refer to the amino acid triad (His, Asp, Ser) involved in catalysis. Open diamonds (◇) refer to N-linked carbohydrate chains.

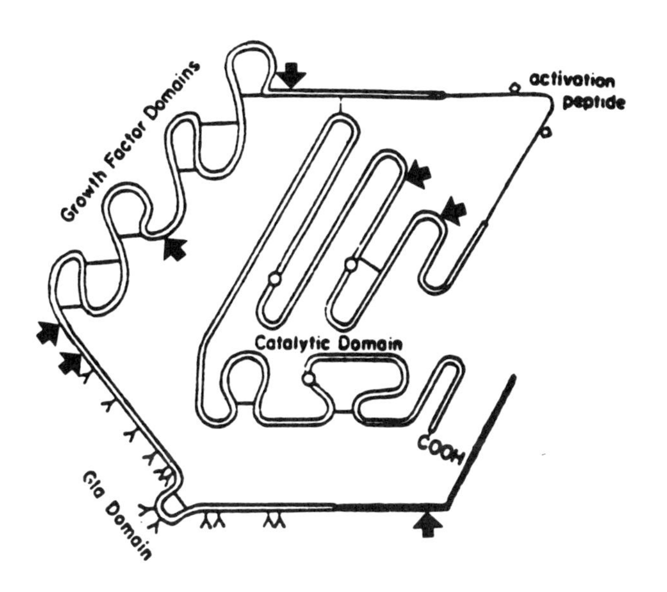

Factor IX

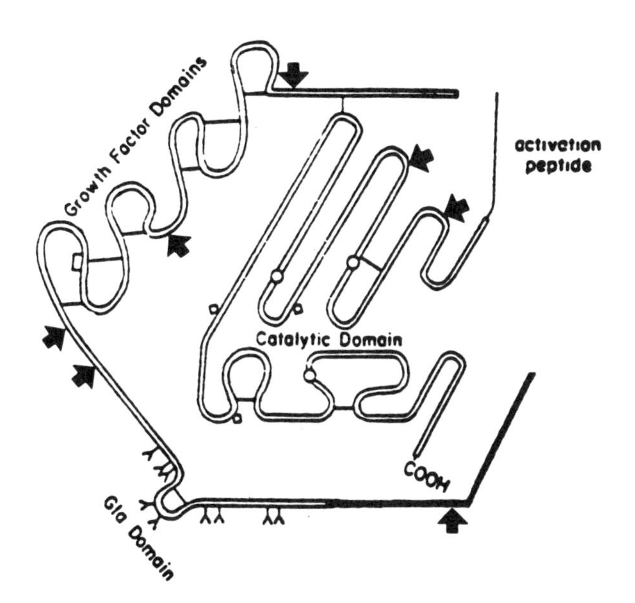

Protein C

FIG. 2. *continued*

sequence within the introns in the genes coding for these proteins vary, however, and this leads to a major difference in the size of each of the four genes (Table 1).

The gene for factor IX (hemophilia B) is located on the tip of the long arm of the X chromosome (15,16) very near the gene for factor VIII (hemophilia A). At present, a large number of abnormal genes coding for these two proteins have been identified. These include a variety of mutations, such as nucleotide insertions, substitutions, deletions, and changes at intron/exon boundaries. Studies of these abnormalities have been particularly important for clinical diagnosis since these two bleeding disorders are the most frequent of all the coagulation abnormalities. The genes for factor VII and factor X are located on chromosome 13 and are also extremely close to each other (17,18). Indeed, the polyadenylation site at the 3′ end of the gene coding for factor VII is only 2.8 kb from the Met start site on the 5′ end of the gene coding for factor X (Leytus SP, Davie EW, unpublished results). The gene for protein C, a vitamin-K-dependent protein involved in the regulation of the clotting cascade, is located on chromosome 2 (19). This gene is the smallest in this family of proteins, being only 9 kb of DNA in size (20,21).

Factor XI (M_r 120,000) is a plasma glycoprotein that participates in the early phase of the intrinsic pathway of blood coagulation (Fig. 1). It is composed of two identical polypeptide chains that are held together by a bond. The amino-terminal region of the protein contains four tandem repeats of 90 (or 91) amino acids called apple domains (22). These domains show considerable amino acid sequence identity ($\approx$ 60%) with four tandem repeats in the amino-terminal region of plasma prekallikrein (23). Factor XI is converted to factor XIa, a serine protease, by minor proteolysis during the blood coagulation cascade. The gene coding for factor XI is 23 kb in length and contains 15 exons (24). It is located on the tip of the long arm of chromosome 4. A deficiency of factor XI results in a mild bleeding disorder that occurs primarily in Ashkenazi Jews. The frequency for homozygotes in Israel is about one in 190 individuals (25). The occurrence of factor XI deficiency in other populations is extremely rare. Approximately 95% of the factor XI deficiency in the Ashkenazi population is due to two point mutations (26). These occur at Glu-117 (GAA) converting it to a stop codon (TAA) and the substitution of Phe-283 (TCC) by Leu (TCT). These two mutations occur in the Ashkenazi population at a ratio of 1:2:1 (Asakai R, unpublished results). Accordingly, these two mutations appear to be due to a typical founder effect. The continued presence of these two mutations in the Ashkenazi population may be due to some close linkage of the factor XI gene to another critical gene(s) on chromosome 4.

Factor XIII (M_r 320,000) circulates in blood as a precursor to a transglutaminase. It is converted to an active enzyme, factor XIIIa, by thrombin in the presence of fibrin (Fig. 1). It is composed of two *a* subunits and two *b* subunits (27,28). The active enzyme requires calcium and catalyzes the formation of ε-(γ-glutamyl)lysine bonds between the γ-carboxyl group of a glutamine residue in one fibrin monomer and the ε-amino group of a lysine residue in a second fibrin monomer (29). Accordingly, it stabilizes the fibrin clot that is generated during the coagulation cascade by the formation of covalent crosslinks between the fibrin monomers.

The gene for the *a* subunit of factor XIII spans more than 160 kb of DNA and includes 15 exons (30). It is located on chromosome 6 at p24-25 (31,32). The gene for the *b* subunit spans 28 kb of DNA and includes 12 exons (32a). It is located on chromosome 1 at q31-32.2 (33). The amino acid sequence of the *b* subunit contains 10 tandem repeats of about 60 amino acids. These repetitive sequences, called sushi structures, have characteristic disulfide bonds between the first and third, and the second and fourth cystine residues in each sushi structure (34). Interestingly, each of the sushi structures are coded by one exon. A large number of proteins have been identified in recent years that contain sushi structures. Indeed, it is one of the largest superfamilies thus far identified and, at present, includes more than 25 different proteins. Accordingly, it appears to be a family that has resulted from exon shuffling during the evolutionary process.

Two of the largest proteins that participate in the coagulation cascade include factor V and factor VIII (35). Both are activated by minor proteolysis and participate as cofactors in the coagulation cascade (Fig. 1). Furthermore, the activated forms of these two proteins, i.e., factor Va and factor VIIIa, are inactivated by activated protein C. Factor VIII (M_r 330,000) greatly increases the V_{max} in the activation of factor X by factor IXa, whereas factor Va greatly increases the V_{max} in the activation of prothrombin by factor Xa (36,37). The gene for factor VIII is approximately 186 kb in size (38) and is located on the long arm of the X chromosome (39). It consists of 26 exons and represents about 0.1% of the X chromosome. Exon 14 in the center of the factor VIII gene is extremely large, containing approximately 31 kb of DNA. Also, the intron between exons 22 and 23 is very large, being approximately 32 kb in size.

Factor V has been cloned with a cDNA of approximately 7 kb (40–42). The gene coding for this protein, however, has not been characterized thus far.

A remarkable feature of the coagulation proteins is the presence of similar domain structures in each of the plasma proteins. These proteins are all members of families that contain common domains, such as kringle structures, growth factor domains, sushi structures, apple domains, or serine protease domains. This suggests that they apparently have evolved in part by exon shuffling (11,43,44), as well as mutations involving insertions, substitutions, and deletions.

REFERENCES

1. Doolittle RF. Structural aspects of the fibrinogen-fibrin conversion. *Adv Protein Chem* 1973;27:1.
2. Francis CW, Marder VJ, Martin SE. Demonstration of a large molecular weight variant of the γ chain of normal human plasma fibrinogen. *J Biol Chem* 1980;255:5599–5604.
3. Wolfenstein-Todel C, Mosesson MW. Human plasma fibrinogen heterogeneity: Evidence for an extended carboxyl-terminal sequence in a normal γ chain variant (γ'). *Proc Natl Acad Sci USA* 1980;77:5069–5073.
4. Chung DW, Harris JE, Davie EW. Nucleotide sequences of the three genes coding for human fibrinogen. *Adv Exp Med Biol* 1990, in press.
5. Henry I, Uzan G, Weil D, et al. The genes coding for A alpha-, B beta-, and gamma-chains of fibrinogen map to 4q2. *Am J Hum Genet* 1984;36:760.
6. Kant JA, Fornace AJ Jr., Saxe D, Simon MI, McBride OW, Crabtree GR. Evolution and organization of the fibrinogen locus on chromosome 4. Gene duplication accompanied by transposition and inversion. *Proc Natl Acad Sci USA* 1985;82:2344–2348.
7. Rixon MW, Chung DW, Davie EW. Nucleotide sequence of the gene for the γ chain of human fibrinogen. *Biochemistry* 1985;24:2077–2086.
8. Doolittle RF. The structure of vertebrate fibrinogen. *Ann NY Acad Sci* 1983;408:13.
9. Crabtree GR, Kant JA. Coordinate accumulation of the mRNAs for the α, β and γ chains of rat fibrinogen following defibrination. *J Biol Chem* 1982;257:7277–7279.
10. Farrell DH, Mulvihill ER, Chung DW, Davie EW. Expression of functional human fibrinogen from cDNA clones in baby hamster kidney cells. *Blood,* Suppl 1 1989;196.
11. Hedner U, Davie EW. Introduction to hemostasis and the vitamin K-dependent coagulation factors. In: Scriver CR, Beaudet AL, Sly WS, Valle D, eds. *The Metabolic Basis of Inherited Disease,* II, Sixth Edition. New York: McGraw-Hill, Inc., 1989;2107–2134.
12. Magnusson S, Petersen TE, Sottrup-Jensen L, Claeys H. In: Reich E, Rifkin DB, Shaw E, eds. *Proteases and Biological Control.* New York: Cold Spring Harbor Laboratories, 1975;123–149.
13. Royle NJ, Irwin DM, Koschinsky ML, MacGillivray RTA, Hamerton JL. Human genes encoding prothrombin and ceruloplasmin map to 11p11-q12 and 3q21-24, respectively. *Somat Cell Mol Genet* 1987;13:285–292.
14. Degen SJF, MacGillivray RTA, Davie EW. Characterization of the complementary deoxyribonucleic acid and gene coding for human prothrombin. *Biochemistry* 1983;22:2087–2097.
15. Camerino G, Grzeschik KH, Jaye M, et al. Regional localization on the human X chromosome and polymorphism of the coagulation factor IX gene (hemophilia B locus). *Proc Natl Acad Sci USA* 1984;81:498–502.
16. Chance PF, Dyer KA, Kurachi K, et al. Regional localization of human factor IX gene by molecular hybridization. *Hum Genet* 1983;65:207.
17. Ott R, Pfeiffer RA. Evidence that activities of coagulation factors VII and X are linked to chromosome 13 (q34). *Hum Hered* 1984;34:123.
18. Gilgenkrantz S, Briquel ME, Andre E, et al. Structural genes of coagulation factors VII and X located on 13q34. *Ann Genet* 1986;29:32.
19. Rocchi M, Roncuzzi L, Santamaria R, Archidiacono N, Dente L, Romeo G. Mapping through somatic cell hybrids and cDNA probes of protein C to chromosome 2, factor X to chromosome 13, and alpha 1-acid glycoprotein to chromosome 9. *Hum Genet* 1986;74:30.
20. Foster DC, Yoshitake S, Davie EW. The nucleotide sequence of the gene for human protein C. *Proc Natl Acad Sci USA* 1985;82:4673–4677.
21. Plutzky J, Hoskins JA, Long GL, Crabtree GR. Evolution and organization of the human protein C gene. *Proc Natl Acad Sci USA* 1986;83:546.
22. Fujikawa K, Chung DW, Hendrickson LE, Davie EW. Amino acid sequence of human factor XI, a blood coagulation factor with four tandem repeats that are highly homologous with plasma prekallikrein. *Biochemistry* 1986;25:2417–2424.
23. Chung DW, Fujikawa K, McMullen BA, Davie EW. Human plasma prekallikrein, a zymogen to a serine protease that contains four tandem repeats. *Biochemistry* 1986;25:2410–2417.

24. Asakai R, Davie EW, Chung DW. Organization of the gene for human factor XI. *Biochemistry* 1987;26:7221–7228.
25. Seligsohn U. Factor XI (PTA) deficiency. In: Goodman RM, Motulsky AG, eds. *Genetic Diseases Among Ashkenazi Jews*. New York: Raven Press, 1979;141–147.
26. Asakai R, Chung DW, Ratnoff OD, Davie EW. Factor XI (plasma thromboplastin antecedent) deficiency in Ashkenazi Jews is a bleeding disorder that can result from three types of point mutations. *Proc Natl Acad Sci USA* 1989;86:7667–7671.
27. Schwartz ML, Pizzo SV, Hill RL, McKee PA. Human factor XIII from plasma and platelets. Molecular weights, subunit structures, proteolytic activation, and cross-linking of fibrinogen and fibrin. *J Biol Chem* 1973;248:1395–1407.
28. Chung SI, Lewis MS, Folk JE. Relationships of the catalytic properties of human plasma and platelet transglutaminases (activated blood coagulation factor XIII) to their subunit structures. *J Biol Chem* 1974;249:940–950.
29. Chen R, Doolittle RF. γ-γ cross-linking sites in human and bovine fibrin. *Biochemistry* 1971;10:4486–4491.
30. Ichinose A, Davie EW. Characterization of the gene for the *a* subunit of human factor XIII (plasma transglutaminase), a blood coagulation factor. *Proc Natl Acad Sci USA* 1988;85:5829–5833.
31. Board PG, Webb GC, McKee J, Ichinose A. Localization of the coagulation factor XIII A subunit gene (F13A) to chromosome bands 6p24→p25. *Cytogenet Cell Genet* 1988;48:25–27.
32. Olaisen B, Gedde-Dahl T, Teisberg P, et al. A structural locus for coagulation factor XIII A (F13A) is located distal to the HLA region on chromosome 6p in man. *Am J Hum Genet* 1985;37:215.
32a. Bottenus RE, Ichinose A, Davie EW. Nucleotide sequence of the gene for the b subunit of human factor XIII. Biochemistry 1990;29:11195–11209.
33. Webb GC, Coggan M, Ichinose A, Board PG. Localization of the coagulation factor XIII B subunit gene (F13B) to chromosome bands 1q31-32.1 and restriction fragment length polymorphism at the locus. *Hum Genet* 1989;81:157–160.
34. Lozier J, Takahashi N, Putnam FW. Complete amino acid sequence of human plasma β_2 glycoprotein I. *Proc Natl Acad Sci USA* 1984;81:3640–3644.
35. Kane WH, Davie EW. Blood coagulation factors V and VIII: Structural and functional similarities and their relationship to hemorrhagic and thrombotic disorders. *Blood* 1988;71:539–555.
36. van Dieijen G, Tans G, Rosing J, Hemker HC. The role of phospholipid and factor VIIIa in the activation of bovine factor X. *J Biol Chem* 1981;256:3433.
37. Rosing J, Tans G, Grovers-Riemslag JWP, Zwaal RFA, Hemker HC. The role of phospholipids and factor Va in the prothrombinase complex. *J Biol Chem* 1980;255:274–283.
38. Gitschier J, Wood WI, Goralka TM, et al. Characterization of human factor VIII gene. *Nature* 1984;312:326–330.
39. Drayna D, White R. The genetic linkage map of the human X chromosome. *Science* 1985;230:753.
40. Kane WH, Davie EW. Cloning of a cDNA coding for human factor V, a blood coagulation factor homologous to factor VIII and ceruloplasmin. *Proc Natl Acad Sci USA* 1986;83:6800–6804.
41. Kane WH, Ichinose A, Hagen FS, Davie EW. Cloning of cDNAs coding for the heavy chain region and connecting region of human factor V, a blood coagulation factor with four types of internal repeats. *Biochemistry* 1987;26:6508–6514.
42. Jenny RJ, Pittman DD, Toole JJ, et al. Complete cDNA and derived amino acid sequence of human factor V. *Proc Natl Acad Sci USA* 1987;84:4846.
43. Gilbert W. Genes-in-pieces revisited. *Science* 1985;228:823–824.
44. Patthy L. Evolution of the proteases of blood coagulation and fibrinolysis by assembly from modules. *Cell* 1985;41:657–663.

Etiology of Human Disease at the DNA Level,
edited by Jan Lindsten and Ulf Pettersson.

7

DNA Markers in Neurogenetic Disorders

James F. Gusella

Molecular Neurogenetics Laboratory, Massachusetts General Hospital and Department of Genetics, Harvard University, Boston Massachusetts 02114

Direct identification of an abnormal protein or a reduced enzyme activity has been the traditional method for defining the primary defects in genetic disorders, but has been applicable mainly to recessive diseases. Once identified as the site of the genetic defect, the gene encoding the candidate protein now usually can be cloned with relative ease. However, most disorders affecting primarily the adult nervous system are inherited in a dominant fashion. Moreover, since the disorders affect a tissue of very high complexity that is prone to secondary alterations and is limited in its accessibility, a direct search for the defective protein is not likely to bear fruit. The advent of polymorphic markers in the form of readily detectable DNA variations throughout the human genome has advanced a potent alternative strategy for attacking such disorders (1). In this case, family studies are performed on pedigrees in which the disorder is segregating in order to identify a marker genetically linked to the disease. The knowledge of the chromosomal location of the genetic defect that is inferred from linked markers can then be employed in designing strategies to isolate the disease gene and thereby determine the nature of the defective protein it encodes.

We and our collaborators have been successful in the first step of this strategy in six distinct disorders. In 1983, in the first assignment of an autosomal disease locus using this strategy, we identified a DNA marker that placed the Huntington's disease defect on chromosome 4 (2). Subsequently, we have localized the respective genetic defects in a form of familial Alzheimer's disease to chromosome 21 (3), von Recklinghausen neurofibromatosis to chromosome 17 (4), bilateral acoustic neurofibromatosis to chromosome 22 (5), von Hippel-Lindau disease to chromosome 3 (6), and torsion dystonia to chromosome 9 (7). In each case, more detailed regional mapping of the disease locus has been undertaken in order to provide the foundation for attempting to clone and characterize the genetic defect. Here I will review the current status of our work on the first four of these inherited neurological disorders.

HUNTINGTON'S DISEASE

Huntington's disease (HD) is a progressive neurodegenerative disorder manifesting motor abnormalities, personality changes, and cognitive decline (8). The tell-tale signs of the autosomal dominant genetic defect are most easily recognized in the uncontrolled dance-like movements or chorea that it induces in all parts of the body. HD displays a distinctive pattern of neuropathology, with a dramatic loss of neurons in the neostriatum, but the biochemical basis for this neuronal death is not known. Symptoms of HD first appear between 35 and 50 years of age in most cases, but can occur as early as age 2 or as late as age 80. Essentially all carriers of the defective gene eventually succumb to this untreatable disorder.

The discovery in 1983 of a DNA marker linked to HD, which opened a new path to research in this disorder, was made possible by the availability of an enormous HD pedigree from Venezuela that was ideal for genetic linkage studies (2,9). The DNA marker D4S10 assigned the HD gene to chromosome 4, but the region of the genetic defect was progressively narrowed to only the 4p16.3 subband by a combination of *in situ* hybridization, dosage studies, somatic cell hybrid mapping, and genetic linkage analysis (10–17). A collaborative effort catalyzed by the Hereditary Disease Foundation and involving a number of different groups (D. Housman, Massachusetts Institute of Technology; A-M Frischauf and H. Lehrach, Imperial Cancer Research Fund; F. Collins, University of Michigan; J. Gusella, Massachusetts General Hospital, and J. Wasmuth, University of California at Irvine) was formed several years ago to foster an accelerated approach to cloning the HD gene through cooperative effort. The net result has been the development of detailed genetic and physical maps of 4p16.3 (18–21) (Fig. 1).

Unfortunately, analysis of recombination events in HD pedigrees has to date failed to confine the genetic defect to a single limited portion of the map (18). Two candidate regions remain. A small candidate region at the very tip of 4p is supported by several apparent recombination events (18,22), but comprises only about 100,000 bp of DNA (23). The region contains a large number of different repetitive sequences, including subtelomeric repeats shared with other chromosomes, and does not seem to be a particularly hospitable location for a protein coding sequence. The alternative candidate region, supported by only a couple of recombination events in HD pedigrees, is located internally in 4p16.3 but is much larger and contains one of the two remaining gaps in the pulsed-field gel restriction map of 4p16.3 (24). A limited number of sites within two distinct portions of this internal candidate region of 2 Mb have displayed linkage disequilibrium with the HD mutation (25,26).

A resolution of which candidate region contains the HD gene will require the availability of a DNA marker at the very telomere of the chromosome

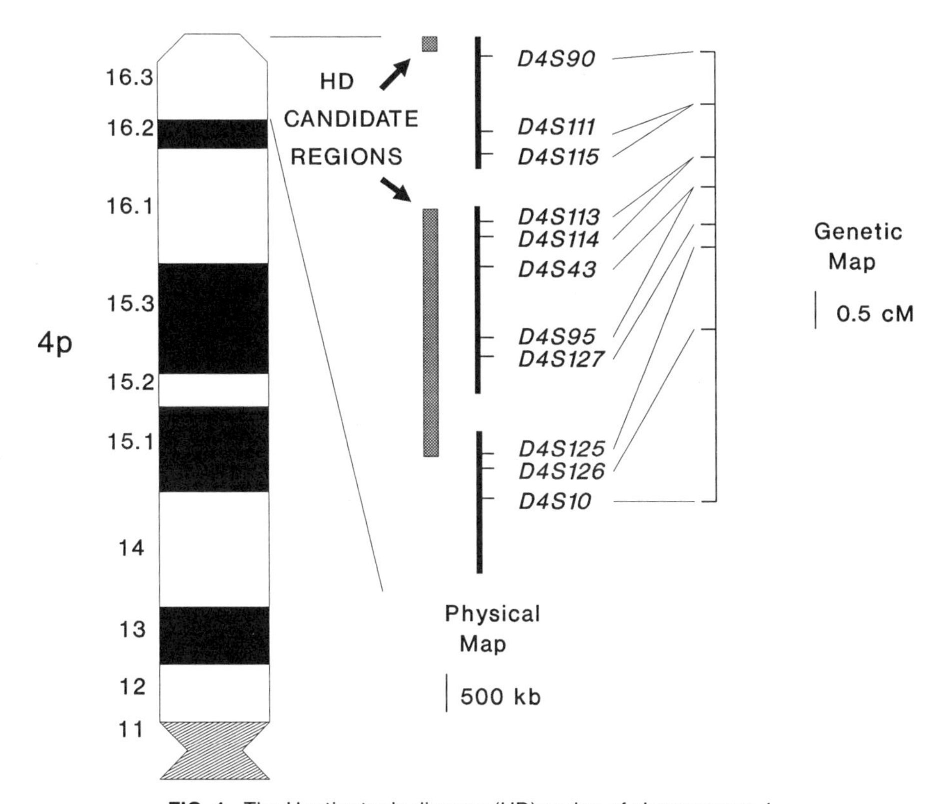

FIG. 1. The Huntington's disease (HD) region of chromosome 4.

to determine which set of apparent recombination events is correct. The others must then be explained as misdiagnoses, double recombinants, or gene conversions. Linkage disequilibrium would tend to support the internal location, but does not hone in on a small area. Consequently, in the absence of additional recombination data to narrow the search, it may be necessary to scan all genes in this internal region for evidence of the HD mutation.

ALZHEIMER'S DISEASE

Alzheimer's disease (AD) has been a much more difficult disorder in which to apply the genetic linkage strategy than HD (27). AD is a late onset disorder involving a progressive dementia that typically begins after age 70. However, as there are several other possible causes of senile dementia, AD can most reliably be identified by a post-mortem neuropathologic examination to observe the neuritic plaques of insoluble amyloid and the intracellular neurofibrillary tangles characteristic of the disorder. Few cases of AD show clear heritability, but a very small minority are represented in large pedigrees typically with early onset of the disorder (<55 years) and an apparent autosomal dominant pattern of inheritance. We have used this form of AD, known as familial Alzheimer's disease (FAD) to apply the genetic linkage strategy.

We targeted chromosome 21 (Fig. 2), because of the similarity in neuropathology observed in AD and Down's syndrome (DS), and found evidence for coinheritance of markers in the proximal portion of 21q with FAD in four large pedigrees (3). Chromosome 21 was also found to encode the precursor of the amyloid found in the senile plaques of both AD and DS, but this gene (*APP*) was not the site of the genetic defect in the FAD pedigrees (28,29). Instead, the FAD defect was located closer to the centromere.

Two subsequent studies using different populations and diagnostic criteria failed to confirm this linkage (30,31) until Goate et al. (32) published an independent confirmation using the same DNA markers. A large collaborative study involving the majority of the groups now working on FAD has pooled all available data and concluded that the etiology of AD is heterogeneous (33). The overall evidence for a genetic locus on chromosome 21 remains solid, but this genetic defect appears to explain the presence of the disease

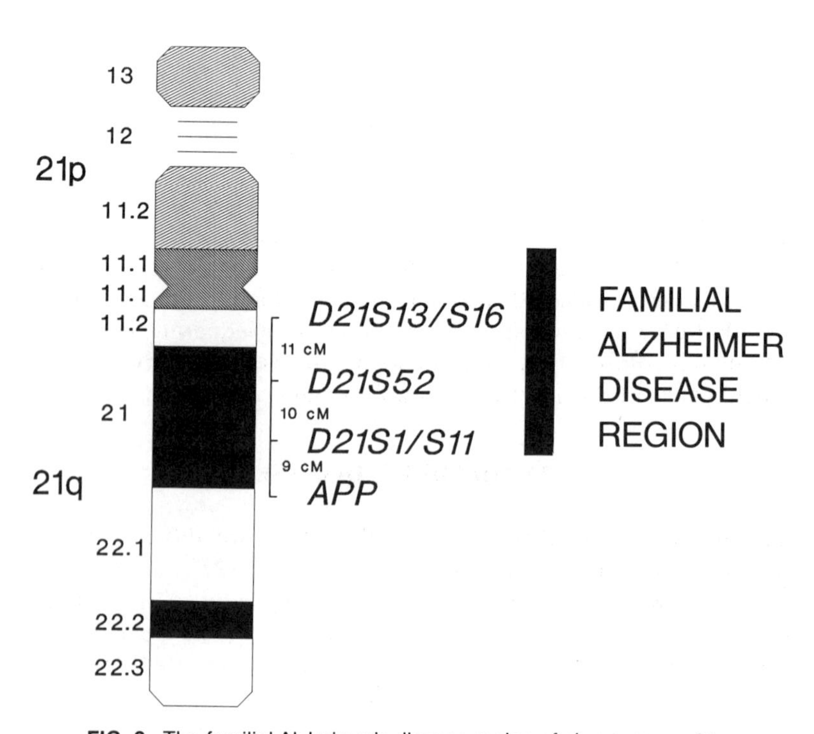

FIG. 2. The familial Alzheimer's disease region of chromosome 21.

in probably only a subset of early onset families. The remaining early onset families and those cases with late onset could represent a number of causes, from a second fully penetrant genetic locus in some cases, to some combination of loci showing reduced penetrance, genetic predisposition, or environmental agents in others. Unfortunately, this multiplicity of causes, and the possibility of more than one being present within a given small pedigree, combined with the difficulty of diagnosis and the late age of onset, make fine localization of the FAD defect on chromosome 21 problematic. Similarly, identification of loci on other chromosomes that could be involved in some families is an equally daunting task, suggesting that completion of the "location cloning" strategy in FAD will be quite difficult.

NEUROFIBROMATOSIS

The term *neurofibromatosis* (NF) has been applied to two disorders that share overlapping phenotypes but are genetically distinct. Both von Recklinghausen's NF (NF1) and bilateral acoustic NF (NF2) involve the disordered propagation of Schwann cells to form primarily benign tumors. However, in general the effects of the former are more readily visible in the periphery while the latter most frequently attacks the central nervous system.

NF1 is very frequent, with an incidence of about 1 in 3,000, but is extremely variable in its expression (34). It is seen as an autosomal dominant disorder in all races, although half of all cases are considered to represent new mutations. The cardinal signs of NF1 are: the formation of benign cutaneous neurofibromas, which may be extremely numerous and are disfiguring in severe cases; the increased frequency of café-au-lait spots in the skin; and the presence of Lisch nodules or raised hamartomas in the iris. Patients with NF1 can also suffer from mental retardation, learning disabilities, macrocephaly, bone abnormalities, etc., as well as being predisposed to malignant tumors such as neurofibrosarcomas and optic gliomas.

NF2 is much less common than NF1 (perhaps 1/50,000) but is much more consistent with its presentation and often more devastating for the patient (35). The hallmark of NF2 is the formation of bilateral acoustic neuromas, Schwann cell tumors of the eighth cranial nerve, but NF2 patients also suffer from multiple recurrent meningiomas and spinal nerve root neurofibromas. The tumor types seen in NF2 are also seen with greater frequency in the general population, but as solitary sporadic tumors, such as unilateral acoustic neuromas or single meningiomas. Patients with NF2 do not show an increase in the number of cutaneous café-au-lait spots and do not possess Lisch nodules.

Our approach to neurofibromatosis was a bipartite one: 1) We searched for consistent chromosomal rearrangements in the tumors associated with each disorder that might denote the presence of the disease gene on a particular chromosome; and 2) We undertook a genetic linkage study using candidate genes as well as highly polymorphic loci throughout the genome as linkage markers. The first route was quickly successful in NF2. We detected specific loss of alleles from chromosome 22 in sporadic acoustic neuromas and in sporadic meningiomas suggesting that the two might have a common etiology and result from loss of a tumor suppressor locus (36,37). Subsequent analysis of bilateral acoustic neuromas from NF2 patients demonstrated the same specific chromosome 22 loss with high frequency, as did meningiomas and neurofibrosarcomas in NF2 (38). Furthermore, multiple independent tumors from the same individual always lost the same allele, suggesting the deletion of a gene whose homologous allele was already inactivated as a result of germ-line mutation (39). As a final test of the tumor suppressor model, we demonstrated that in a large pedigree with NF2, the genetic defect shows linkage to DNA markers in the same chromosomal region showing most frequent loss in the tumors (5). Thus, all evidence is consistent with the existence on chromosome 22 of a tumor suppressor gene analogous to that in retinoblastoma, but effective primarily in neural crest cells. We have undertaken a combination of tumor deletion studies and further genetic linkage analyses to identify recombination events, with the results summarized in Fig. 3. The *NF2* gene has been confined to a region of

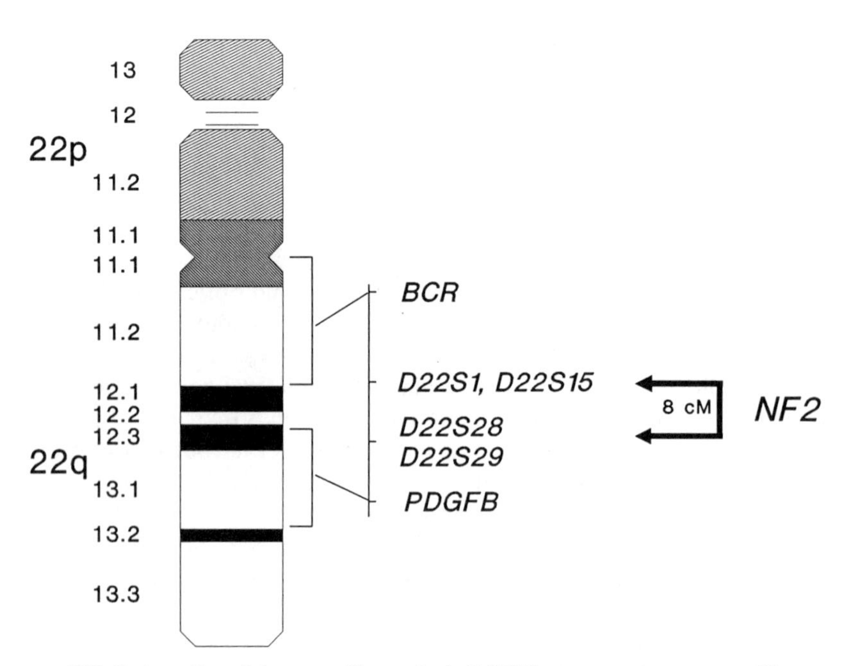

FIG. 3. Location of the neurofibromatosis 2 (NF2) gene on chromosome 22.

8 cM in the central portion of the long arm of chromosome 22, and our efforts are now focused on the saturation of this region with cloned DNA (39).

NF1 was ideally suited to the genetic linkage route because of its high frequency in the population and relative ease of early diagnosis. The tumor analysis strategy was less applicable to NF1 because the benign neurofibromas contain a mixture of cell types that might mask any chromosomal alteration in a subpopulation, such as Schwann cells. A large number of groups set out to localize the *NF1* gene, and after excluding much of the genome, two reported linkage to DNA markers on chromosome 17 (4,40). In our case, the linked marker was the candidate gene encoding the nerve growth factor receptor (*NGFR*), but while it assigned the defect to chromosome 17, a number of recombinants were detected indicating that *NGFR* was not the site of the NF1 mutation.

Subsequent to the initial reports of linkage, *NF1* has been narrowed to a stretch of DNA located between flanking markers on the proximal long arm of chromosome 17. Attention has focused on the analysis of two inherited translocation chromosomes that cosegregate with NF1 in specific pedigrees and appear to mark the site of the disease gene (40,41) (Fig. 4). An intensive analysis of genes in the vicinity of these translocations is now being carried out, but as yet no definitive *NF1* candidate has emerged. However, since the site of the *NF1* gene has apparently been pinpointed, the isolation and characterization of the defect appears imminent.

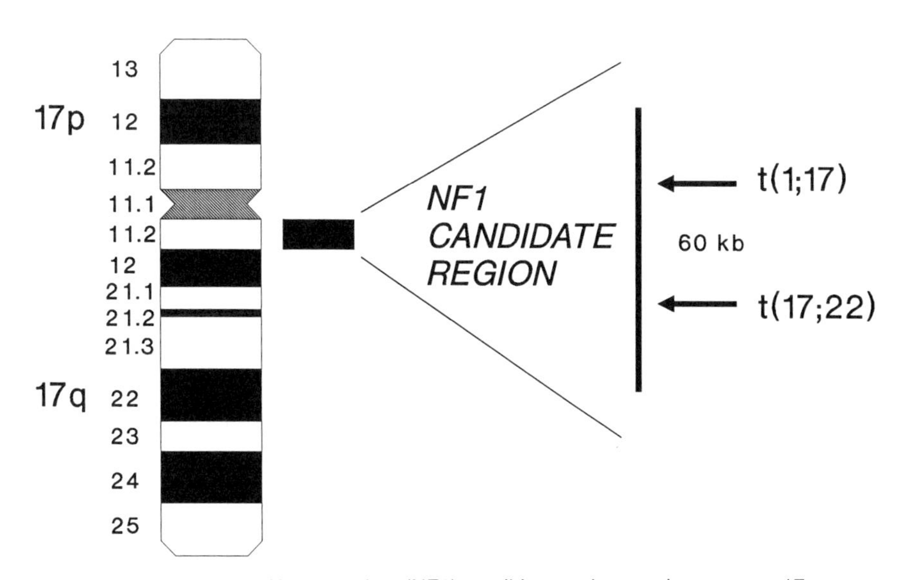

FIG. 4. The neurofibromatosis 1 (NF1) candidate region on chromosome 17.

Once *NF1* had been mapped on 17q, we also undertook a more concerted effort to analyze *NF1*-related tumors for loss of alleles on this chromosome that might support the notion that *NF1* is a tumor suppressor gene. The benign neurofibromas consistently showed retention of both copies of chromosome 17, but since these represent mixed cell populations no firm conclusion was possible. However, malignant neurofibrosarcomas from NF1 that comprise a pure population of transformed cells showed frequent loss of chromosome 17 alleles (43). Analysis of a number of these tumors indicated that although some displayed loss of the *NF1* region, others showed loss of alleles only on 17p. In fact, the most common region of loss centered on the region of 17p containing the known tumor suppressor gene *TP53*. DNA sequence analysis of the remaining copy of the *TP53* gene in these tumors has revealed that some have undergone point mutation, suggesting that alteration or inactivation of the *TP53* gene is an important step in the progression of neurofibrosarcomas. The question of whether the *NF1* gene is also a tumor suppressor gene that must be inactivated for tumor formation to occur remains an open question.

SUMMARY

The chromosomal localization of inherited neurological diseases has become straightforward and has been highly successful. Effort in many disorders is now concentrating on the steps necessary to move from linked markers to the disease gene itself. Each disorder presents a different problem, depending on its chromosomal location, the nature of the disorder and its phenotype, and the biological and pedigree resources available to support the search for the defect. In some cases, the task of identifying the disease gene will soon be successful while in others many years may be necessary. In almost all cases, however, the directed approach of aiming at the primary cause of the disorder is eventually likely to provide a much deeper understanding of the fundamental disease process than complete reliance on investigation of the diseased tissue, which must typically be limited to descriptive studies.

ACKNOWLEDGMENTS

This work is supported by NIH grants NS16367, NS22224, NS24279, and NS20012, by the National Neurofibromatosis Foundation, the Hereditary Disease Foundation, and the American Health Assistance Foundation.

REFERENCES

1. Gusella JF. DNA polymorphism and human disease. *Annu Rev Biochem* 1986;55: 831–854.

2. Gusella JF, Wexler NS, Conneally PM, et al. A polymorphic DNA marker genetically linked to Huntington's disease. *Nature* 1983;306:234–238.
3. St George-Hyslop PH, Tanzi RE, Polinsky RJ, et al. The genetic defect causing familial Alzheimer's disease maps on chromosome 21. *Science* 1987;235:885–890.
4. Seizinger BR, Rouleau GA, Ozelius LJ, et al. Genetic linkage of von Recklinghausen neurofibromatosis to the nerve growth factor receptor gene. *Cell* 1987;49:589–594.
5. Rouleau GA, Wertelecki W, Haines JL, et al. Genetic linkage of bilateral acoustic neurofibromatosis to a DNA marker on chromosome 22. *Nature* 1987;329:246–248.
6. Seizinger BR, Rouleau GA, Ozelius LJ, et al. Von Hippel-Lindau disease maps to the region of chromosome 3 associated with renal cell carcinoma. *Nature* 1988;332:268–269.
7. Ozelius L, Kramer PL, Moskowitz CB, et al. Human gene for torsion dystonia located on chromosome 9q32-q34. *Neuron* 1989;2:1427–1434.
8. Martin JB, Gusella JF. Huntington's disease: Pathogenesis and management. *N Engl J Med* 1986;315:1267–1276.
9. Gusella JF, Tanzi RE, Anderson MA, et al. DNA markers for nervous system diseases. *Science* 1984;225:1320–1326.
10. Gusella JF, Tanzi RE, Bader PI, et al. Deletion of Huntington's disease-linked G8(*D4S10*) locus in Wolf-Hirschhorn syndrome. *Nature* 1985;318:75–78.
11. Zabel BU, Naylor SL, Sakaguchi AY, Gusella JF. Mapping of the DNA locus *D4S10* and the linked Huntington's disease gene to 4p16-p15. Cytogenet Cell Genet 1986;42: 187–191.
12. Magenis RE, Gusella J, Weliky K, et al. Huntington disease-linked restriction fragment length polymorphism localized within band p16.1 of chromosome 4 by in situ hybridization. *Am J Hum Genet* 1986;39:383–392.
13. Wang HS, Greenberg CR, Hewitt J, Kalousek D, Hayden MR. Subregional assignment of the linked marker G8 (D4S10) for Huntington's disease to chromosome 4p16.1-16.3. *Am J Hum Genet* 1986;39:392–396.
14. Landegent JE, Jansen in de Wal N, Fisser-Groen YM, Bakker E, Van der Ploeg M, Pearson PL. Fine mapping of the Huntington disease linked D4S10 locus by non-radioactive in situ hybridization. *Hum Genet* 1986;73:354–357.
15. MacDonald ME, Anderson MA, Gilliam TC, et al. A somatic cell hybrid panel for localizing DNA segments near the Huntington's disease gene. *Genomics* 1987;1:29–34.
16. Smith B, Skarecky D, Bengtsson U, Magenis RE, Carpenter N, Wasmuth JJ. Isolation of DNA markers in the direction of the Huntington disease gene from the G8 locus. *Am J Hum Genet* 1988;42:335–344.
17. Gilliam TC, Tanzi RE, Haines JL, et al. Localization of the Huntington's disease gene to a small segment of chromosome 4 flanked by D4S10 and the telomere. *Cell* 1987;50:565–571.
18. MacDonald ME, Haines JL, Zimmer M, et al. Recombination events suggest possible locations for the Huntington's disease gene. *Neuron* 1989;3:183–190.
19. Youngman S, Sarafarazi M, Bucan M, et al. A new DNA marker [D4S90] is terminally located on the short arm of chromosome 4 close to the Huntington's disease gene. *Genomics* 1989;5:802–809.
20. Bucan M, Zimmer M, Whaley WL, et al. Physical maps of 4p16.3, the area expected to contain the Huntington's disease mutation. *Genomics* 1990;6:1–15.
21. Allitto BA, MacDonald ME, Bucan M, et al. Increased recombination adjacent to the Huntington's disease-linked *D4S10* marker. *Genomics*, in press.
22. Robbins C, Theilman J, Youngman S, et al. Evidence from family studies that the gene causing Huntington disease is telomeric to *D4S95* and *D4S90*. *Am J Hum Genet* 1989;44:422–425.
23. Bates GP, MacDonald ME, Baxendale S, et al. A YAC telomere clone spanning a favoured location of the Huntington's disease gene. *Am J Hum Genet* 1990;46:762–775.
24. Whaley L, Bates G, Novelletto A, et al. Mapping of cosmid clones in the Huntington disease region of chromosome 4. *Somatic Cell and Molecular Genetics*, in press.
25. Snell RG, Lazarou L, Youngman S, et al. Linkage disequilibrium in Huntington's disease: An improved localization for the gene. *J Med Genet* 1989;26:673–675.
26. Theilman J, Kanani S, Shiang R, et al. Non-random association between alleles detected at D4S95 and D4S98 and the Huntington's disease gene. *J Med Genet* 1989;26:676–681.
27. St George-Hyslop PM, Myers RH, Haines JL, et al. Familial Alzheimer's disease: Progress and problems. *Neurobiol Aging* 1989;10:417–425.

28. Tanzi RE, St. George-Hyslop PH, Haines JL, et al. The genetic defect in familial Alzheimer disease in not tightly linked to the amyloid beta protein gene. *Nature* 1987;329:156–157.
29. Von Broeckhoven C, Genthe AM, Vandenberghe A, et al. Failure of familial Alzheimer's disease to segregate with the A4-amyloid gene in several European families. *Nature* 1987;329:153–155.
30. Pericak-Vance MA, Yamaoka LH, Haynes CS, et al. Genetic linkage studies in Alzheimer's disease families. *Exp Neurol* 1988;102:271–279.
31. Schellenberg GD, Bird TD, Wijsman EM, et al. Absence of linkage of chromosome 21q21 markers to familial Alzheimer's disease. *Science* 1988;241:1507–1510.
32. Goate AM, Haynes AR, Owen MJ, et al. Predisposing locus for AD on chromosome 21. *Lancet* 1989;1:352–355.
33. St George-Hyslop PH, Haines JL, Farrer LA, et al. Genetic linkage studies suggest that Alzheimer's disease is not a single homogeneous entity. *Nature,* 1990;347:194–197.
34. Riccardi VM, Eichner JE. *Neurofibromatosis: Phenotype, Natural History, and Pathogenesis*. Johns Hopkins University Press, Baltimore, MD 1986.
35. Martuza RL, Eldridge R. Neurofibromatosis 2 (bilateral acoustic neurofibromatosis). *N Engl J Med* 1988;318:684–688.
36. Seizinger BR, Martuza RL, Gusella JF. Loss of genes on chromosome 22 in tumorigenesis of human acoustic neuroma. *Nature* 1986;322:644–647.
37. Seizinger BR, de la Monte S, Atkins L, Gusella JF, Martuza RL. A molecular genetic approach to human meningioma: Loss of genes on chromosome 22. *Proc Natl Acad Sci USA* 1987;84:5419–5423.
38. Seizinger BR, Rouleau G, Ozelius LJ, et al. Common pathogenetic mechanism for three tumor types in bilateral acoustic neurofibromatosis. *Science* 1987;236:317–319.
39. Rouleau GA, Seizinger BR, Wertelecki W, et al. Flanking markers bracket the neurofibromatosis type 2 (NF2) gene on chromosome 22. *Am J Hum Genet* 1990;46:323–328.
40. Barker D, Wright E, Nguyen K, et al. Gene for von Recklinghausen neurofibromatosis is in the pericentric region of chromosome 17. *Science* 1987;236:1100–1102.
41. Ledbetter DH, Rich DC, O'Connell P, Leppert M, Baty BJ, Carey JC. Precise localization of NF1 to 17q11.2 by balanced translocation. *Am J Hum Genet* 1989;44:20–24.
42. Menon AG, Ledbetter DH, Rich DG, et al. Characterization of a translocation within the von Recklinghausen neurofibromatosis region of chromosome 17. *Genomics* 1989;5:245–249.
43. Menon AG, Anderson KM, Riccardi VM, et al. The common region of deletions on chromosome 17 in neurofibrosarcomas is distinct from the region containing the NF1 gene. *Proc Natl Acad Sci USA,* 1990;87:5435–5439.

Etiology of Human Disease at the DNA Level,
edited by Jan Lindsten and Ulf Pettersson.

8

Genetic Predisposition to Type I Diabetes Mellitus: The Role of Gene Products of the Major Histocompatibility Complex

Hugh O. McDevitt

Departments of Microbiology and Immunology, and Medicine, Stanford University School of Medicine, Stanford, California 94305

Type I insulin-dependent diabetes mellitus (IDDM) manifests in childhood and early adolescence, and is characterized primarily by destruction of the β cells in the islets of Langerhans, resulting in complete insulin deficiency. The disease occurs sporadically in the population, and ultimately affects 0.5–0.6% of the population. There is abundant evidence suggesting that the disease is autoimmune in nature (1). At the time of onset, more than 90% of patients have autoantibodies to human insulin, or to one of several poorly characterized islet-cell-specific antigens. Initiation of the autoimmune process is multifactorial, controlled by several genes, and results in clinically asymptomatic destruction of the insulin-producing β cells in the islets of Langerhans.

There are several reasons to suspect that the disease is clinically heterogeneous (1). Because the process results in no symptoms until all of the insulin-producing cells have been destroyed, it is clear that effective prevention of the disease depends upon identifying susceptible individuals in the population, and initiating treatment prior to the completion of the pathogenic process.

It is apparent that environmental factors (e.g., infectious agents) play a major role in the initiation of IDDM. Thus, the concordance rate for type I diabetes in monozygotic twin pairs is only 30–40%. Furthermore, since major histocompatibility complex (MHC) identical siblings show a concordance rate of only 5–10% (in contrast to the rate in monozygotic twins), it is clear that several genes other than genes in the human leukocyte antigen (HLA) complex—the human MHC—are also involved in susceptibility to this disease. Evidence from the non-obese diabetic (NOD) mouse strain, an excellent spontaneous animal model of human type I diabetes, indicates that there are at least three, and possibly as many as six independent recessive genes that predispose to type I diabetes (2).

Following earlier findings showing that 95% of patients possess either the HLA-DR3 or -DR4 class II MHC alleles (see references in ref. 3), extensive sequence analysis of the HLA-DR and -DQ genes in both normal and diabetic individuals, as well the I-A genes in the NOD mouse, has suggested that susceptibility is determined to a large extent by alleles of the HLA-DQ molecule, and by its murine homologue, the I-A molecule. Moreover, the residue in position 57 of the DQβ chain is believed to make a major contribution to susceptibility or resistance to type I diabetes (4). These observations confirmed earlier studies (5) using restriction fragment length polymorphism analysis, which indicated that the HLA class II MHC DQ genotype was the major MHC-linked factor in determining susceptibility to IDDM.

The acute insulitis seen at onset in NOD mice and humans, the presence of a variety of autoantibodies specific for β-cell-expressed molecules, and the strong association with particular HLA-DQ alleles all indicate that the destruction of β cells is due to an autoimmune process. Although the precise events initiating this autoimmune process remain obscure, there is extensive evidence, particularly in the case of congenital rubella in man, and lymphocytic choriomeningitis virus infection in the mouse, that specific viral infections may greatly increase or decrease the incidence of type I diabetes in genetically susceptible individuals (6–8).

ROLE OF MHC CLASS II MOLECULES IN IMMUNITY AND AUTOIMMUNITY

A number of recent discoveries have shed light upon the critical factors involved in the initiation of a normal immune response, and presumably, an autoimmune response (9–11). To initiate a $CD4^+$ helper T-cell response, the T-cell receptor must recognize a peptide fragment of a foreign or self protein antigen found in the binding groove of a self class II MHC molecule. The peptide fragment bound by self class II MHC is derived from intracellular proteolytic digestion of native foreign or self proteins. The clarification of the molecular structure of class I (12,13) and class II (14) MHC molecules has greatly enhanced our understanding of the peptide-binding capabilities of MHC molecules.

It is now apparent that one major goal in unraveling the molecular pathogenesis of autoimmunity is to obtain complete characterization of the MHC molecules responsible for presenting critical self peptides to specific helper T cells that initiate the autoimmune response. This task requires isolation of T-cell clones, which are capable of transferring the disease in question, in this case IDDM, determination of the target antigens and the immunodominant peptide epitope of that antigen, and identification of the MHC molecules that present the critical peptides to the T-cell clones. With this information, the molecular steps involved in the initiation of autoimmune damage can be elucidated.

Unfortunately, for most autoimmune diseases, including IDDM, rheumatoid arthritis, multiple sclerosis, and many others, the target antigen(s) has not yet been identified. As a result, attempts to characterize in detail T-cell receptor repertoire utilization in these diseases remains difficult. In the absence of knowledge of the target autoantigen, the first step in analysis must focus on the MHC molecules, which mediate genetic susceptibility to a specific autoimmune disease.

CLASS II MHC ALLELES PREDISPOSING TO IDDM IN MAN

As noted above, MHC class II allelic sequencing of HLA-DR and HLA-DQ gene products from normal and diabetic individuals demonstrates that the alleles found in patients were the same as those in the general population (4). When all of the available sequence data for HLA-DRβI, -DRβIII, -DQαI, and -DQβI loci were compared with respect to an increase, a decrease, or no change in the frequency of that allele in IDDM patients versus normals, residue 57 of the DQβ chain correlated consistently with susceptibility (4). Among the subtypes of the DR2,-4, and -6 haplotypes, the DQ alleles found to be positively associated with IDDM encoded a non-charged amino acid (alanine, serine, or valine) at position 57. In contrast, the DQ alleles found to be decreased in incidence in IDDM patients, or to be positively associated with *resistance* to IDDM encoded an aspartic acid residue at DQβ 57. Furthermore, Acha-Orbea (15) had earlier shown that the β-chain sequence of the NOD mouse is unique among inbred murine strains in possessing serine at position 57 instead of aspartic acid, and histidine at position 56.

Although the lack of an aspartic acid residue at DQβ 57 correlated well with susceptibility to type I diabetes (4,16), a number of exceptions were noted. One such exception is the DQβ1 allele, DQw2, which on the DR3 haplotype is associated with IDDM susceptibility, but on the DR7 haplotype in Caucasians is neutral or negative with respect to susceptibility. It was observed that the α chain of the DQw2 allele on the Caucasian DR7 haplotype was different from that seen on the DR3 haplotype. An explanation for these findings came from the work of Todd et al. (17), who demonstrated that among Black IDDM patients, the DR7 haplotype identical to that seen in Caucasians was decreased in incidence. However, a recombinant DR7 haplotype was also found in the Black IDDM population that carried the same DQw2β1 allele, but carried the DQα allele (DQA3) from the Caucasian DR4 haplotype.

This new DR7 haplotype apparently arose by recombination, and was positively associated with susceptibility to IDDM in Blacks, with a relative risk of 13. This result indicates that while the residue at position 57 in the DQβ1 allele is a major factor in susceptibility to IDDM, the sequence in the DQα1 allele is also important, and in some cases, can override the susceptibility mediated by a nonaspartic acid residue at DQβ57. Further analysis of HLA-D region associations with IDDM in Black and Japanese populations has shown that the DQA3 allele of the DQα1 locus is most strongly associated with susceptibility to IDDM in Japanese patients (most of whom have aspartic acid at DQβ 57), and in Black patients carrying a DQβ Asp 57^+ allele.

It should be noted that the incidence of IDDM in Japanese is the lowest in the world, and is from 1/10th to 1/30th of that seen in Caucasians. Furthermore, the incidence of diabetes in Blacks is also much lower than in Caucasians. These DQβ Asp 57^+ diabetic patients are apparently similar to the 10–15% of Caucasian diabetics who also carry a DQβ Asp 57^+ allele. Thus, in Caucasians, the amino acid at DQβ 57 correlates most consistently with susceptibility or resistance to IDDM, while in Japanese patients, the DQα1 correlates more consistently with susceptibility to diabetes. It is entirely possible that these different associations may be due to heterogeneity within the disease, or to recognition of a separate target autoantigen or a separate peptide epitope on the same autoantigen in the two populations.

In any event, it seems clear that the amino acid sequence of the DQ molecule is the best candidate for mediating susceptibility or resistance to IDDM. In the majority of Caucasian patients, the amino acid sequence at DQβ 57 is a critical factor, while in DQβ Asp 57^+ Japanese patients and Black patients, the allele at the DQα locus is of major importance.

The mechanism by which HLA-DQ molecules mediate susceptibility or resistance to IDDM is speculative (3,18,19). Since most patients are homozygous for a nonaspartic residue at DQβ 57, inheritance of the ability to mount an autoimmune response in IDDM is recessive. If, as postulated above, HLA-DQ is acting as a restriction element for the presentation of a β-cell-specific peptide to T cells, the recessive inheritance pattern is not typical of the immune response function of MHC class II molecules. Dominant nonresponsiveness may be due to active suppression induced by the self peptide bound in a DQβ Asp 57^+ peptide binding site, or might be due to tolerance. Tolerance could conceivably take two forms. In the first, the sequence of a DQβ Asp 57^+ chain may be similar or identical to the critical peptide epitope of an islet cell autoantigen. In the second, some other self peptide in the DQβ Asp 57^+ binding site may delete that population of T cells required for recognition of an islet cell autoantigen in the peptide binding site of a DQβ Asp 57^- molecule. There is some evidence for a suppressor mechanism in the NOD mouse (20). Analysis of the precise mechanism would be greatly facilitated by identification of the target autoantigen.

CLASS II MOLECULES MEDIATING SUSCEPTIBILITY TO IDDM IN THE NOD MOUSE

As alluded to earlier, NOD mice express the murine counterpart of the HLA-DQ molecule, the I-A molecule, and lack the homologue of HLA-DR, the I-E molecule. The α chain in the NOD mouse is identical to I-A^d, while the β chain is unique (15). The NOD β chain is most similar to I-A^f, but has two striking sequence differences that distinguish it from all other inbred I-Aβ chain sequences. Histidine is found in place of proline at position 56, and serine is found in place of aspartic acid at position 57.

This observation was one of several that led to the suggestion that the residue at position 57 in the I-A/DQβ chain was an important factor in determining susceptibility or resistance to IDDM. A gene in the MHC, presumably I-A, is one of three recessive genes required for susceptibility to IDDM in the NOD mouse (2). The NOD mouse thus provides a spontaneous experimental animal model of IDDM susceptibility in which the effect of specific amino acid sequence changes in the I-Aβ chain on susceptibility to IDDM can be assessed directly.

Initially, experiments (Acha-Orbea and McDevitt et al., unpublished results) were carried out in which an I-$A\beta^d$ genomic transgene was introduced into NOD embryos. I-$A\beta^d$ has several sequence differences from I-$A\beta^{NOD}$ and encodes proline and aspartic acid at positions 56 and 57, respectively. Unexpectedly, these animals became immunodeficient due to a block in B-cell development that was attributed to extremely high levels of I-$A\beta^d$ mRNA expression. Consequently, these mice could not be used to assess the effect of an Asp 57^+ β chain on diabetes susceptibility. Interestingly, this phenotype was essentially identical with that of another transgenic line carrying 60–120 copies of I-$A\beta^k$, and also suffering from immunodeficiency (21,22). In the mean time, several other authors have reported the effects of a variety of transgenes on the development of insulitis and/or IDDM in the NOD mouse (summarized in Table 1).

Studies in which an I-$A\beta^k$ transgene or an I-$A\alpha^k$ transgene alone were introduced in NOD mice demonstrated that the expression of these molecules failed to prevent insulitis or diabetes. The lack of a noticeable affect

TABLE 1. *Effect of I-A transgenes on insulitis or diabetes in the NOD mouse*

Transgene(s)	Insulitis (%)	Diabetes (no.)	Reference
$A\beta^d$	B-cell deficiency		Acha-Orbea et al., unpublished
$A\beta^k$ or $A\alpha^k$	83	—	23
$A\alpha^k/A\beta^k$	10–20		24
$A\alpha^k/A\beta^k$	31	—	23
$A\alpha^k/A\beta^k$; 57 Ser	17	—	23
$A\beta^{NOD}$; 56 His→Pro	↓	0/8	25

on disease by these transgene products may be due to the relatively low expression of the I-Aα^d/I-Aβ^k heterodimer resulting from poor interchain pairing (23). Two laboratories have recently demonstrated that the introduction of an I-Aα^k and an I-Aβ^k transgene together results in a striking decrease in the incidence of insulitis, and subsequently in the incidence of diabetes (23,24). This finding, in itself, provides convincing evidence that the I-A molecule is the MHC-linked gene that contributes to susceptibility to IDDM in the NOD mouse.

Miyazaki et al. (23), have recently introduced into embryos the I-Aα^k gene and a mutated I-Aβ^k gene in which residue 57 had been changed from aspartic acid to serine via site-specific mutagenesis. Although expressed at roughly 25% of endogenous levels, the heterodimer containing the mutated Aβ^k chain protected against insulitis as well or better than the native I-Aα^k/I-Aβ^k transgene. This finding would argue that while the I-A^k molecule is capable of preventing the development of IDDM, residue 57 is not essential for its ability to prevent disease. However, since I-A^k differs from I-A^{NOD} at a number of residues, this molecule may not be the most appropriate to test the effect of residue 57 alone on susceptibility.

Clearly, the most direct approach to test the role of residue 57 in the Aβ chain in mediating susceptibility to IDDM is to mutate residue 57 in the Aβ^{NOD} chain from serine to aspartic acid. Since the immediately adjacent residue in position 56 in the Aβ^{NOD} chain is histidine—an unusual residue at this position—it may also be necessary to mutate this histidine back to proline. Interestingly, Lund et al. (25) have shown (Table 1) that an Aβ^{NOD} gene in which position 56 was mutated from histidine to proline is sufficient to protect the NOD transgenics from insulitis and diabetes. While only eight mice were assessed for the development of diabetes, a total of 330 islets from these mice were found to have a marked decrease in both insulitis and diabetes. Additional experiments are still required—most notably, mutating Aβ^{NOD} at position 57 alone and changing Aβ^{NOD} at both positions 56 and 57 to resemble the sequence found in Aβ^d.

Nonetheless, these findings provide further evidence in support of a key role of this region of the Aβ chain, and, by analogy, the same region of the DQβ chain, in generating a class II molecule with a peptide binding site that is critical for mediating the autoimmune response.

Several authors (summarized in ref. 25) have reported that expression of an I-E molecule in the NOD mouse following the introduction of an Eα^k transgene can also prevent diabetes. The mechanism by which an I-E molecule can suppress the development of IDDM mediated by an I-A molecule is not yet apparent. One possibility is that the I-E molecule functions by deleting T cells expressing a critical set of Vβ receptor genes required for recognition of the I-A-islet cell peptide complex. However, recent experiments in which Eα^k transgenes containing mutated promoter sequences were introduced into the NOD mouse strain suggest otherwise (26). Mice carrying transgenes in which the promoter had been altered to direct I-E expression to specific subsets of immunocompetent cells exhibited intrathymic elimination of specific T-cell receptor Vβ subsets typically seen with the wild-type I-Eα transgene. However, only the wild-type I-Eα transgene was capable of preventing insulitis and diabetes, whereas the promoter-altered transgenes could not. These results rule out the possibility that I-E may delete specific T-cell receptor subsets, although the exact mechanism by which I-E mediates protection is still uncertain.

COMBINED GENETIC AND IMMUNOLOGIC APPROACHES TO THE PREVENTION OF TYPE I DIABETES

The evidence presented above for both human and murine IDDM indicates that susceptibility to this disease is due to the interaction of multiple genes with a marked influence by as yet unidentified environmental factors. HLA-DQ in humans, and I-A in the mouse, appear to be the most important MHC-linked loci determining susceptibility or resistance to IDDM. However, other genes in the MHC complex, such as TNFα and lymphotoxin genes, may also have a role. The amino acid at position 57 in the β chain and surrounding residues appear to be critical for determining whether a particular DQ/I-A molecule mediates susceptibility or resistance. At the same time, it is clear that residues in the α chain also may have a key function in certain α chain sequences.

The importance of residue 57 in influencing immune responsiveness and autoimmunity suggests that this residue can play a critical role in the structure of the class II MHC peptide binding site. This suggestion is borne out by the occurrence of several alleles in the human population that differ primarily at position 57. Thus, DQw7 differs from DQw8 at only four positions, one of which is position 57. Similarly, DQw9 is identical in α and β chain sequence, except at residue 57 in the β chain where DQw9 encodes an aspartic acid, while DQw8 encodes a serine. The relatively frequent occurrence in the population of similar or identical alleles, differing only at residue 57 of the β chain, indicates that both kinds of molecules—Asp 57^+ and Asp 57^-—are selected for and may have a significant survival value. Analysis of peptide MHC interactions with Asp 57^+ and Asp 57^- alleles in both mouse and man, as well as other structural studies, may explain the importance of this residue in the function of the molecule (3).

In lieu of the fact that the target autoantigen and the characteristics of the responding T-cell receptor population remain unknown, strategies for the prevention of type I diabetes must rely on the currently available information. Determination of DQ genotype at birth by allele-specific oligonucleotide hybridization would permit an identification of the DQβ non-Asp/non-Asp homozygotes in the population that comprise 85–90% of type I diabetics with an onset prior to the age of 17. Preliminary studies suggest that DQβ non-Asp/non-Asp homozygotes comprise approximately 20% of the general population. The risk of this subset of the population for IDDM is approximately 3–5%.

The ability to identify the most susceptible individuals in the population would be greatly enhanced if methods were available to detect one or two other recessive genes influencing susceptibility, of the type already described in the NOD mouse (2). Ability to type for two or three recessive genes determining susceptibility would permit a high degree of accuracy in ascertaining susceptibility. Clearly, absolute accuracy can never be achieved, since the concordance rate in monozygotic twin pairs for IDDM is only 30–50%.

Once a genetically susceptible population has been identified, these individuals can be followed periodically for the development of autoantibodies (anti-insulin, anti-islet cell cytoplasmic antibody, anti-polar antigen antibody, and anti-64Kd antibody) that are known to appear in diabetes months or years prior to the completion of the β-cell destructive process (1). Individuals who develop antibodies to one or more islet cell components can be used as candidates to initiate an immunosuppressive therapy program once signs of diminished insulin reserve become apparent, but well in advance of complete β-cell destruction. Such an approach at the present time is entirely experimental. Successful trials have been conducted with cyclosporin A, and trials are currently being contemplated with other less toxic immunosuppressive compounds such as azathioprine. In the future, in may be possible to take advantage of the evidence implicating the DQ molecule as the restricting element that mediates the autoimmune recognition of an islet cell peptide. For instance, blocking peptides or blocking drugs may be developed that have a high affinity for the DQβ 57 aspartic acid negative class II molecules, which in turn may prevent continued presentation of the islet cell peptide, and thereby interrupt and suppress the autoimmune process.

Use of such blocking peptides has resulted in the ability to *prevent* autoimmunity in the experimental allergic encephalomyelitis model (EAE) (16,27). However, it has not yet been possible to use blocking peptides to treat and suppress an ongoing autoimmune response in the EAE model. A great deal of experimentation is still required to explore completely and test this approach to allele-specific immunosuppression.

CONCLUSIONS

Although type I insulin-dependent diabetes is a sporadic disease in the population, and one in which only 10–15% of new cases occur in families with a previous case, the evidence from a large number of serologic, immunologic, and DNA sequencing studies of MHC alleles in patients and in experimental animals clearly shows that susceptibility is genetically determined by multiple genes, as well as by environmental factors. The allelic forms of one of these genes, HLA-DQ, have been characterized in sufficient detail to permit the design of preliminary approaches to identifying susceptible individuals, and developing methods for suppressing the autoimmune process prior to the development of the full clinical picture of IDDM. In the future, the characterization of other genes involved in susceptibility, the identification of the target autoantigen, and the development of HLA-DQ blocking peptides or drugs offers the possibility of preventing 80–90% of new cases of IDDM—a disease that in the past 50 years has been increasing in frequency in developed countries.

ACKNOWLEDGMENTS

These studies were supported by grants from the National Institutes of Health.

REFERENCES

1. Eisenbarth GS. Type I diabetes mellitus: a chronic autoimmune disease. *N Engl J Med* 1986;314:1360.
2. Leiter EH. The genetics of diabetes susceptibility in mice. *FASEB J.* 3:2231, 1989.
3. Todd JA. Genetic control of autoimmunity in type I diabetes. *Immunol. Today.* 1990;11:122–129.
4. Todd JA, Bell JI, and McDevitt HO. HLA-DQβ gene contributes to susceptibility and resistance to insulin-dependent diabetes mellitus. *Nature* 1987;329:599.
5. Nepom BS, Palmer J, Kim SJ, et al. Specific genomic markers for the HLA-DQ subregion discriminate between $DR4^+$ IDDM and $DR4^+$ seropositive juvenile rheumatoid arthritis. *J Exp Med* 1986;164:354.
6. Rubinstein P, Walker ME, Fedun B, Witt ME, and Cooper LZ. The HLA system in congenital rubella patients with and without diabetes. Ginsberg-Fellner F. *Diabetes* December 1982.
7. Oldstone MBA. Viruses as therapeutic agents. (Treatment of non-obese insulin-dependent mice with virus prevents insulin-dependent diabetes while maintaining general immune competence). *J Exp Med* 1990 (I);171:2077–2089.
8. Oldstone MBA, Ahmed R, and Salvato M. Viruses as therapeutic agents. (Viral reassortants map prevention of insulin-dependent diabetes mellitus to the small RNA of lymphocytic choriomeningitis virus.) 1990 (II); *J Exp Med* 171:2091–2100.
9. Babbitt BP, Allen PM, Matsueda G, Haber E, and Unanue ER. Binding of immunogenic peptides to Ia histocompatibility molecules. *Nature* 1985;317:359.
10. Grey HM, Sette A, and Buus S. How T cells see antigen. *Sci. Am.* 1989;261:56.
11. Schwartz RJ. Immune response (Ir) genes of the murine major histocompatibility complex. *Adv Immunol* 1986;38:31.
12. Bjorkman PJ, Saper MA, Samraoui B, Bennett WS, Strominger JL, and Wiley DC. Structure of the human class I histocompatibility antigen, HLA-A2. *Nature* 1987(a);329:506.
13. Bjorkman PJ, Saper MA, Samraoui B, Bennett WS, Strominger JL, and Wiley DC. The foreign antigen binding site and T cell recognition regions of class I histocompatibility antigens. *Nature* 1987(b);329:512.

14. Brown JH, Jardetzky T, Saper MA, et al. A hypothetical model of the foreign antigen binding site of class II histocompatibility molecules. *Nature* 1988;332:845.
15. Acha-Orbea H, and McDevitt HO. The first external domain of the non-obese diabetic mouse class II I-Aβ chain is unique. *Proc Natl Acad Sci USA*. 1987;84:2435.
16. Morel PA, Dorman JS, Todd JA, McDevitt HO, and Trucco M. Aspartic acid at position 57 of the HLA-DQβ chain protects against type I diabetes: a family study. *Proc Natl Acad Sci USA*. 1988;85:8111.
17. Todd JA, Mijovic C, Fletcher J, Jenkins D, Bradwell AR, and Barnett AH. Identification of susceptibility loci for insulin-dependent diabetes mellitus by trans-racial gene mapping. *Nature* 1989;338:587–589.
18. Sinha AA, Lopez TM, and McDevitt HO. Autoimmune diseases: the failure of self tolerance. *Science* 1990;248:1380–1388.
19. Smilek DE, Lock CB, and McDevitt HO. Antigen recognition and peptide-mediated immunotherapy in autoimmune disease. Immunol Rev 1990; (in press).
20. Boitard C, Yasunami R, Dardenne M, and Bach JF. T cell-mediated inhibition of the transfer of autoimmune diabetes in NOD mice. *J Exp Med* 1989;169:1669–1680.
21. Gilfillan S, Aiso S, Michie SA, and McDevitt HO. Immune deficiency due to high copy numbers of an $A\beta^k$ transgene. *Proc Natl Acad Sci USA* 1990;(in press).
22. Gilfillan S, Aiso S, Michie SA, and McDevitt HO. The effect of excess β chain synthesis on cell surface expression of allele mismatched class II heterodimers in vivo. *Proc Natl Acad Sci USA* 1990;(in press).
23. Miyazaki T, Unot M, Uehira M, et al. Direct evidence for the contribution of the unique I-A NOD to the development of the insulitis in non-obese diabetic mice. *Nature* 1990;345:722.
24. Slattery RM, Kjer-Nielsen L, Allison J, Charlton B, Mandel TE, and Miller JFAP. Prevention of diabetes in non-obese diabetic $I\text{-}A^k$ transgenic mice. *Nature* 1990;245:724.
25. Lund T, O'Reilly L, Hutchings P, et al. Prevention of insulin-dependent diabetes mellitus in non-obese diabetic mice by transgenes encoding modified I-Aβ chain or normal I-Eα chain. *Nature* 1990;345:727.
26. Bohme J, Schuhbaur JB, Kanagawa O, Benoist C, and Mathis D. MHC-linked protection from diabetes dissociated from clonal deletion of T cells. *Science* 1990;49:293–295.
27. Wraith DC, Smilek DE, Mitchell DJ, Steinman L, and McDevitt HO. Antigen recognition in autoimmune encephalomyelitis and the potential for peptide-mediated immunotherapy. *Cell* 1989;59:247.

Multigenic Disorders

Etiology of Human Disease at the DNA Level,
edited by Jan Lindsten and Ulf Pettersson.

9

Diabetes Mellitus: Identification of Susceptibility Genes

*Graeme I. Bell, *Song-hua Wu, *Marsha Newman,
**Stefan S. Fajans, *Mitsuko Seino, *Susumu Seino,
*Nancy J. Cox

**Howard Hughes Medical Institute and Departments of Biochemistry and Molecular Biology and of Medicine, The University of Chicago, Chicago, Illinois 60637; **Department of Internal Medicine, Division of Endocrinology and Metabolism, University of Michigan Medical Center, Ann Arbor, Michigan 48109*

Diabetes mellitus is a clinical condition characterized by elevated blood glucose levels and, in its fully developed state, by passage of large volumes of urine, excessive thirst and hunger, weight loss and weakness, which can lead to coma and even death (1). It derives its name from the passage of large volumes of sweet-tasting urine: Greek, *diabetes,* a siphon, and *mellitus,* honey. Diabetes mellitus represents a major public health problem in industrialized countries. It is the leading cause of adult blindness and amputations and a major cause of renal failure. In addition, it is a major risk factor for cardiovascular disease.

In normal healthy individuals, plasma glucose levels are maintained between approximately 65 and 140 mg/dL (3.6–7.8 m*M,* respectively) through a balance between cellular glucose uptake and utilization and hepatic glucose production. The diagnosis of diabetes is made when:

1. Fasting plasma glucose concentrations (venous) are ≥ 140 mg/dL (7.8 m*M*) on at least two separate occasions.
2. Plasma glucose concentrations (venous) are ≥ 200 mg/dL (11.1 m*M*) at 2 h and one other intervening point after oral ingestion of 75 g of glucose.

There are two major forms of diabetes mellitus: insulin-dependent (IDDM or type I) and non-insulin-dependent (NIDDM or type II). The hyperglycemia of IDDM results from an absolute deficiency of insulin due to specific autoimmunological destruction of the insulin-secreting β cells of the pancreatic islets of Langerhans. These patients require therapy with exogenous insulin to avoid severe hyperglycemia and ketoacidosis, which if untreated can result in diabetic coma and death. By contrast, in NIDDM patients, the β cells are still able to synthesize and secrete insulin. The hyperglycemia of NIDDM results from reduced insulin levels in some patients and from a relative deficiency of insulin in others. Paradoxically, patients with mild NIDDM may be hyperinsulinemic as well as hyperglycemic; however, although the insulin levels may be high, they are inappropriately low for the degree of hyperglycemia. Patients with NIDDM frequently do not require insulin therapy and their hyperglycemia is usually controlled by diet or drugs that improve insulin secretion and possibly insulin responsiveness of target tissues.

The prevalence of diabetes mellitus varies widely in different parts of the world with the highest frequencies in industrialized countries and in populations participating in industrialized lifestyles (1,2). In the United States, the prevalence may be ~5% with 10–15% of affected individuals having IDDM and the remainder having NIDDM (2). However, in some groups such as Mexican-Americans and Pima Indians (a group of American Indians residing in Arizona), the frequency of NIDDM is much higher than in the general population, ~15% and 40%, respectively. The incidence of IDDM and NIDDM varies in different age groups (2,3). The incidence of IDDM is greatest in adolescents and declines with increasing age. By contrast, the incidence of NIDDM is very low in individuals <20 years of age and steadily increases being >10-fold higher in individuals older than 70 years than those 20–29 years of age.

GENES AND DIABETES MELLITUS

Genetic factors contribute to the development of both IDDM and NIDDM (1–6). However, they are not sufficient and overt diabetes develops only when other genetic or environmental factors supervene, i.e., one does not inherit diabetes mellitus per se but rather a genetic predisposition. In addition to genetic and environmental factors, stochastic factors may influence its development, e.g., T-cell receptors that recognize β-cell antigens and mediate the autoimmunological destruction of the β cells thereby leading to IDDM are products of genes whose germ-line sequences have been altered by somatic rearrangement that may be a stochastic process.

Evidence for genetic factors in the development of both forms of diabetes mellitus has come from studies of concordance (i.e., both twins affected) of diabetes mellitus in identical (monozygotic) and nonidentical (dizygotic) twin pairs and its frequency in first-degree relatives of affected individuals (Table 1) (1–4,6). If IDDM and NIDDM were due entirely to genetic factors, concordance rates of 100% would be expected. The extent to which these rates fall short of 100% is an indication of the importance of nongenetic/environmental factors in disease etiology. The concordance in monozygotic twin pairs is <100% for both IDDM and NIDDM indicating that nongenetic

TABLE 1. *Empirical risk for diabetes according to affected members of family and concordance for monozygotic twins*

Type of diabetes	Risk or concordance (%)
Insulin-dependent diabetes mellitus	
HLA-identical sibling	15
Non-HLA-identical sibling	1
One parent	3
Both parents	20
Monozygous twin	35
Non-insulin-dependent diabetes mellitus	
First-degree relative	10–40
Both parents	<30
Monozygous twin	55–90
Maturity-onset diabetes of the young	
First-degree relative	50

Adapted from Foster, ref. 1; Bennett, ref. 2; Kobberling and Tillil, ref. 3; Rotter et al., ref. 4; Fajans, refs. 5 and 5a; Kingston, ref. 47.

factors play an important role in their etiology. The higher concordance of disease in NIDDM twin pairs (up to >90%) compared to those with IDDM (~35%) and the greater relative risk of NIDDM compared to IDDM in first-degree relatives suggests that the genetic component has a more prominent role in the etiology of NIDDM than it does in IDDM.

Although genetic susceptibility is an important determinant of disease status, studies of the segregation of IDDM and NIDDM in families have indicated that neither disorder has a simple autosomal dominant (with the possible exception of maturity-onset diabetes of the young or MODY) or recessive mode of inheritance. Nor is inheritance sex-linked. Several models have been proposed to explain the complex, nonmendelian mode of inheritance of diabetes mellitus and the contribution of genetic factors to its etiology:

1. Diabetes mellitus is a multifactorial trait with contributions from many unknown genes each with a small effect (polygenes) acting together with environmental (nongenetic) factors. If the contribution of the genetic and nongenetic factors exceeds a threshold, disease occurs.
2. A small number of primary susceptibility genes are responsible for most of the risk and only one or two of these genes contribute to the development of diabetes mellitus in each family. The phenotypic expression of these genes is modified by genetic background and/or environmental factors.

There is support for a model of diabetes susceptibility in which one or a few primary susceptibility genes contribute most of the genetic liability in each family. A major fraction of the genetic susceptibility for IDDM is contributed by a gene in the major histocompatibility complex (MHC) (7,8).

Studies in populations with unusually high frequencies of NIDDM such as the Pima Indians and Micronesians have revealed that fasting plasma glucose levels are bimodal, i.e., distributed around two distinct mean values. This bimodal distribution has been interpreted to mean that codominant inheritance of a single major gene influences glucose tolerance (2). Mutations in the insulin (*INS*) or insulin receptor (*INSR*) genes are the primary genetic lesion contributing to the development of NIDDM in a small subset of diabetic patients (9,10).

Although one or a few major genes may be responsible for a significant fraction of diabetes susceptibility in many families, the mode of inheritance of diabetes is complex and does not conform to a simple dominant or recessive mode of transmission possibly because different susceptibility loci have different modes of transmission, e.g., in general, disorders resulting from mutations in enzymes often have a recessive mode of inheritance whereas disorders arising from defects in structural proteins are dominant. Different mutations in the same gene may also act in a dominant or recessive fashion depending upon their effects on function. For example, extreme insulin-resistant syndromes resulting from mutations in the extracellular portion of the INSR are inherited as a recessive trait whereas mutations in the tyrosine kinase domain are dominant (10–12). In addition to mutations in the primary susceptibility gene, the genetic background and environmental factors will have a variable role in converting genetic susceptibility into overt disease.

Approaches that have been used to identify diabetes-susceptibility genes include: (a) population-based association studies and (b) family-based studies including linkage studies in families as well as analysis of allele sharing in affected sib pairs. In population-based association studies, the frequency of a marker, often a candidate gene, is compared between affected and normal healthy subjects (13,14). If the disorder is found to be associated with a marker more frequently than expected by chance, this may suggest a causal relationship between genetic variation in the marker, or a locus in linkage disequilibrium with the marker, and the disease. However, there are also other explanations for the difference in the frequency of the marker between affected and normal subjects including population stratification and epistasis (i.e., functional association between two loci). In addition, genetic linkage between the marker and disease cannot be presumed from their association; conversely, lack of association does not preclude linkage between the marker and the disease. While these restrictions may reduce the utility of population association studies for identifying disease-susceptibility genes, it should be noted that under some conditions association studies have good power to detect loci with even minor effects on disease susceptibility and in addition may detect loci that are unlikely to be identified in family studies (14).

Two classes of family-based studies may be considered for the identification of diabetes-susceptibility genes: linkage studies, and studies focusing on affected relative pairs. Linkage analysis can be a powerful tool for the identifying disease loci. In a linkage study, the cosegregation of markers (DNA or protein) and disease is assessed within families, and the calculation of a lod ($\log_{10}$ odds ratio) score within each family facilitates the accumulation of evidence for or against linkage across families (15). This approach has been successful for a growing number of single gene disorders (16), including disorders that are inherited in a recessive fashion such as cystic fibrosis as well as dominantly inherited disorders such as adult polycystic kidney disease. The number of mapped disease loci is likely to continue to increase as the human genetic map becomes saturated with informative marker loci. A linkage study also provides an estimate of the recombination frequency between the marker and disease gene, thereby facilitating the generation of a fine-structure genetic map of the region around the disease gene that can be used in molecular studies to facilitate its isolation.

Linkage studies are not as easily applied in the analysis of the genetics of complex disorders such as IDDM or NIDDM as in a simple monogenic disorder. Calculation of lod scores requires that the mode of transmission of the disease and allele frequencies for the disease-susceptibility loci be specified; these are unlikely to be known for a complex trait. One method often used for linkage analyses of complex disorders is to allow for incomplete penetrance and to test a variety of models using different modes of inheritance and allele frequencies for the susceptibility loci. Computer simulation studies suggest a true linkage would not be missed using such a conservative approach (17). The effects of multiple analyses on the criteria to establish linkage (a lod score $\geqslant 3.0$ or an odds ratio of 1,000 to 1) are uncertain. However, it seems likely that there is increased risk of obtaining a false linkage. Consequently it may be necessary to increase the lod score that will be accepted as proof of linkage. If evidence for linkage is obtained, the estimate of the recombination fraction between the marker and disease gene will be uncertain, which increases the difficulty of precisely mapping and ultimately finding the disease locus. In addition to these problems, complex disorders may be genetically heterogeneous and allowing for incomplete penetrance decreases the power to detect heterogeneity. Although it has been suggested that studies of families that are large enough to allow detection or rejection of linkage may reduce the problem of genetic heterogeneity, there are also difficulties with studies of large pedigrees. If the frequency of the disorder in the population is relatively high, choosing large pedigrees with many affected individuals may increase the probability of genetic heterogeneity, e.g., due to marrying in of asymptomatic but genetically-susceptible individuals. In addition, the susceptibility gene in the large pedigree may not be important for the development of the disease in the general population. The evolution of the evidence for linkage between manic-depressive illness and chromosome 11 markers, initially for and then against, provides a cogent example of the difficulties of performing linkage studies for complex disorders (18). Lander and Botstein have recently suggested an approach for mapping complex genetic traits in humans using a complete restriction fragment length polymorphism (RFLP) linkage map (19,20) in which a collection of individual families is typed for a panel of RFLPs that span the entire genome. The simultaneous search method is then used to identify a marker tightly linked to the disease locus.

It is possible to simplify studies of the genetics of complex disorders by examining only affected members of families (21–23). Thus, problems in analysis due to confounding factors such as incomplete penetrance and variable age-of-onset may be eliminated. Several methods employing this strategy have been suggested, the most widely used of which is the analysis of affected sib pairs. Briefly, a pair of affected sibs would be expected to inherit 2, 1 or 0 markers identical by descent from their parents with probabilities of 1/4, 1/2 and 1/4, respectively, if the marker locus is not linked to a locus affecting disease susceptibility. If the marker is linked to a locus increasing susceptibility to disease, the probabilities will be distorted from their expected values towards increased sharing in affected sib pairs. The precise level of distortion depends upon the magnitude of the contribution of that disease susceptibility locus, its inheritance and the recombination frequency between the marker and disease susceptibility locus. In general, this method has been used only to test for evidence of linkage and this is easily accomplished by comparing the observed proportion of affected sib pairs sharing 2, 1 or 0 parental marker alleles to that expected under the hypothesis of no linkage between marker and disease. Although this method can provide evidence for linkage, it provides no estimate of the recombination frequency between the marker and disease loci. In addition, the affected sib pair and related approaches using only affected individuals do not utilize the full information available in the family. Although uncertainties regarding penetrance reduce the amount of information that can be gained from studying unaffected individuals, this loss of information is a complex function of the penetrance and mode of inheritance. For example, there is more information in the unaffected members when a disease is dominantly inherited than when recessively inherited. Thus, even with the same degree of incomplete penetrance, there will be a differential loss of information depending upon the mode of transmission. Recent studies assessing the power of affected relative pairs have indicated that other pairs of relatives may be preferable to sibs depending upon the mode of transmission, penetrance and marker informativeness (22,23). However, some of the relative pairs that would be most useful (e.g., grandparent-grandchild) are difficult to collect in a late-onset disease like NIDDM. Although only information on affected individuals is analyzed, unaffected individuals in the family may have to be typed to establish inheritance of marker alleles. Thus, the savings in effort are not as great as might be expected and great deal of potentially useful information is not used in the final analyses. In addition, some of the affected relative pair approaches allow for no more than two affected members of each family to be included in the analysis, leading to further loss of potential information, although corrections that permit analysis of multiple pairs of affected relatives may ultimately be offered.

Because of inherent problems associated with both lod score and affected relative pairs strategies, it may be prudent at this time to analyze family data using both methods.

IDDM SUSCEPTIBILITY GENES

Association studies in populations and analysis of allele sharing in affected sib pairs have identified both major and minor susceptibility genes that contribute to the development of IDDM (4). A gene(s) in the vicinity of the MHC, class II, DQ β locus (HLA-DQβ—human chromosome 6p21.3) contributes a significant fraction of the genetic liability for the development of IDDM (7,8). However, this gene is not sufficient and other genetic and/or nongenetic factors must be present for IDDM to develop. Rotter and Landaw have estimated that the MHC may be responsible for 60–70% of the overall genetic susceptibility to IDDM (24). Studies of the nonobese diabetic (NOD) mouse, a murine model of IDDM, suggest that three recessive genes are required to cause diabetes (25). The major susceptibility gene in the NOD mouse has been identified [I-A^{NOD} (Idd-1^{S}), the murine gene corresponding to human HLA-DQβ]; however, the identities of the other two genes (Idd-2^{S}, Idd-3^{S}), which represent modifying or minor susceptibility genes, are unknown. In humans, a number of genes not linked to the HLA complex have been suggested as candidate loci whose products might play a role in the pathogenesis of IDDM. These include the insulin gene (*INS*—chromosome 11p15.5), immunoglobulin heavy- and light-chain genes (chromosomes 14q32 and 22q11, respectively) and T-lymphocyte receptor β-chain gene (*TCRB*—chromosome 7q35) (4,26–28). Genetic variation in the region of these genes may be associated with a modest (2–4-fold) increased risk of developing IDDM. Although it has been possible to demonstrate linkage between the MHC and IDDM, it has not been possible to demonstrate linkage between these other genes and IDDM suggesting that they represent minor susceptibility determinants. Moreover, it is unlikely that they are human homologues of Idd-2^{S} or Idd-3^{S}, the non-MHC-linked susceptibility loci in the NOD mouse.

NIDDM SUSCEPTIBILITY GENES

Twin and family studies have shown that genetics play a more prominent role in the development of NIDDM than in IDDM (1–5). However, the genes contributing to NIDDM remain poorly characterized. Association studies in populations have identified several minor NIDDM susceptibility genes. Although HLA alleles are not associated with NIDDM in most populations, statistically significant positive associations with NIDDM have been observed for HLA alleles in the Xhosa of Southern Africa, Pima Indians, Micronesians and Finns (2). Rhesus factor (*RH*—chromosome 1p36.2-p34), haptoglobin (*HP*—chromosome 16q22.1) and insulin receptor (*INSR*—chromosome 19p13.3-p13.2) polymorphisms have been associated with NIDDM in one Mexican-American population but not in a second (29–31). *RH* has been associated with gestational diabetes in Italians (32). Apolipoprotein A1 (*APOA1*—chromosome 11q23-q24) RFLPs have been associated with NIDDM in Polish (33) and Chinese populations (34); Xiang et al. (34) have estimated that the *APOA1* locus may account for ~8% of the difference between baseline and total possible risk of NIDDM in overweight Chinese. Apolipoprotein B (*APOB*—chromosome 2p24-p23) RFLPs are positively associated with diabetes in Chinese (the *APOB* locus accounts for ~1% of total possible risk in individuals who are lean or normal weight) (34) and *INSR* RFLPs are associated with decreased risk in this population (34) whereas they have been positively associated with NIDDM in Mexican Americans (31) and in one study of Caucasians (35). In other studies, *INSR* RFLPs are not associated with NIDDM (4,6,10). *INS* RFLPs have been extensively examined in many populations and variation at this locus is no longer believed to be associated with NIDDM (27). In Caucasians, *INSR* and *IGF2* alleles interact to confer additional risk for gestational diabetes mellitus (36). Genetic variation in/near the *GLUT1* facilitative glucose transporter gene (*GLUT1*—chromosome 1p33) has been associated with NIDDM in Northern and Southern Europeans and Japanese (37) but not in Caucasians residing in the United States, nor in African-Americans, Chinese or Micronesians (34,38,39).

INSULIN AND INSULIN RECEPTOR MUTATIONS AND NIDDM

Mutations in the *INS* and *INSR* genes have been reported in NIDDM patients (9,10). Five mutations that result in the synthesis of an abnormal proinsulin molecule have been described (Fig. 1). All the patients expressing an abnormal proinsulin/insulin are heterozygous and express both mutant and normal proteins. Three of the *INS* mutations, $B24^{Ser}$, $B25^{Leu}$ and $A3^{Leu}$, result in a molecule having reduced receptor binding and biological activity with values that are 0.2–5% of those of insulin. The codominant expression of these variants with normal insulin appears to increase the risk of developing NIDDM. However, expression of a single allele coding for a structurally abnormal insulin is not always associated with NIDDM and the expression of NIDDM may vary between and within families having the same *INS* mutation depending upon the presence of other diabetogenic factors, either

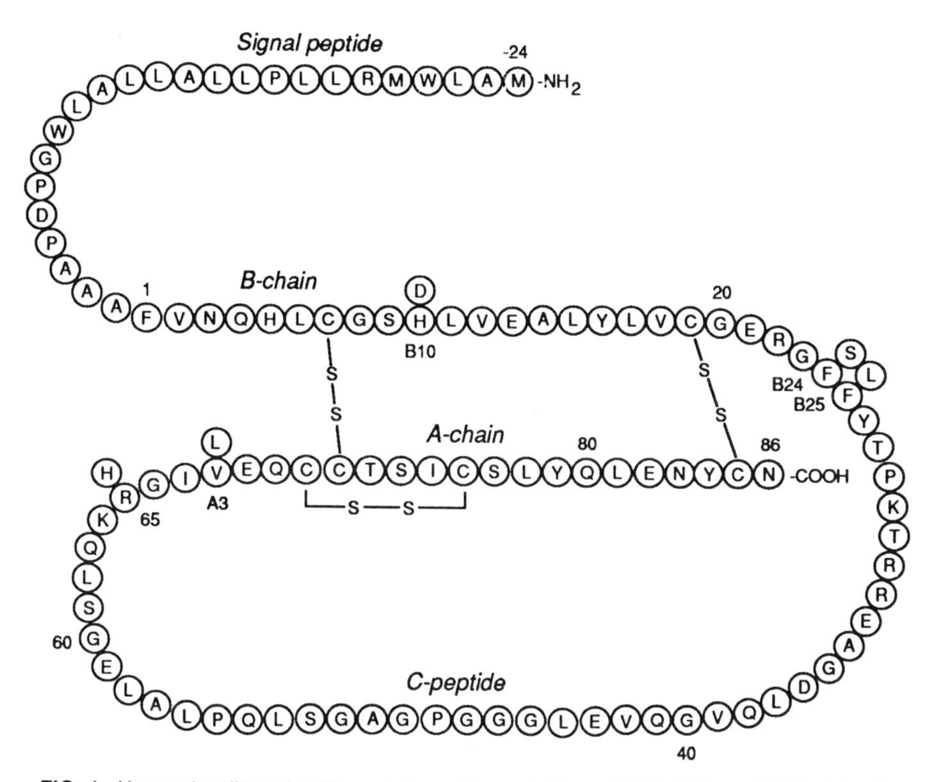

FIG. 1. Human insulin/proinsulin mutations. The mutations at B10, B24 and B25 of the B-chain, A3 of the A-chain and residue 65 of proinsulin are indicated. The domains of the 110 amino acid preproinsulin molecule are noted. The amino acid residues are indicated using their single-letter abbreviations.

genetic or acquired (Table 2). The frequency of *INS* mutations in most populations seems to be quite low and they are a rare cause of NIDDM. However, in French Canadians and Japanese (populations in which two or more unrelated individuals with the same mutation, $B24^{Leu}$ and $A3^{Leu}$, respectively, have been identified presumably due to a founder effect), they may be responsible for NIDDM in a measurable subgroup of NIDDM patients. It is now possible to determine the frequency of the $B24^{Leu}$ and $A3^{Leu}$ mutations in these populations through use of the polymerase chain reaction and hybridization with allele-specific oligonucleotides (40) or by allele-specific amplification (41).

Syndromes of extreme insulin resistance such as leprechaunism, Rabson-Mendenhall syndrome, type A severe insulin resistance with acanthosis nigricans, and congenital lipodystrophy are associated with the expression of mutated insulin receptors (10). The mutations in 14 patients with these syndromes have been identified (Fig. 2). Studies of these patients and their het-

TABLE 2. *Insulin mutations and NIDDM*

Mutation	Family	No. of members with mutation	No. of susceptible members with NIDDM	Ref.
A3^{Leu}	1	5	2	48
	2	4	0	49
	3	5	3	50,51
B24^{Ser}		6	2	52

erozygous parents have revealed that insulin resistance/hyperinsulinemia is inherited as an autosomal dominant trait with a partial gene dosage effect (10,42). Although insulin resistance is inherited as an autosomal dominant trait, the syndromes associated with extreme insulin resistance have both recessive and dominant modes of inheritance. The patients with extreme insulin resistance due to mutations in the extracellular portion of the *INSR* all appear to be genetic compounds or are homozygous for the same mutation (Fig. 2); thus, the mode of inheritance of the syndrome is recessive. By contrast, patients with extreme insulin resistance due to mutations in the intracellular tyrosine kinase domain express both normal and mutant *INSR* proteins and thus the mode of inheritance is dominant. In fact, mutations in the tyrosine kinase domain appear, in many instances, to function as dominant-negative mutations since the function of the normal protein also seems to be impaired. The expression of a mutant *INSR* increases the risk of developing NIDDM and 9 of the 14 individuals expressing a mutant insulin receptor have diabetic glucose tolerance (Fig. 2). Insulin resistance is a risk factor contributing to the development of NIDDM; however, the contribution of genetic variation in the *INSR* to the development of insulin resistance associated with NIDDM is unknown. Taylor and his colleagues (10) suggest that the frequency of *INSR* mutations may be ~0.05%, although this is likely to be a minimum estimate. The frequency of heterozygous carriers, i.e., individuals having a normal and mutant *INSR* gene, would then be ~0.1%. If heterozygous carriers have an increased risk of developing NIDDM, they may represent a small but measurable subgroup of patients. It may be possible to identify such individuals and thereby establish the contribution of *INSR* mutations to the development of NIDDM using new screening procedures such as GC-clamped polymerase chain reaction together with denaturing gradient gel electrophoresis, a strategy that is able to identify more than 90% of the nucleotide substitutions in segments of amplified genomic DNA (43). We are using this approach to screen a cohort of 50 NIDDM patients for mutations that alter the amino acid sequence of the *INSR*. The analysis of exon 17, the exon encoding the ATP-binding region of the tyrosine kinase (Fig. 2), has revealed a silent nucleotide substitution in the codon

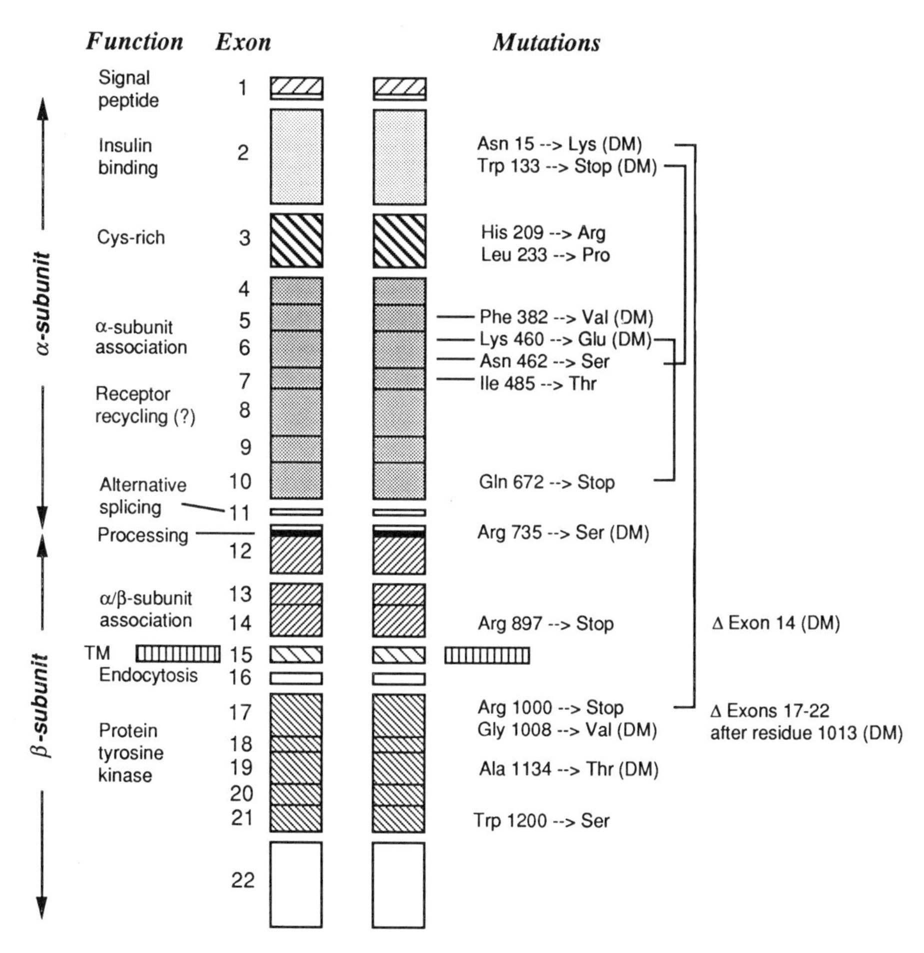

FIG. 2. Human insulin receptor mutations. Seventeen mutations in the *INSR* genes of 14 subjects with syndromes of extreme insulin resistance are shown. The functional regions of the INSR encoded by each exon are indicated. DM indicates that the patient also has NIDDM: Δ indicates a deletion in the *INSR* gene. Five patients with mutations in the INSR external to the plasma membrane (TM) are homozygous for the same mutation identical by descent (Arg^{209}/Arg^{209}, Pro^{233}/Pro^{233}, Val^{382}/Val^{382}, Thr^{485}/Thr^{485} and Ser^{735}/Ser^{735}), and five are genetic compounds [$Lys^{15}/Stop^{1000}$, $Stop^{133}/Ser^{462}$, $Glu^{460}/Stop^{672}$, $Stop^{897}$/? (the identity of the mutation in the second allele has not been identified), Δ exon 14/?]. Four subjects with mutations in the tyrosine kinase domain of the receptor are heterozygotes and express both mutant and normal INSRs: Gly^{1008}/Val^{1008}, Ala^{1134}/Thr^{1134}, Trp^{1200}/Ser^{1200} and normal/Δ exons 17–22. These data are from references 10, 55–61 and references therein.

for Tyr[984] (TAC or TAT) described previously (10) as well as one in the codon for His[1046] (CAC or CAT). The identification of a subgroup of NIDDM patients in whom the primary genetic lesion is known would represent a significant advance in our understanding of the etiology of NIDDM and would allow us to begin to try to identify the nongenetic factors that convert susceptibility into overt disease.

The *INS* and *INSR* mutations may be useful models for studying the etiology of NIDDM. These genes have features expected for NIDDM susceptibility genes including being incompletely penetrant. As a consequence, they increase the risk of developing NIDDM and overt disease only develops in those individuals having other genetic or acquired defects, i.e., the *INS* or *INSR* mutation is necessary but not sufficient to cause NIDDM. The *INS* and *INSR* mutations also illustrate the potential genetic heterogeneity of NIDDM in that very different types of genes (i.e., hormone or hormone receptor) expressed in different tissues (the insulin-secreting β cell as well as insulin-responsive cells) can result in hyperglycemia and NIDDM.

LINKAGE STUDIES IN NIDDM

Studies of mutant *INS* and *INSR* proteins have identified primary susceptibility genes in a subgroup of NIDDM patients. Population association studies have identified components of the genetic background whose variation may modify the expression of the diabetic phenotype. However, the major gene(s) responsible for NIDDM susceptibility in most subjects has not been identified.

Recent successes with a number of different simple monogenic disorders (16) indicate that a DNA marker linked to a disease locus can be identified provided that the disorder is common enough that families with multiple affected members can be collected for study. Identifying a DNA marker(s) linked to NIDDM will be more difficult than for a simple monogenic disorder such as cystic fibrosis because NIDDM does not exhibit simple mendelian recessive or dominant inheritance. In addition, it is likely to be genetically heterogeneous with mutations in several different genes being the cause of hyperglycemia. It may also be difficult to obtain large, multigenerational families because of its late age of onset. The analyses, whether using lod scores and testing a variety of modes of transmission or concentrating on affected sib (or other relative) pairs, also present difficulties and traditional criteria for significance may not be adequate to establish linkage.

These problems notwithstanding, investigators have begun to use linkage approaches to identify NIDDM susceptibility genes. However, initial studies have only examined candidate genes (*INS, INSR* and *GLUT1*) in selected families and have provided no evidence that these genes are linked to NIDDM (44,45). In the most comprehensive study to date, Cox et al. (44) examined the linkage of *INS* and *INSR* markers in 20 African-American families. The lod score (using several different models of transmission) and affected sib pair approaches both gave negative results for linkage of *INS* or *INSR* with NIDDM. However, it was not possible to exclude either of these loci as being major susceptibility loci in a small proportion of families. Moreover, since a relatively small number of families were examined, it was not possible to exclude these genes as minor susceptibility or modifying loci.

It is important in evaluating strategies for identifying disease markers in complex disorders to consider the likelihood of success under various conditions. Analysis of 50–100 nuclear families with at least four or more sibs, at least two of whom have NIDDM, may be sufficient to detect linkage if there is a single major locus for NIDDM segregating in the majority of families, even allowing for reduced penetrance (19,20, NJC, unpublished data). However, if there are five or six major diabetogenic loci, only one or two of which are likely to be important in any individual family, a sample of 50–100 families may be too small to allow detection of linkage. Moreover, attempts to confirm a linkage could lead to conflicting and confusing results if the frequencies of the diabetogenic genes vary between study groups. Linkage studies using a single pedigree that is large enough to provide unambiguous evidence for or against linkage may be preferred at this time since it is likely that a diabetes-susceptibility gene segregating in such a pedigree could be identified using current methods. However, since NIDDM is quite common, even studies of a single pedigree could be confounded due to intrapedigree heterogeneity.

Multigenerational pedigrees with the MODY form of NIDDM have been described (5). MODY is defined as "NIDDM in the young and autosomal dominant inheritance." It can be diagnosed at a young age (<25 yr) and frequently in early adolescence (9–14 yr) if sought by routine plasma glucose testing. MODY is usually asymptomatic in younger age groups although some patients may have symptoms if stressed by infection. The prevalence of MODY among NIDDM patients is uncertain and estimates range from 0.2 to 18% in different populations. The phenotypic (hormonal and metabolic) characteristics of MODY subjects suggest that MODY, like NIDDM, is heterogeneous and likely to have different causes in each MODY family although it is a homogeneous trait within a family.

The RW family is a MODY family that has been studied since 1958 (5). It is a five-generation pedigree of over 250 individuals, 47 of whom have MODY. Fibroblast and/or lymphoblastoid cell lines are available for 116 individuals. Diabetic subjects in the RW pedigree have a reduced insulin secretory response to glucose and it has been proposed that the delayed and decreased insulin secretory response is the manifestation of the basic genetic defect that leads to diabetes. In this and other regards, the phenotype of diabetic subjects in the RW pedigree is similar to that of many subjects with the more common form of NIDDM suggesting that elucidation of the defect in the RW family could lead to an understanding of these forms as well.

We have initiated studies to identify a marker that is linked to MODY using a segment of the RW pedigree (Fig. 3). Of the 49 individuals being studied, 19 have MODY, 4 have impaired glucose tolerance (prediabetic?), 1 has IDDM and 25 have normal glucose tolerance. Together with Dr. Richard Spielman and his colleagues, University of Pennsylvania, we have analyzed over 75 polymorphic DNA markers for linkage to MODY including genes involved in carbohydrate and lipid metabolism as well as random DNA markers. Markers were selected primarily on the basis of heterozygosity of >30% and a genetic distance between markers of ~20 centimorgans (cM). None of the markers that have been tested shows linkage to MODY. In most instances, it has been possible to exclude the disease gene from the marker with a lod score of −2 at a recombination fraction of >0.1. We estimate that we have excluded the MODY gene from 30–40% of the genome including chromosomes X and Y (inheritance is autosomal) as well as more than 75% and 50% of chromosomes 4 and 19, respectively. Exclusion data for several candidate genes as well as several other markers that we have examined are summarized in Table 3.

The linkage analysis program LIPED (15) was used for all two-point linkage analyses. In these analyses, MODY was assumed to be homogeneous within the RW pedigree and transmitted as an autosomal dominant trait. We used a straight-line age-of-onset correction and allowed for separate penetrances in males and females. In one model, MODY was assumed to be completely penetrant in males (the probability that a susceptible heterozygote or homozygote develops MODY increases linearly from 0 at age 8 to 1.0 at age 25) and incompletely penetrant in females (the probability that a susceptible heterozygote or homozygote develops MODY increases linearly from 0 at age 8 to 0.95 at age 25). This model is consistent with the presence of two possible female carriers (unaffected females older than 25 years having an affected child: individuals 8072 and 8364, Fig. 3) (5). In a second model, we allowed for reduced penetrance in both males and females (the probability that a heterozygote or homozygote develops MODY increases linearly from 0 at age 8 to 0.8 at age 25 in males and to 0.6 at age 25 in females). The rationale for including a model allowing for reduced penetrance was to minimize the risk of missing a true linkage. If genetic susceptibility to MODY is not expressed in some individuals, a model assuming complete penetrance could lead to rejection of a true linkage. Since there are two possible female carriers in this pedigree, we made a careful examination of the proportion of at-risk individuals who are affected. While the overall proportion of at-risk individuals who are affected is not significantly different from 50%, when the sexes are considered separately, significantly less than 50% of the at-risk females have MODY, while the proportion of at-risk males who have MODY is not significantly different from 50%. Since the high proportion of affected individuals in generation II may have contributed to the ascertainment of this pedigree, we estimated male and female penetrances by determining the proportion of affected individuals in the at-

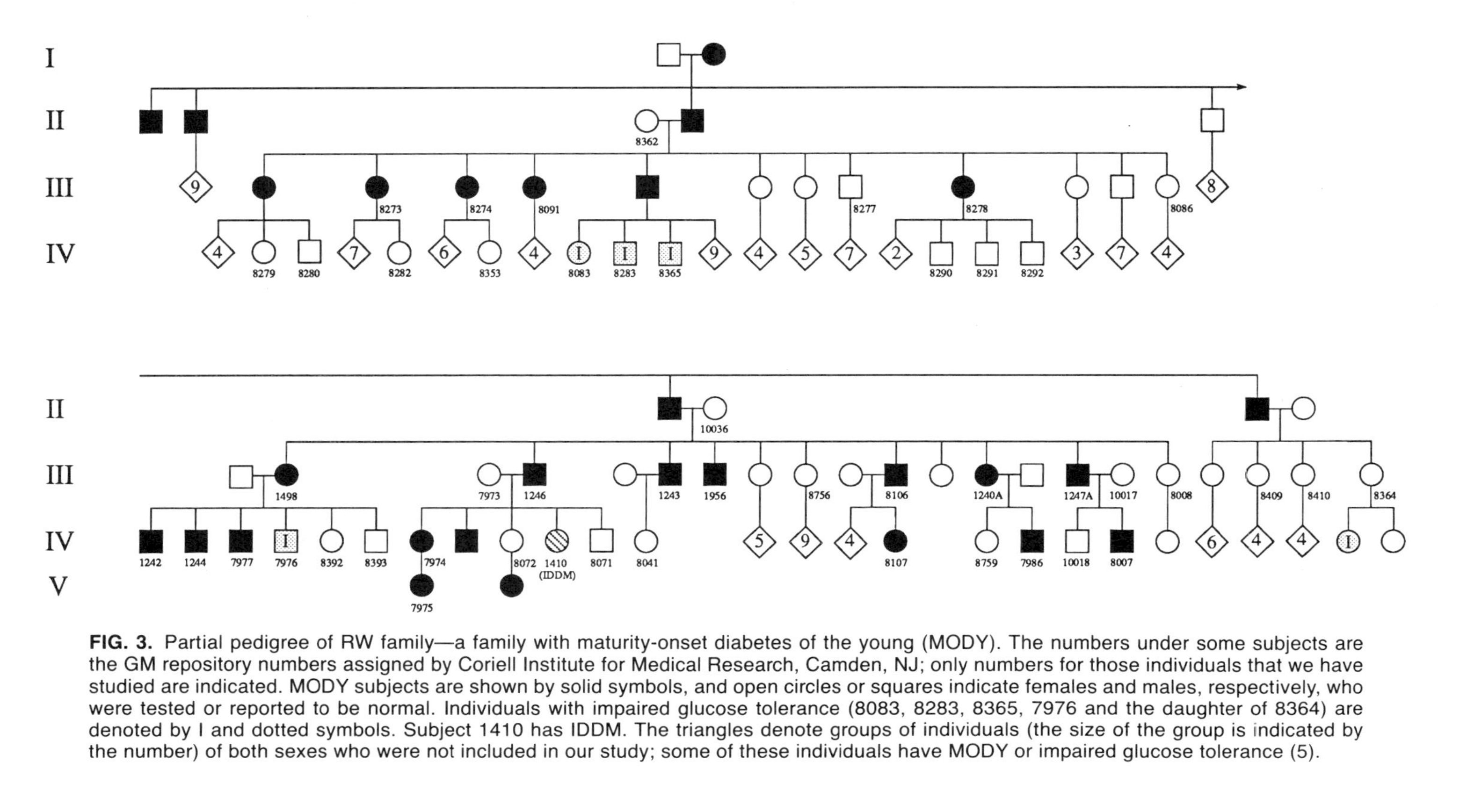

FIG. 3. Partial pedigree of RW family—a family with maturity-onset diabetes of the young (MODY). The numbers under some subjects are the GM repository numbers assigned by Coriell Institute for Medical Research, Camden, NJ; only numbers for those individuals that we have studied are indicated. MODY subjects are shown by solid symbols, and open circles or squares indicate females and males, respectively, who were tested or reported to be normal. Individuals with impaired glucose tolerance (8083, 8283, 8365, 7976 and the daughter of 8364) are denoted by I and dotted symbols. Subject 1410 has IDDM. The triangles denote groups of individuals (the size of the group is indicated by the number) of both sexes who were not included in our study; some of these individuals have MODY or impaired glucose tolerance (5).

TABLE 3. *Pairwise lod scores between selected markers and MODY*

		Recombination fraction (θ)				
Marker[a]	Chromosome	0.00	0.10	0.20	0.30	0.40
GLUT1	1p33	−∞	−2.6	−0.6	0.0	0.1
		−5.0	−1.4	−0.1	0.2	0.2
AMY2B	1p21	−∞	−7.3	−3.4	−1.4	−0.4
		−20.0	−4.1	−2.1	−0.9	−0.2
APOA2	1q21–q23	−∞	−1.7	−0.4	0.1	0.1
		−4.3	−1.5	−0.8	−0.3	0.0
APOB	2p24–p23	−∞	−7.0	−2.9	−1.0	−0.2
		−∞	−4.2	−1.9	−0.7	−0.1
GLUT2	3q26	−∞	−2.4	−1.1	−0.5	−0.2
		−2.4	−1.2	−0.5	−0.2	−0.1
D4S10	4p16.3	−∞	−6.1	−3.2	−1.6	−0.6
		−7.5	−3.5	−1.9	−1.0	−0.4
D4S144	4p	−∞	−1.7	−0.5	−0.1	0.0
		−5.6	−1.5	−0.5	−0.1	0.0
RAF1P1	4p16.1	−17.4	−2.7	−1.5	−0.7	−0.3
		−2.7	−1.7	−1.0	−0.5	−0.2
D4S35	4p11–q11	−∞	−3.0	−1.3	−0.5	−0.2
		−5.4	−1.6	−0.7	−0.3	−0.9
D4S1	4cen–q21	−10.6	−2.3	−1.1	−0.5	−0.2
		−3.4	−1.5	−0.7	−0.4	−0.1
ADH3/EGF	4q25	−∞	−4.9	−2.1	−0.9	−0.5
		−∞	−2.4	−1.6	−0.9	−0.5
FGB	4q28	−∞	−2.2	−0.5	0.0	0.1
		−1.7	−1.0	−0.1	0.2	0.2
D4S139	4q	−∞	−8.0	−4.1	−2.0	−0.2
		−8.3	−4.6	−2.8	−1.5	−0.6
HLA-DR	6p21.3	−∞	−3.5	−1.5	−0.6	−0.2
		−9.7	−2.4	−1.1	−0.5	−0.1
LPL	8p22	−∞	−7.3	−3.7	−1.8	−0.7
		−∞	−3.8	−2.5	−1.3	−0.6
HK1	10q22	−14.4	−6.1	−3.0	−1.5	−0.6
		−9.0	−4.0	−2.4	−1.2	−0.5
APOA1	11q23–q24	−∞	−2.5	−1.1	−0.5	−0.1
		−6.5	−2.3	−1.1	−0.5	−0.1

risk members of generations III and IV. In addition, only individuals who had undergone a glucose tolerance test were included in these calculations. In both models, the penetrance for the normal homozygote was assumed to be 0 throughout life and the MODY susceptibility allele frequency was assumed to 0.0001, consistent with a lifetime prevalence for MODY of 0.02%. Reducing the penetrance in the linkage analysis decreased the region of exclusion, sometimes considerably. This effect varied with different markers, being most pronounced when apparent recombinations occurred more frequently among unaffected individuals in the pedigree. We also examined the effect of increasing the MODY susceptibility allele frequency. In general, the region of exclusion was only slightly reduced when the MODY allele

TABLE 3. *Continued.*

Marker[a]	Chromosome	Recombination fraction (θ) 0.00	0.10	0.20	0.30	0.40
GLUT3/A2M	12p13.3	−∞	−1.4	−0.6	−0.3	−0.1
		−5.8	−0.9	−0.4	−0.3	−0.1
IGF1	12q23	−∞	−2.4	−0.8	−0.1	0.1
		−4.8	−0.9	−0.2	−0.1	−0.1
LIPC	15q21–q23	−∞	−5.1	−2.2	−0.8	−0.2
		−9.7	−3.1	−1.7	−0.8	−0.2
GLUT4/D17S1	17p13	−∞	−1.1	−0.5	−0.2	0.0
		−3.2	−0.8	−0.4	−0.1	0.0
INSR	19p13.3–p13.2	−∞	−7.9	−3.9	−1.9	−0.7
		−7.9	−4.7	−2.4	−1.2	−0.4
LDLR	19p13.2–p13.1	−∞	−4.3	−1.7	−0.6	−0.2
		−13.7	−2.2	−0.9	−0.3	−0.1
D19S7	19cen–q12	−∞	−4.3	−2.1	−1.0	−0.4
		−5.2	−2.8	−1.5	−0.7	−0.3
D19S9	19q12–q13.3	−11.6	−2.5	−1.4	−0.8	−0.3
		−4.1	−1.7	−1.0	−0.6	−0.2
APOC2	19q13.2	−∞	−2.0	−0.8	−0.3	−0.1
		−5.0	−1.1	−0.6	−0.2	−0.1
D19S36	19q13.2	−∞	−4.3	−1.8	−0.7	−0.2
		−6.2	−1.4	−0.3	0.1	0.1

[a]The INS and INSR genes have previously been shown not to be linked to MODY (53,54). At each marker, the first line of data corresponds to results obtained with the highly penetrant model and the second with the model with reduced penetrance. The gene symbols are: *GLUT1,* GLUT1/erythrocyte glucose transporter; *AMY2B,* pancreatic amylase, alpha 2B; *APOA2,* apolipoprotein A-II; *APOB,* apolipoprotein B; *GLUT2,* GLUT2/liver glucose transporter; *RAF1P1,* raf-1 oncogene pseudogene 1; *ADH3,* alcohol dehydrogenase (class I), gamma polypeptide; *EGF,* epidermal growth factor; *FGB,* fibrinogen, B beta polypeptide; *HLA-DR,* major histocompatibility complex, class II, DR; *LPL,* lipoprotein lipase; *HK1,* hexokinase I; *APOA1,* apolipoprotein A-I; *GLUT3,* GLUT3/brain glucose transporter; *A2M,* alpha-2-macroglobulin; *IGF1,* insulin-like growth factor I; *LIPC,* hepatic lipase; *GLUT4,* GLUT4/muscle-fat glucose transporter; *INSR,* insulin receptor; *LDLR,* low density lipoprotein receptor; and APOC2, apolipoprotein C-II.

frequency was increased to 0.001 (lifetime prevalence 0.2%). The magnitude of this effect was most pronounced when apparent recombination with the marker occurred in affected individuals.

It seems likely that it is only a matter of time until a marker for MODY in the RW family will be identified and then studies directed towards the isolation and characterization of the RW MODY gene can begin. It will then be possible to assess the contribution of this gene, perhaps in the form of alternative alleles, to the development of other forms of MODY as well as the common form(s) of NIDDM. Since MODY is likely to be genetically heterogeneous, linkage studies in other MODY families may lead to the identification of other primary diabetes-susceptibility genes.

CONCLUSION

Rotter and Rimoin have described over 60 different syndromes associated with glucose intolerance (4). Most are rare and although they serve to illustrate the different mechanisms that can lead to hyperglycemia provide few clues to the chromosomal location or identity of the major susceptibility genes that may be responsible for NIDDM. Although there are examples of chromosomal abnormalities associated with glucose intolerance such as Down's, Klinefelter's and Turner's syndromes, there are no case reports of individuals with chromosomal deletions having NIDDM that might provide a clue as to the location of the gene responsible for some forms of NIDDM. However, just as the follow-up of a case report of an interstitial deletion of chromosome 5 in a mentally retarded individual with multiple developmental abnormalities and familial adenomatous polyposis (FAP) led to the identification of DNA markers on the long arm of chromosome 5 that were linked to the FAP gene (46), the identification of developmentally disabled individuals with NIDDM may lead to the identification of markers for NIDDM. In addition, since a comprehensive human genetic linkage map will be available in the near future, it is important to begin to select and characterize clinically NIDDM families for linkage studies.

The identification of the genes that contribute to IDDM or NIDDM susceptibility will have profound implications for the treatment and prevention of these major public health problems since preventative medicine or therapy could be directed to the unique genotype of the individual rather than to the entire population.

ACKNOWLEDGMENTS

The studies from our laboratory were supported by the Howard Hughes Medical Institute, the Diabetes Research and Training Center of The University of Chicago (DK-20595), and a gift from Sandoz Research Institute.

NOTE ADDED IN PROOF

A DNA polymorphism that cosegregates with MODY in the RW family (Fig. 3) has been identified in the adenosine deaminase gene (*ADA*) on the long arm of chromosome 20 (62). The maximum lod score for linkage between MODY and *ADA* was 5.25 at a recombination fraction of 0.00. These results indicate that the odds are greater than 178,000:1 that the gene responsible for MODY in this family is tightly linked to *ADA* on chromosome 20q.

REFERENCES

1. Foster DW. Diabetes mellitus. In: Scriver CR, Beaudet AL, Sly WS, Valle D, eds. *Metabolic basis of inherited disease,* 6th ed, vol 1, New York: McGraw-Hill, 1989:375–397.
2. Bennett PH. Epidemiology of diabetes mellitus. In: Rifkin D, Porte D, eds. *Diabetes mellitus: theory and practice,* New York: Elsevier, 1990:357–377.
3. Kobberling J, Tillil H. Empirical risk figures for first degree relatives of non-insulin dependent diabetics. In: Kobberling J, Tattersall R, eds. *The genetics of diabetes mellitus.* London: Academic Press, 1982:201–209.
4. Rotter JI, Vadheim CM, Rimoin DL. Genetics of diabetes mellitus. In: Rifkin D, Porte D, eds. *Diabetes mellitus: theory and practice,* New York: Elsevier, 1990:378–413.
5. Fajans SS. Maturity-onset diabetes of the young (MODY). *Diabetes/Metab Rev* 1989;5:579–606.

5a. Fajans SS. Scope and heterogeneous nature of MODY. *Diabetes Care* 1990;13:49–64.

6. O'Rahilly S, Wainscoat JS, Turner RC. Type 2 (non-insulin-dependent) diabetes mellitus: new genetics for old nightmares. *Diabetologia* 1988;31:407–414.
7. Todd JA. Genetic control of autoimmunity in type 1 diabetes. *Immunol Today* 1990;11:122–129.
8. Baisch JM, Weeks T, Giles R, Hoover M, Stastny P, Capra JD. Analysis of HLA-DQ genotypes and susceptibility in insulin-dependent diabetes mellitus. *N Engl J Med* 1990;322:1836–1841.
9. Steiner DF, Tager HS, Chan SJ, Nanjo K, Sanke T, Rubenstein AH. Lessons learned from molecular biology of insulin-gene mutations. *Diabetes Care* 1990;13:600–609.
10. Taylor SI, Kadowaki T, Kadowaki H, Accili D, Cama A, McKeon C. Mutations in insulin-receptor gene in insulin-resistant patients. *Diabetes Care* 1990;13:257–279.
11. Taira M, Taira M, Hashimoto N, et al. Human diabetes associated with a deletion of the tyrosine kinase domain of the insulin receptor. *Science* 1989;245:63–66.
12. Odawara M, Kadowaki T, Yamamoto R, et al. Human diabetes associated with a mutation in the tyrosine kinase domain of the insulin receptor. *Science* 1989;245:66–68.
13. Emery AEH. *Methodology in medical genetics: an introduction to statistical methods.* Edinburgh: Churchill Livingston, 1986:114–125.
14. Cox NJ, Bell GI. Disease associations: chance, artifact, or susceptibility genes? *Diabetes* 1989;38:947–950.
15. Ott J. *Analysis of human genetic linkage.* Baltimore: The Johns Hopkins University Press, 1985.
16. Cooper DN, Schmidtke J. Analysis and diagnosis of human inherited disease by recombinant DNA methods. In: Scriver CR, Beaudet AL, Sly WS, Valle D, eds. *Metabolic basis of inherited disease,* 6th ed, vol 1, New York: McGraw-Hill, 1989:55–72.
17. Cox NJ, Hodge SE, Marazita ML, Spence MA, Kidd KK. Some effects of selection strategies on linkage analysis. *Genet Epidemiol* 1988;5:289–297.
18. Kelsoe JR, Ginns EI, Egeland JA, et al. Re-evaluation of the linkage relationship between chromosome 11p loci and the gene for bipolar affective disorder in the Old Order Amish. *Nature* 1989;342:238–243.
19. Lander ES, Botstein D. Mapping complex genetic traits in humans: new methods using a complete RFLP linkage map. *Cold Spring Harbor Symp Quant Biol* 1986;51:49–62.
20. Lander ES. Mapping complex genetic traits in humans. In: Davies KE, ed. *Genome analysis: a practical approach.* Oxford: IRL Press, 1988:171–189.
21. Weeks DE, Lange K. The affected-pedigree-member method of linkage analysis. *Am J Hum Genet* 1988;42:315–326.
22. Risch N. Linkage strategies for genetically complex traits. II. The power of affected relative pairs. *Am J Hum Genet* 1990;46:229–241.
23. Risch N. Linkage strategies for genetically complex traits. II. The effect of marker polymorphism on analysis of affected relative pairs. *Am J Hum Genet* 1990;46:242–253.
24. Rotter JI, Landaw EM. Measuring the genetic contribution of a single locus to a multilocus disease. *Clin Genet* 1984;26:529–542.
25. Prochazka M, Leiter EH, Serreze DV, Coleman DL. Three recessive loci required for insulin-dependent diabetes in nonobese diabetic mice. *Science* 1987;237:286–289.

26. Concannon P, Wright JA, Wright LG, Sylvester DR, Spielman RS. T-cell receptor genes and insulin-dependent diabetes mellitus (IDDM): no evidence for linkage from affected sib pairs. *Am J Hum Genet* 1990;47:45–52.
27. Permutt MA, Elbein SC. Insulin gene in diabetes: analysis through RFLP. *Diabetes Care* 1990;13:364–374.
28. Spielman RS, Baur MP, Clerget-Darpoux F. Genetic analysis of IDDM: summary of GAW5 IDDM results. *Genet Epidemiol* 1989;6:43–58.
29. Stern MP, Ferrell RE, Rosenthal M, Haffner SM, Hazuda HP. Association between NIDDM, RH blood group, and haptoglobin phenotype: results from the San Antonio heart study. *Diabetes* 1986;35:387–391.
30. Iyengar S, Hamman RF, Marshall JA, Baxter J, Majumder PP, Ferrell FE. Genetic studies of Type 2 (non-insulin-dependent) diabetes mellitus: lack of association with seven genetic markers. *Diabetologia* 1989;32:690–693.
31. Raboudi SH, Mitchell BD, Stern MP, et al. Type II diabetes mellitus and polymorphism of insulin-receptor gene in Mexican Americans. *Diabetes* 1989;38:975–980.
32. Bottini E, Gerlini G, Pascone R, Gori MC, Gloria-Bottini F. Is there a role of chromosome 1 in the clinical expression of diabetes mellitus? *Am J Hum Genet* 1988;43:217–219.
33. Buraczynska M, Hanzlik J, Grzywa M. Apolipoprotein A-I gene polymorphism and susceptibility of non-insulin-dependent diabetes mellitus. *Am J Hum Genet* 1985;37:1129–1135.
34. Xiang K-S, Cox NJ, Sanz N, Huang P, Karam JH, Bell GI. Insulin-receptor and apolipoprotein genes contribute to development of NIDDM in Chinese Americans. *Diabetes* 1989;38:17–23.
35. McClain DA, Henry RR, Ullrich A, Olefsky JM. Restriction-fragment-length polymorphism in insulin-receptor gene and insulin resistance in NIDDM. *Diabetes* 1988;37: 1071–1075.
36. Ober C, Xiang K-S, Thisted RA, Indovina KA, Wason CJ, Dooley S. Increased risk for gestational diabetes mellitus associated with insulin receptor and insulin-like growth factor II restriction fragment length polymorphisms. *Genet Epidemiol* 1989;6:559–569.
37. Li SR, Baroni MG, Oelbaum RS, Stock J, Galton DJ. Association of genetic variant of the glucose transporter with non-insulin-dependent diabetes mellitus. *Lancet* 1988;2: 368–370.
38. Cox NJ, Xiang K-S, Bell GI, Karam JH. Glucose transporter gene and non-insulin-dependent diabetes. *Lancet* 1988;2:793–794.
39. Serjeantson SW, White B, Bell GI, Zimmet P. The glucose transporter gene and type 2 diabetes in the Pacific. In: Larkins RG, Zimmet PZ, Chisholm DJ, eds. *Diabetes 1988,* Amsterdam: Excerpta Medica, 1988:329–333.
40. Miyano M, Nanjo K, Chan SJ, Sanke T, Kondo M, Steiner DF. Use of in vitro DNA amplification to screen family members for an insulin gene mutation. *Diabetes* 1988;37:862–866.
41. Newton CR, Graham A, Heptinstall LE, et al. Analysis of any point mutation in DNA. The amplification refractory mutation system (ARMS). *Nucleic Acids Res* 1989;17: 2503–2516.
42. Lekanne Deprez RH, Potter van Loon BJ, van der Zon GCM, et al. Individuals with only one allele for a functional insulin receptor have a tendency to hyperinsulinaemia but not to hyperglycaemia. *Diabetologia* 1989;32:740–744.
43. Sheffield VC, Cox DR, Lerman LS, Myers RM. Attachment of a 40-base-pair G+C-rich (GC-clamp) to genomic DNA fragments by the polymerase chain reaction results in improved detection of single-base changes. *Proc Natl Acad Sci USA* 1989;86:232–236.
44. Cox NJ, Epstein PA, Spielman RS. Linkage studies on NIDDM and the insulin and insulin-receptor genes. *Diabetes* 1989;38:653–658.
45. O'Rahilly S, Patel P, Wainscoat JS, Turner RC. Analysis of the HepG2/erythrocyte glucose transporter locus in a family with Type 2 (non-insulin-dependent) diabetes and obesity. *Diabetologia* 1989;32:266–269.
46. Bodmer WF, Bailey CJ, Bodmer J, et al. Localization of the gene for familial adenomatous polyposis on chromosome 5. *Nature* 1987;328:614–616.
47. Kingston HM. Genetics of common disorders. *Br Med J* 1989;298:949–952.

48. Nanjo K, Sanke T, Miyano M, et al. Diabetes due to secretion of a structurally abnormal insulin (insulin Wakayama): clinical and functional characterization of [Leu^{A3}] insulin. *J Clin Invest* 1986;77:514–519.
49. Nanjo K, Miyano M, Kondo M, et al. Insulin Wakayama: familial mutant insulin syndrome in Japan. *Diabetologia* 1987;30:87–92.
50. Iwamoto Y, Sakura H, Ishii Y, et al. A new case of abnormal insulinemia with diabetes: reduced insulin values determined by radioreceptor assay. *Diabetes* 1986;35:1237–1242.
51. Awata T, Iwamoto Y, Matsuda A, Kuzuya T. Identification of nucleotide substitution in gene encoding [Leu^{A3}] insulin in third Japanese family. *Diabetes* 1988;37:1068–1070.
52. Haneda M, Polonsky KS, Bergenstal RM, et al. Familial hyperinsulinemia due to a structurally abnormal insulin: definition of an emerging new clinical syndrome. *N Engl J Med* 1984;310:1288–1294.
53. Andreone T, Fajans S, Rotwein P, Skolnick M, Permutt MA. Insulin gene analysis in a family with maturity-onset diabetes of the young. *Diabetes* 1985;34:108–114.
54. Elbein SC, Borecki I, Corsetti L, et al. Linkage analysis of the human insulin receptor gene and maturity onset diabetes of the young. *Diabetologia* 1987;30:641–647.
55. Yoshimasa Y, Seino S, Whittaker J, et al. Insulin-resistant diabetes due to a point mutation that prevents insulin proreceptor processing. *Science* 1988;240:784–787.
56. Kadowaki T, Kadowaki H, Taylor SI. A nonsense mutation causing decreased levels of insulin receptor mRNA: detection by a simplified technique for direct sequencing of genomic DNA amplified by the polymerase chain reaction. *Proc Natl Acad Sci USA* 1990;87:658–662.
57. Shimada F, Taira M, Suzuki Y, et al. Insulin-resistant diabetes associated with partial deletion of insulin-receptor gene. *Lancet* 1990;1:1179–1181.
58. Kadowaki T, Kadowaki H, Rechler MM, Roth J, Gorden P, Taylor SI. Five mutant alleles of the insulin receptor gene in patients with genetic forms of insulin resistance. *J Clin Invest* 1990;44:14979–14985.
59. Moller DE, Yokota A, White MF, Pazianos AG, Flier JS. A naturally occurring mutation of receptor Ala^{1134} impairs tyrosine kinase function and is associated with dominantly inherited insulin resistance. *J Biol Chem* 1990;265:in press.
60. Moller DE, Yokota A, Ginsberg-Fellner F, Flier JS. Functional properties of a naturally occurring Trp^{1200}$\rightarrow$Ser^{1200} mutation of the insulin receptor. *Mol Endocrinol* 1990;4:in press.
61. Yokota A, Moller DE, Flier JS. Homozygous mutation of position 485 of the insulin receptor α-subunit in a patient with lipodystrophy and severe insulin resistance. *Diabetes* 1990;1839:235A.
62. Bell GI, Xiang K-S, Newman MV, et al. Gene for non-insulin-dependent diabetes mellitus (MODY subtype) is linked to DNA polymorphism on human chromosome 20q. *Proc Natl Acad Sci USA* 1991;88:in press.

Etiology of Human Disease at the DNA Level,
edited by Jan Lindsten and Ulf Pettersson.

10

Genetic Control of Plasma Fibrinogen Levels: An Example of Gene–Environment Interaction in the Etiology of a Multifactorial Disorder

*Steve E. Humphries, *Fiona R. Green, *Angela E. Thomas,
**Cecily H. Kelleher, and **Tom W. Meade

**Arterial Disease Research Unit, Charing Cross Sunley Research Centre, Hammersmith, London W6 8LW, England, U.K. **MRC Epidemiology and Medical Care Unit, Northwick Park Hospital, Harrow HA1 3UJ, England, U.K.*

HIGH PLASMA FIBRINOGEN AND ISCHAEMIC HEART DISEASE

Several prospective studies have now shown a direct association between plasma fibrinogen concentration and the subsequent incidence of ischaemic heart disease (IHD) (1–4) and stroke (2). In two of these studies (1,2) the association between plasma fibrinogen and IHD incidence is at least as strong if not stronger than that between blood cholesterol and IHD. In men in the Northwick Park Heart Study (NPHS) (1), an elevation of 1 standard deviation in fibrinogen (about 0.6 g/L) was associated with an 84% increase in the risk of IHD within the next 5 years. It is well known that individuals who smoke have a higher risk of both IHD and stroke (5,6) and many studies have observed that individuals who smoke have elevated levels of plasma fibrinogen (1–8). It is very likely that a substantial proportion of the association between smoking and IHD is mediated through the plasma fibrinogen concentration (1,6).

There are several mechanisms whereby elevated levels of fibrinogen may be involved in IHD. High levels of fibrinogen increase blood viscosity, which itself may partly be involved. Individuals with elevated fibrinogen levels may have an increased propensity for coagulation, and thrombus formation in an artery that is already narrowed by atherosclerosis is a frequent cause of acute symptoms such as myocardial infarction or stroke (9). Alternatively, elevated levels of fibrinogen may be having a direct effect on the development of the atherosclerotic lesion. Fibrinogen interacts with specific receptors on activated platelets and increases platelet aggregability in vitro (10,11). Fibrinogen has a mitogenic effect on haemopoetic cells, again through apparently specific cell surface receptors (12), but these studies have not been extended to endothelial cells. It is, however, clear that fibrinogen and fibrinogen degradation products can be detected histochemically and immunologically in the intima of diseased artery walls, and within atherosclerotic plaques (13). Animal studies have also shown that intravascular fibrin deposition is related to plasma fibrinogen levels (14).

Elevated plasma fibrinogen could be caused either by increased synthesis or by reduced removal via the fibrinolytic system. Fibrinolysis occurs as a result of the action of a number of plasma proteins (15) and in vivo studies have shown that the degradation products of injected radiolabelled fibrinogen accumulate in low levels in all the organs and tissues of the body (reviewed in 16). Fibrinogen is synthesized in the liver, and since it is an acute phase protein, its plasma level is raised following infection or injury. Levels are raised by the use of oral contraceptives and following the menopause (7). In both men and women, the fibrinogen level increases with age and obesity and is higher in diabetics than nondiabetics. There is an inverse association between alcohol intake and the plasma fibrinogen concentration. The known individual and environmental factors that influence fibrinogen level account for about 20% of the population variance in fibrinogen (7), although because there is considerable within-individual variability in fibrinogen levels, this is probably a conservative estimate. The extent to which genetic factors may determine the plasma fibrinogen level is of obvious interest.

Each plasma fibrinogen molecule is composed of two each of the Aα-, Bβ- and gamma-fibrinogen polypeptide chains. The complex is held together by a number of inter- and intrachain disulphide links, and the steps in assembly of the complex protein have been elucidated by pulse chase studies in vivo (17) and in vitro in the hepatoma cell line HepG2 (17). Fibrinogen secreted immediately after a pulse of ^{35}S-methionine contains label only in the Bβ-chain, with labelled Aα- and gamma-chains appearing some time later. This demonstrates that there are intracellular pools of Aα- and gamma-chains and that the rate-limiting step in the assembly of plasma fibrinogen is the synthesis of Bβ-chains (18). We could therefore speculate that an alteration in the level of synthesis of the Bβ-chain may have an effect on the amount of fibrinogen secreted by the liver.

GENETIC CONTRIBUTION TO DETERMINATION OF FIBRINOGEN LEVELS

A priori it is likely that variation at a number of different unlinked loci will be involved in determining between-individual differences in plasma fibrinogen levels. Data from path analysis have suggested an estimate for her-

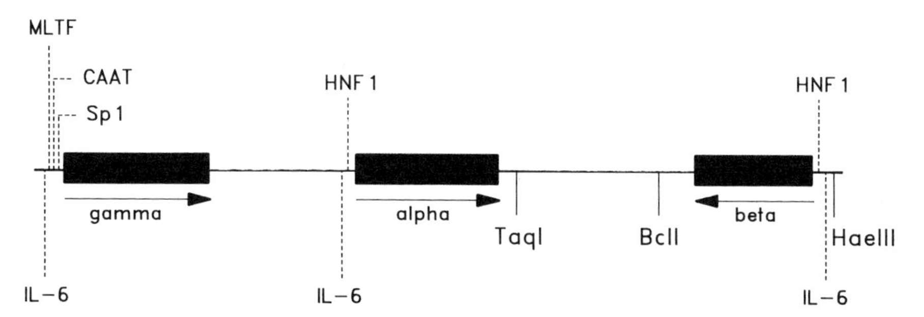

FIG. 1. Fibrinogen gene locus showing proposed binding sites for transcription factors (data from 30,40). MLTF, adenovirus major late transcription factor; CAAT, CAAT binding factor; Sp 1, transcription factor Sp 1; HNF 1, hepatic nuclear factor 1; 1L-6, interleukin-6-responsive element. Horizontal arrows indicate direction of transcription of genes. The positions of the variable site for the enzymes *Taq*I, *Bcl*I, and *Hae*III are shown.

itability of fibrinogen levels of 0.5 (19) but there have as yet been no reports of biometrical analysis to investigate the possibility of a major gene determining fibrinogen levels. Candidate genes that may be involved in determining between-individual differences in plasma fibrinogen include the fibrinogen genes themselves, and those involved in fibrin generation and fibrinolysis such as thrombin and plasminogen.

The cDNA sequence and gene structure for all three fibrinogen genes have now been determined. The genes are in a cluster of less than 50 kb (Fig. 1) on the long arm of chromosome 4 (20,21). Northern blot experiments using cDNA probes have shown that each chain is synthesized as a separate mRNA, and that the levels of all three mRNAs are coordinately controlled to some extent (22).

In our initial studies, we used three two-allele restriction fragment length polymorphisms (RFLPs) of the fibrinogen genes to investigate the role of genetic variation at this locus in the determination of plasma fibrinogen levels (23). In a sample of 91 healthy individuals, statistically significant differences in mean fibrinogen levels were found in groups of individuals with different RFLP genotypes (Table 1). The largest effect was associated with the polymorphism detected with the β-fibrinogen probe and the enzyme *Bcl*I, explaining 9.5% of the sample variance. We obtained similar results, but with a smaller effect on fibrinogen levels, using a *Taq*I RFLP detected with the α-fibrinogen probe. These data suggest that variation at the fibrinogen gene locus does determine in part the between-individual differences in fibrinogen levels seen in healthy individuals in the UK. A recent study in twins in Norway failed to find an association between these same polymorphisms and fibrinogen levels (24). The reason for this is unclear and may reflect genetic differences between Norway and the UK, or be a result of differences in the fibrinogen assay used or study design.

No large studies have yet been reported on the frequency of fibrinogen

TABLE 1. *Mean (and approximate SD) of plasma fibrinogen levels in 91 healthy individuals from North London, grouped according to* Bcl*I (B)* Taq *I (T), and* Hae*III (H) RFLP genotype*

Genotype	No.	Fibrinogen g/L	$R^2 \times 100$
B1B1	50	2.74 (0.47)*	
B1B2	37	2.98 (0.58)	9.0%
B2B2	4	3.69 (0.67)	
T1T1	47	3.00 (0.60)	
T1T2	37	2.74 (0.50)	4.2%
T2T2	7	2.76 (0.27)	
H1H1	47	2.8 (0.48)	
H1H2	33	3.1 (0.50)	5.0%
H2H2	8	3.3 (0.69)	

*$p = 0.025$.
(Data from 23 and unpublished).

RFLPs between patients with coronary artery disease (CAD) and controls, to determine if variation associated with the *Bcl*I RFLP contributes significantly to CAD risk. By comparison with data from the NPHS (1), 45–55 year old men with the genotype *B2B2* (mean fibrinogen 3.69 g/L) would experience a roughly twofold increase in risk of an IHD event over 5 years compared to those with the genotype *B1B1* (mean fibrinogen 2.7 g/L). It is thus possible that the frequency of the *B2* allele would be higher in individuals suffering a myocardial infarction (MI). However, the development of CAD is multifactorial involving interaction between many genetic and environmental factors, and variation at any one gene locus is unlikely to be having a major impact on CAD risk in the general population. Power calculations indicate that if an effect associated with the *Bcl*I RFLP was to increase risk twofold, it would require a sample of 300 patients and 300 controls to detect a difference in frequency with 90% power at 0.05% level of significance. This RFLP frequency approach has been used extensively for the apoprotein genes but results have been conflicting, at least in part because the samples used have been too small, but also because of problems of ethnic heterogeneity (25).

THE FIBRINOGEN GENES ARE A POLYGENIC SYSTEM

In disorders caused by single gene defects, the rare mutations in the gene that cause clinical consequences have a drastic effect on the function or level of expression of the protein. However, there are many potential sites in a gene where a single base change may have only a *small* effect on gene expression. For example, some sequence changes in the promoter or enhancer region may alter the rate of transcription by only 10–20% (e.g., 26). Sequence changes may affect the efficiency of correct splicing of the RNA in the nucleus or stability of the mRNA, and result in reduced mRNA levels in the cytoplasm. In addition, changes in the amino acid sequence of the protein may alter its function or its rate of removal. The expression of the candidate gene will therefore be the result of the summation of the effects of sequence variation in and around the gene, with each variation having an additive and interchangeable or possibly an interactive effect. In the general population, such sequence variations will have occurred at different times in evolutionary history, and by chance some will have reached polymorphic frequencies. These variations may occur in several different combinations in the population, with the number of combinations depending in part on the rate of recombination that is causing the decay of the evolutionary association between the different mutations at the particular gene locus. Each gene thus represents a "polygenic system," with overall expression being determined by effects of variation at a number of sites. This concept is particularly relevant in the genetic analysis of a quantitative trait. For a protein like fibrinogen comprised of several polypeptides these possibilities are increased by the fact that there are three genes involved, requiring coordinate regulation.

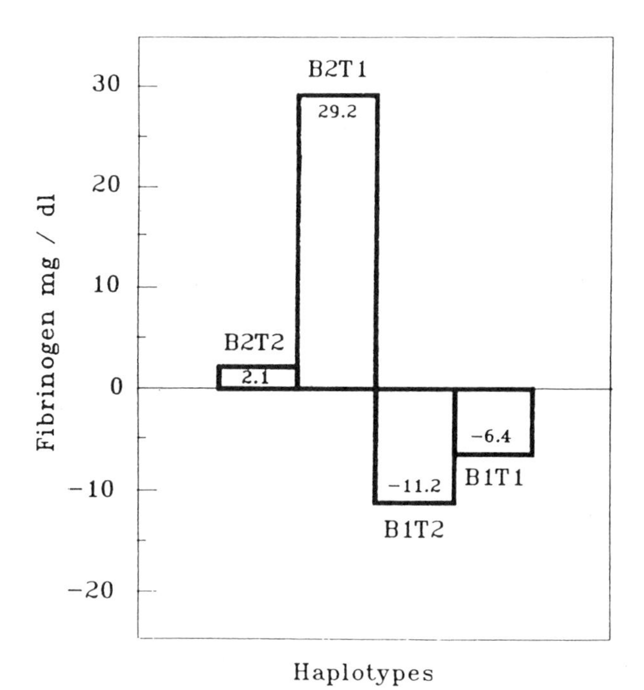

FIG. 2. Average excess of fibrinogen levels associated with the *BclI* and *TaqI* haplotypes observed in 91 healthy individuals from North London (data from 22). Frequencies of the four haplotypes are *B1-T1* = 0.54; *B1-T2* = 0.21; *B2-T1* = 0.18; *B2-T2* = 0.071.

The concept is illustrated by the data shown in Table 1, where the *Taq*I *T1* allele and the *Bcl*I *B2* allele are both associated with an elevation of plasma fibrinogen levels. In this small sample there is no detectable linkage disequilibrium between these two variable sites (delta $= 0.16$ $p > 0.1$), and all four possible 2-RFLP haplotypes were observed unambiguously (23). As expected from an additive model, the group of individuals with the genotype *B1B1/T2T2* had amongst the lowest levels of fibrinogen (2.58 g/L $n = 21$) and the two individuals with the genotype *B2B2/T1T1* amongst the highest (4.44 g/L). Taken together, variation associated with the *Bcl*I and *Taq*I RFLPs explained 15% of the sample variance in fibrinogen levels (23). We have used these data to estimate the "average excess" (27), associated with the four haplotypes in this sample. As can be seen in Fig. 2, the *B2-T1* haplotype is associated with a large positive effect on fibrinogen levels, the *B1-T2* and *B1-T1* haplotypes with a negative effect of similar size, and the *B2-T2* haplotype with an effect close to zero. This suggests the presence of at least *three* functionally distinct haplotypes in this sample, and the use of further polymorphisms and a larger sample should allow a more precise definition of this.

MECHANISM OF ASSOCIATION

The DNA sequence differences that create all three of the polymorphisms map outside the gene-coding region (19 and our unpublished observations) with the *Bcl*I variant site mapping in the α-β intergenic region. Therefore, it is most likely that these sequence differences themselves do not alter the function or level of production of fibrinogen, but are in population association, because of their evolutionary history, with some other functionally significant differences. Since the synthesis of the Bβ-protein appears to be the rate-limiting step in production of fibrinogen, the promoter of the β-gene is a prime candidate for the location of such sequence differences. In the whole population we predict that several such differences will exist, and will be present, singly or in combination, in different individuals who have high or low plasma fibrinogen levels.

The sequence of the β-gene promotor was published in 1987 (28) and the region from −150 bp to the start of transcription contains a number of elements of interest (Fig. 3). Only in this region is there significant sequence homology between the rat and human genes suggesting that this sequence has an important conserved function (28). This region also has homology with other "acute phase" genes such as α-1-antitrypsin (29). In addition, this region has been reported to contain all the information required to act as a promoter in HepG2 cells and has been shown to bind proteins from a HepG2 cell nuclear extract (30). The sequence from −89 to −76 contains a conserved liver-specific transcription element which binds hepatocyte nuclear

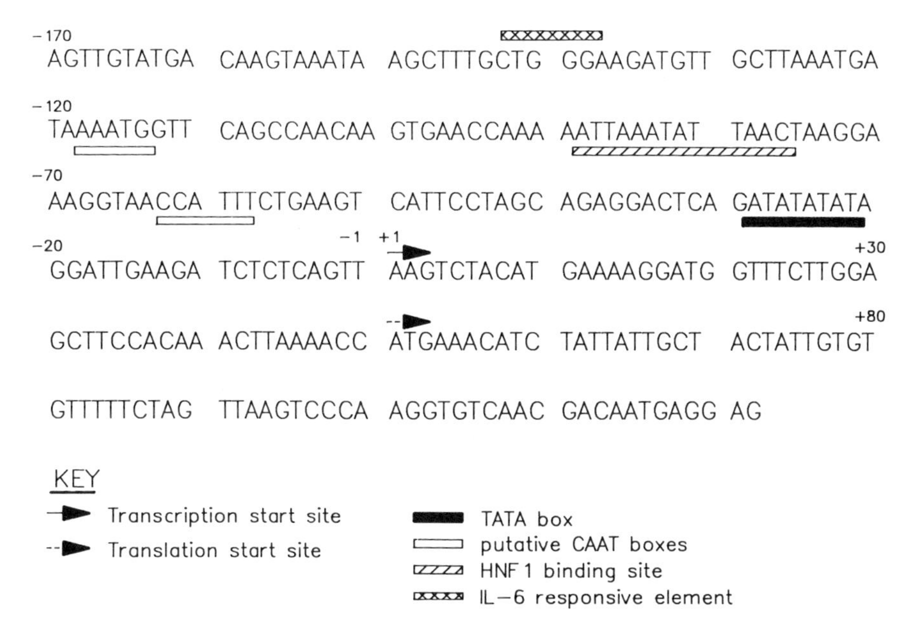

FIG. 3. Nucleotide sequence of the 5′ flanking region and first exon of the human β-fibrinogen gene (28), showing *cis*-acting elements homologous to those found to be involved in transcriptional control of rat β-fibrinogen gene expression. HNF 1, hepatic nuclear factor 1; IL-6, interleukin 6.

factor 1 (HNF1) (30,31), and deletion mapping shows that this region also contains the interleukin 6 (IL-6)-responsive element (31), which has been identified in other genes as the motif CTGGGA (32,33).

To screen for sequence differences, in vitro gene amplification was used to synthesise a 500-bp fragment covering this region of the gene, from individuals of different genotype. Using a combination of oligonucleotide melting and digestion with frequent-cutting restriction enzymes, no differences were detected in the promoter-proximal 150 bp. However a common variation was detected using the enzyme *Hae*III, with a loss of the predicted *Hae*III site, at position −453, in roughly 20% of alleles examined. Using the 91 individuals from our initial study we observed that this *Hae*III site was associated with small differences in plasma fibrinogen levels explaining 5% of the sample variance, with individuals homozygous for the lack of the site (genotype *H2H2*) having fibrinogen levels 14% higher than the mean of the sample (Table 1). Although this site is outside the region of the reported promoter sequence it is possible that it is having a direct effect on transcription. Preliminary analysis has demonstrated that a hepatic nuclear protein does bind to this region of the gene and that binding is altered by the *Hae*III sequence change (Green F, Cortese R, Humphries SE, unpublished observations). In order to test whether this is also causing a change in transcription of the gene, experiments are in progress to insert this fragment of the gene into the appropriate vector to test promoter strength.

There is almost complete linkage disequilibrium between the *Hae*III variable site and the *Ava*II variable site used previously and strong linkage disequilibrium with the *Bcl*I variable site (unpublished data). However, the association between the *Hae*III site and fibrinogen levels we have observed is weaker than that for the *Bcl*I site. The data thus suggest that we have not yet identified the functionally important sequence difference originally detected with the *Bcl*I RFLP, and it is likely that there are other functionally important sequence elements that are involved in transcriptional control located outside the 500-bp upstream-region of the β-gene transcriptional start. There may be transcription control elements in the introns of the β-gene, enhancer elements at a distance from the gene, or important elements in the α- or γ-genes that affect plasma fibrinogen level. In particular it is known that dexamethasone stimulates plasma fibrinogen secretion by both liver cells and HepG2 cells, at least partly through an effect on transcription. Recently, the dexamethasone-responsive element for fibrinogen has been identified, by deletion mapping, to be located between 2,900 and 1,500 bp upstream from the start of transcription of the β-gene (31). The sequence for this region of the gene has not been reported, but it represents another potential area for detailed analysis. Since linkage disequilibrium is more dependent on evolutionary history and less on physical distance, (at least over short genetic distances—see 34), it is possible that the functionally important sequence change detected by the *Bcl*I RFLP is at considerable distance away (say 10–50 kb). Thus sequencing studies in these other more distant regions of the gene cluster are warranted to search for the presence of distant enhancer elements.

WITHIN-INDIVIDUAL VARIATION IN FIBRINOGEN LEVELS MAY BE INFLUENCED BY VARIATION AT THE FIBRINOGEN GENE LOCUS

Fibrinogen is an acute phase protein and after infection or injury, plasma levels rise by up to twofold within 24–48 hours. Because of its sensitivity to these and other environmental factors, the within-individual variation of fibrinogen levels is high, accounting for up to 26% of the sample variance in standardised assays (35). One result of this is that when based on a single measure, estimates of the strength of the association between fibrinogen levels and CAD risk or with RFLP genotype will be an underestimate by as much as 20–30% (35). The extent of this within-individual variation is likely to have clinical consequences. An individual whose mean fibrinogen level is low but whose levels fluctuate over a wide range may sometimes be at a greater risk of an acute thrombotic event when at a "peak" value, compared

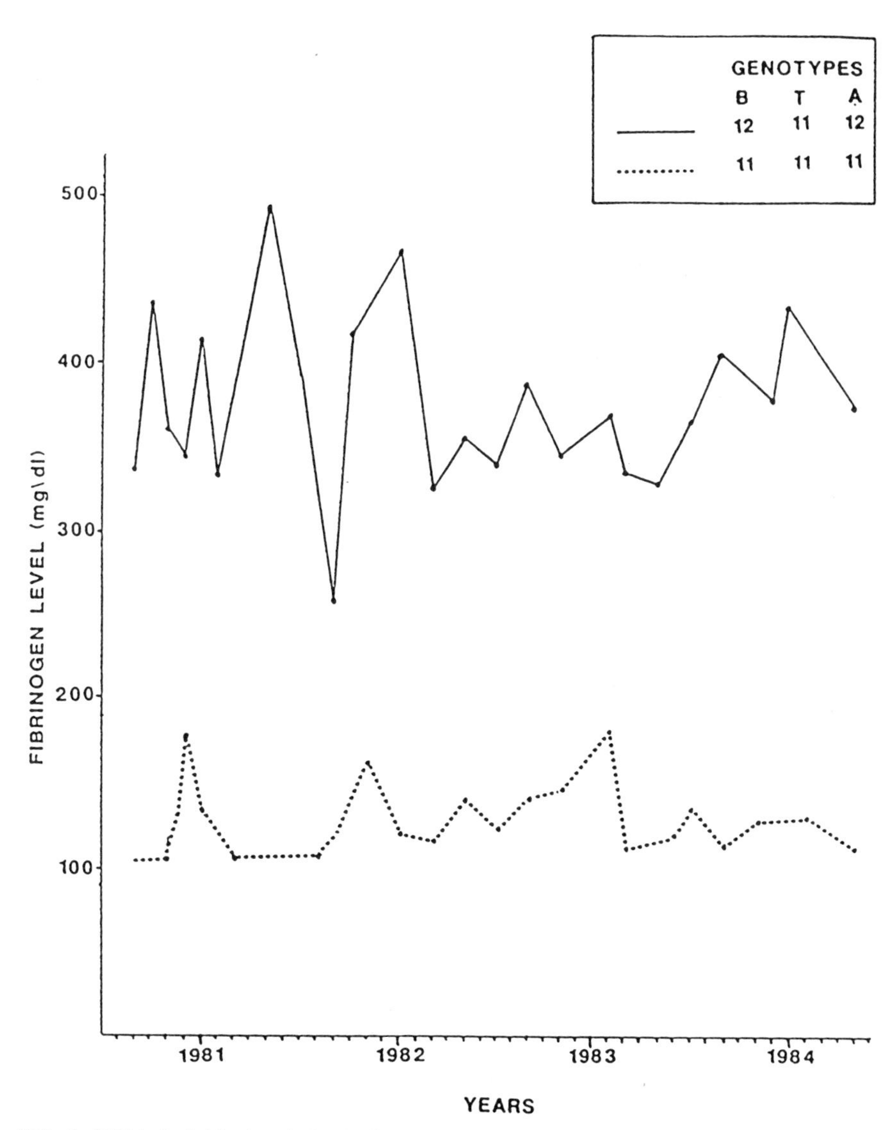

FIG. 4. Within-individual variation in fibrinogen levels over time. Unadjusted fibrinogen values over a 3-year period for individual 12 (64 year old male, mean level 3.9 ± 0.65 g/L) genotype *B1B2,* and individual 9 (32 year old male, mean level 1.3 ± 0.21 g/L) genotype *B1B1* (age at start of study; data from 35,37). B = *BclI*; T = *TaqI*; A = *AraII.*

to an individual whose mean level is higher, but in whom the fluctuations are less. This "variability" concept has been developed by Berg, particularly in relation to genes determining lipid levels (36).

As before, there are a number of candidate genes in which DNA sequence variation may determine the extent of within-individual variability in fibrinogen levels. However, we have recently obtained preliminary evidence that variation at the fibrinogen gene locus is involved, in part, in determining this within-individual variability. We have studied a panel of 14 individuals on whom an average of 20 fibrinogen determinations have been carried out over a period of 3 years (35). In some of these individuals fibrinogen levels vary widely over time, whereas in others they appear more stable. We have determined the genotype of 13 of these individuals, and preliminary analysis suggests that the four individuals with the genotype *B1B1* may have a significantly lower degree of variation overall, compared to the nine with the genotype *B1B2* (37). As an example of this, Fig. 4 shows the fluctuation in fibrinogen levels over time in two of the male individuals from the study, selected to have different *Bcl*I genotype and the highest and lowest within-individual variance estimates. These findings, although only in a small group of subjects, suggest that variation at the fibrinogen gene cluster may interact with an as yet unidentified environmental factor, to determine, in part, an individual's variability in fibrinogen levels over time. One possibility is that variation in sequences such as the IL-6-responsive element proximal to the β-promoter may affect the affinity of the *trans*-acting protein that interacts with the element. This could result in an altered transcription rate of the gene in different individuals in response to a similar environmental stimulus such as infection or smoking. From the point of view of identifying individuals at risk of CAD because of increased thrombotic potential, this is one example in which a single determination of an individual's genotype may carry as much information as multiple determination of fibrinogen levels over several months.

IS THERE INTERACTION BETWEEN SMOKING AND GENOTYPE AT THE FIBRINOGEN LOCUS TO DETERMINE AN INDIVIDUAL'S PLASMA FIBRINOGEN LEVELS?

Besides cross-sectional studies showing a significant gradient of plasma fibrinogen levels between nonsmokers, exsmokers, and current smokers, prospective data show a clear relationship between a reduction of fibrinogen levels associated with stopping smoking, and an elevation of fibrinogen levels upon starting or resuming smoking (6–8,38). There is also good evidence for a dose response between daily use of cigarettes and fibrinogen levels (8). However, although the relationship between smoking and raised plasma fibrinogen is well established, the mechanism whereby smoking has this effect is unclear. One possibility is that macrophages in the lung, recruited as a response to damage induced by smoking, may release low levels of IL-6 into the blood. This IL-6 would result in a chronic stimulation of fibrinogen synthesis through the binding of positive transcription factors to IL-6-responsive elements in the fibrinogen gene promoters.

However, it is possible that the elevation in fibrinogen experienced in response to a certain degree of smoking may vary between different individuals. Information about the genetic determinants of such response would be useful since individuals who are predisposed to this enhanced response are at particular risk of thrombotic events and it may be that for these individuals advice on avoiding smoking may be particularly important. This is another example in which, compared to a single protein measurement, an individual's genotype may contain additional information for predicting future risk of subsequent high fibrinogen levels, and thus the development of IHD.

At the present time, the genes determining response to smoking are unknown. One possibility that could be explored is that variation associated with RFLPs at the fibrinogen gene locus may determine the elevation in plasma fibrinogen seen in response to smoking. Using RFLPs, this would be seen as a greater than expected elevation of fibrinogen levels associated with a certain genotype class, which would indicate that smokers with this genotype may be particularly susceptible to develop elevated levels of fibrinogen because of genetic variations at the fibrinogen gene locus.

We have recently analysed data from a group of 292 healthy men, of whom 41% are current smokers, and determined genotype for the *Hae*III RFLP (39). The results showed that both smoking status and genotype strongly and independently predicted an individual's fibrinogen levels and as expected, the worst possible combination was the possession of the *H2* allele and smoking. There was no evidence for significant interaction between these two factors, but it is possible that the use of other RFLPs in larger samples, and a more precise definition of smoking habit, may allow the detection of smoking-genotype interaction using this approach.

CONCLUSIONS

It is evident that the control of plasma fibrinogen levels is complex, involving not only many environmental factors such as alcohol intake, smoking habit, age, obesity, and the acute phase response, but also genetic factors, as shown by the association of the *Bcl*I RFLP of the β-fibrinogen gene with plasma fibrinogen levels. The development of recombinant DNA technology has made the dissection of the different factors controlling plasma fibrinogen levels a valid proposition and great progress is already being made. Our studies have suggested that variation at the fibrinogen gene locus determines in part not only an individual's plasma level of fibrinogen, but also the extent to which that level varies over time. If eventually established, any such interaction between genetic predisposition and environmental factors will be particularly important in understanding the etiology of this multifactoral disease. It may be possible to develop DNA tests to identify individuals who, on the basis of their genotype, are at high risk of thrombotic ischaemic heart disease. The challenge for the next few years will be to move from the imprecise use of RLPS to the detection and characterisation of the underlying variability in the different genes, which collectively predispose to thrombosis. This should allow the development of precise tests that will have a high degree of accuracy and diagnostic potential. Once identified, the subsequent risk of these individuals can be reduced by modifying life-style or by drug therapy to reduce other known risk factors such as cholesterol levels. Once the mechanisms controlling fibrinogen levels are better understood at the molecular level, it may also be possible to develop directed therapeutic strategies that will reduce fibrinogen synthesis in a specific manner, an approach which is not possible at present. In the future such pharmacological agents may have as wide an impact on reducing IHD as cholesterol-lowering drugs do today.

ACKNOWLEDGEMENTS

The work in the author's laboratory was supported by the British Heart Foundation (RG5 and 87/78), the Medical Research Council, the Charing Cross Sunley Research Trust, and the Tobacco Products Research Trust. We thank Springer-Verlag Publishers for permission to reproduce material from several published papers, and Elaine Osman for assistance in preparation of the manuscript.

REFERENCES

1. Meade TW, Mellows S, Brozovic M, et al. Haemostatic function and ischaemic heart disease: principal results of the Northwick Park Heart Study. *Lancet* 1986;ii:533–537.
2. Wilhelmsen L, Svärdsudd K, Korsan-Bengtsen K, Larsson B, Welin L, and Tibblin G. Fibrinogen as a risk factor for stroke and myocardial infarction. *N Engl J Med* 1984;311:501–505.
3. Kannell WB, Wolf PA, Castelli WP, and D'Agostino RB. Fibrinogen and risk of cardiovascular disease: The Framingham Study. *JAMA* 1987;258:1183–1186.
4. Stone MC, and Thorpe JM. Plasma fibrinogen—a major coronary risk factor. *JR Coll Gen Pract* 1985;35:565–569.
5. Abbot RD, Yin Yin, Reed DM, Yano K. Risk of stroke in male cigarette smokers. *N Engl J Med* 1986;315:717–720.
6. Meade TW, Imeson J, and Stirling Y. Effects of changes in smoking and other characteristics on clotting factors and risk of ischaemic heart disease. *Lancet* 1987;ii:986–988.
7. Meade TW, Chakrabarti R, Haines AP, North WRS and Stirling Y. Characteristics affecting fibrinolytic activity and plasma fibrinogen concentrations. *Br Med J* 1979;i:153–156.

8. Wilkes HC, Kelleher C and Meade TW. Smoking and plasma fibrinogen. *Lancet* 1988;i:307–308.
9. Davies MJ and Thomas A. Thrombosis and acute coronary-artery lesions in sudden cardiac ischaemic death. *N Engl J Med* 1984;310:1137–40.
10. Meade TW, Vickers MV, Thompson SG and Seghatchian MJ. The effects of physiological levels of fibrinogen on platelet aggregation. *Thromb Res* 1985;38:527–34.
11. Marguerie G, Ginsberg MH and Plow EF. The role of fibrinogen in platelet aggregation. In: *Fibrinogen-Fibrin Formation and Fibrinolysis,* Vol. 4, (eds. Lane, D.A., Henschen, A. and Jasani, M.K.), Walter de Gruyter & Co., Berlin, pp. 175–183, 1986.
12. Levesque JP, Hatzfeld A and Hatzfeld J. Fibrinogen mitogenic effect on hemopoietic cell lines: Control via receptor modulation. *Proc Natl Acad Sci USA* 1986;83:6494–6498.
13. Smith EB and Staples EM. Haemostatic factors in human aortic intima. *Lancet* 1981;i:1171–1174.
14. Gurewich V, Lipinski B and Hyde F. The effect of the fibrinogen concentration and the leucocyte count on intravascular fibrin deposition from soluble fibrin monomer complexes. *Thromb Haemost* 1976;36:605–614.
15. Lijnen HR. (1986) On the role of fibrin in the fibrinolytic system. In: *Fibrinogen-Fibrin Formation and Fibrinolysis.* Vol. 4, (eds. Lane, D.A., Henschen, A., and Jasani, M.K.), Walter de Gruyter & Co., Berlin, pp. 121–136.
16. Lane DA. (1986) Clearance and catabolism of fibrinogen and its derivatives—a review. In: *Fibrinogen and Its Derivatives,* (eds. V. Schaefers-Borchel, E. Selmagi and A. Henschen), Elsevier Science Publishers, B.V. (Biomedical Division), Amsterdam, pp. 181–193.
17. Alving BM, Chung S, Murano G, Tang DB and Finlayson JS. Rabbit fibrinogen: Time course of constituent chain production *in vivo. Arch Biochem Biophys* 1982;217:1–9.
18. Yu S, Sher B, Kudryk B and Redman CM. Intracellular assembly of human fibrinogen. *J Biol Chem* 1984;259:10574–10581.
19. Hamsten A, Iselius L, de Faire U and Blomback M. Genetic and cultural inheritance of plasma fibrinogen concentration. *Lancet* 1987;ii:988–990.
20. Humphries SE, Imam AMA, Robbins TP, et al. The identification of a DNA polymorphism of the α-fibrinogen gene and the regional assignment of the human fibrinogen genes to 4q26-qter. *Hum Genet* 1984;88:148–153.
21. Kant JA, Fornace AJ Jr., Saxe D, et al. Evolution and organization of the fibrinogen locus on chromosome 4: Gene duplication accompanied by transposition and inversion. *Proc Natl Acad Sci USA* 1985;82:2344–2348.
22. Crabtree GR and Kant JA. Coordinate accumulation of the mRNAs for the α, β and gamma chains of rat fibrinogen following defibrination. *J Biol Chem* 1982;257:7277–7279.
23. Humphries SE, Cook M, Dubowitz M, Stirling Y and Meade TW. Role of genetic variation at the fibrinogen locus in determination of plasma fibrinogen concentrations. *Lancet* 1987;i:1452–1455.
24. Berg K and Kierulf P. DNA polymorphisms at fibrinogen loci and plasma fibrinogen concentration. *Clin Genet* 1989;36:229–235.
25. Humphries SE. DNA polymorphisms of the apoliprotein genes—their use in the investigation of the genetic component of hyperlipidaemia and atherosclerosis. *Atherosclerosis* 1988;72:89–108.
26. Treisman R, Orkin SH and Maniatis T. Specific transcription and RNA splicing defects in five cloned β-thalassaemia genes. *Nature* 1983;302:591–596.
27. Templeton AR. The general relationship between average effect and average excess. *Genet Res* 1987;49:69–70.
28. Huber P, Dalmon J, Courtois G, et al. Characterization of the 5′-flanking region for the human fibrinogen β gene. *Nucleic Acid Res* 1987;15:1615–1625.
29. Kugler W, Wagner U, and Ryffel GU. Tissue-specificity of liver gene expression: A common liver-specific promoter element. *Nucleic Acids Res* 1988;16:3165–3174.
30. Courtois G, Morgan JG, Campbell LA, Fourel G and Crabtree GR. Interaction of a liver-specific nuclear factor with the fibrinogen and α-1-antitrypsin promoters. *Science* 1987;238:688–692.
31. Huber P, Laurent M and Dalmon J. Human beta-fibrinogen gene expression: Upstream

sequences involved in its tissue specific expression and its dexamethasone and interleukin 6 stimulation. *J Biol Chem* 1990;265:5695–5701.

32. Hattori M, Northemann W and Fey GH. Acute-phase reaction induces a specific complex between hepatic nuclear proteins and the interleukin 6-responsive element of the rat α-2-macroglobulin gene. *Proc Natl Acad Sci USA* 1990;87:2364–2368.
33. Ito T, Tanahashi H, Misumi Y and Sakaki Y. APRF-1. Nuclear factors interacting with an interleukin-6 responsive element of rat α-2-macroglobulin gene. *Nucleic Acids Res* 1989;17:9425–9435.
34. Hegele RA, Plaetke R and Lalouel JM. Linkage disequilibrium between DNA markers at the low-density lipoprotein receptor gene. *Genet Epidemiol* 1990;7:68–81.
35. Thompson SG, Martin JC and Meade TW. Sources of variability in coagulation factor assays. *Thromb Haemost* 1987;58:1073–1077.
36. Berg K. Predictive genetic testing to control coronary heart disease and hyperlipidaemia. *Arteriosclerosis supplement* I. 1989;9:1-51–1-58.
37. Cook M, Godenir N, Green FR, et al. Genetic variation at the fibrinogen locus is involved in determining fibrinogen levels: Evidence from thirteen individuals sampled on repeated occasions. *Biochem Soc Trans* 1988;16:541–542.
38. Balleisen L, Bailey J, Epping PH, et al. Epidemiological study on factor VII, factor VIII and fibrinogen in an industrial population: I. Baseline data on the relation to age, gender, body-weight, smoking, alcohol, pill-using, and the menopause. *Thromb Haemost* 1985;54:475–479.
39. Thomas A, Kelleher C, Green F, Meade TW and Humphries SE. Variation in the promoter region of the β-fibrinogen gene is associated with plasma fibrinogen levels in smokers and non-smokers (in preparation).
40. Morgan JG, Courtois G, Fourel G, et al. Sp1, a CAAT-binding factor, and the adenovirus major late promoter transcription factor interact with functional regions of the gamma-fibrinogen promoter. *Mol Cell Biol* 1988;8:2628–2637.

Etiology of Human Disease at the DNA Level,
edited by Jan Lindsten and Ulf Pettersson.

11

The Low-Density Lipoprotein Receptor:

A Key for Unlocking a Multifactorial Disease

Joseph L. Goldstein and Michael S. Brown

Department of Molecular Genetics, University of Texas Southwestern Medical Center, Dallas, Texas 75235

The chronic degenerative diseases, such as cancer, heart disease, and arthritis, are caused by multiple subtle predisposing genetic differences among individuals that dictate a deleterious response to similar environmental insults. In the past the ability to unravel this complex causal chain has been limited to simple quantitative estimates of heritability. The power of molecular genetics may now make it possible to identify the genetic loci involved and to pinpoint the specific predisposing mutant alleles. This in turn should allow a more precise delineation of the specific environmental agents. This new information should improve our ability to predict the occurrence of a degenerative disease in an individual and hopefully to take appropriate steps to alter the environment, thereby obviating the disease.

One set of diseases that may yield to this approach are the atherosclerotic vascular diseases, whose hallmark is a deposition of blood-borne cholesterol in thickened plaques that narrow the blood vessels. The rate of vascular buildup is in rough proportion to the level of cholesterol in the blood, particularly the level of cholesterol contained in low-density lipoprotein (LDL).

One approach to unraveling atherosclerosis is to understand the factors that dictate the level of LDL cholesterol in the blood. In recent years a major gene controlling LDL levels has been identified and major genetic differences at this locus have been traced. The gene encodes the LDL receptor, which removes LDL from the bloodstream (1). In this chapter we review progress in understanding the LDL receptor and introduce this receptor as a potential key to unlocking the multifactorial nature of atherosclerosis.

LDL RECEPTORS: A MAJOR DETERMINANT OF PLASMA CHOLESTEROL LEVELS

The concentration of LDL in blood of normal individuals varies over a 3-fold range. Every large-scale epidemiologic study ever conducted has shown that the risk of coronary atherosclerosis increases progressively as the plasma LDL level rises. In industrialized societies such as those of the United States and Western Europe, the mean LDL cholesterol level in the population is above the threshold for risk of developing atherosclerosis. In the past these concentrations of LDL have been considered "normal" in the statistical sense that they are the usual values found in such populations. However, they seem not to be normal for the human species in the biologic sense that they lead to accelerated atherosclerosis.

What determines the concentration of LDL cholesterol in the bloodstream? And why do half of all Americans and Europeans have concentrations of plasma LDL that place them at high risk for developing atherosclerosis? Answers are emerging from studies of a class of membrane proteins, called LDL receptors, that were discovered 17 years ago (1). These receptors are present in varying amounts on the surface of most mammalian cells, where they mediate the uptake of plasma LDL, thereby providing the cholesterol that cells need for growth and membrane synthesis. In the body, most LDL receptors are expressed in the liver where they supply cholesterol for secretion into bile, for conversion to bile acids, and for resecretion into the plasma in newly synthesized lipoproteins. LDL receptors are also present in high concentrations in the adrenal cortex and the ovarian corpus luteum, where they provide cholesterol for steroid hormone formation.

The total number of LDL receptors in the body is a major determinant of plasma LDL levels. When the number of LDL receptors is high, LDL is rapidly removed from the plasma and circulating levels are kept low. When LDL receptors are diminished as a result of genetic or acquired defects, LDL levels rise and atherosclerosis is accelerated (2).

All tissues that express LDL receptors contribute to the removal of LDL from plasma and thereby help to determine plasma LDL levels. The liver makes an especially important contribution for two reasons (Fig. 1A). First, the liver expresses the largest number of LDL receptors in the body, approximately 70% of the total. Second, the liver uses its LDL receptors not only for the uptake of LDL, but also for the uptake of intermediate-density lipoprotein (IDL), the precursor of LDL. When LDL receptors are diminished, the liver fails to take up IDL normally. The IDL remains in plasma where it is converted to LDL. Thus, hepatic LDL receptors determine the rate of LDL production as well as its degradation (1,2).

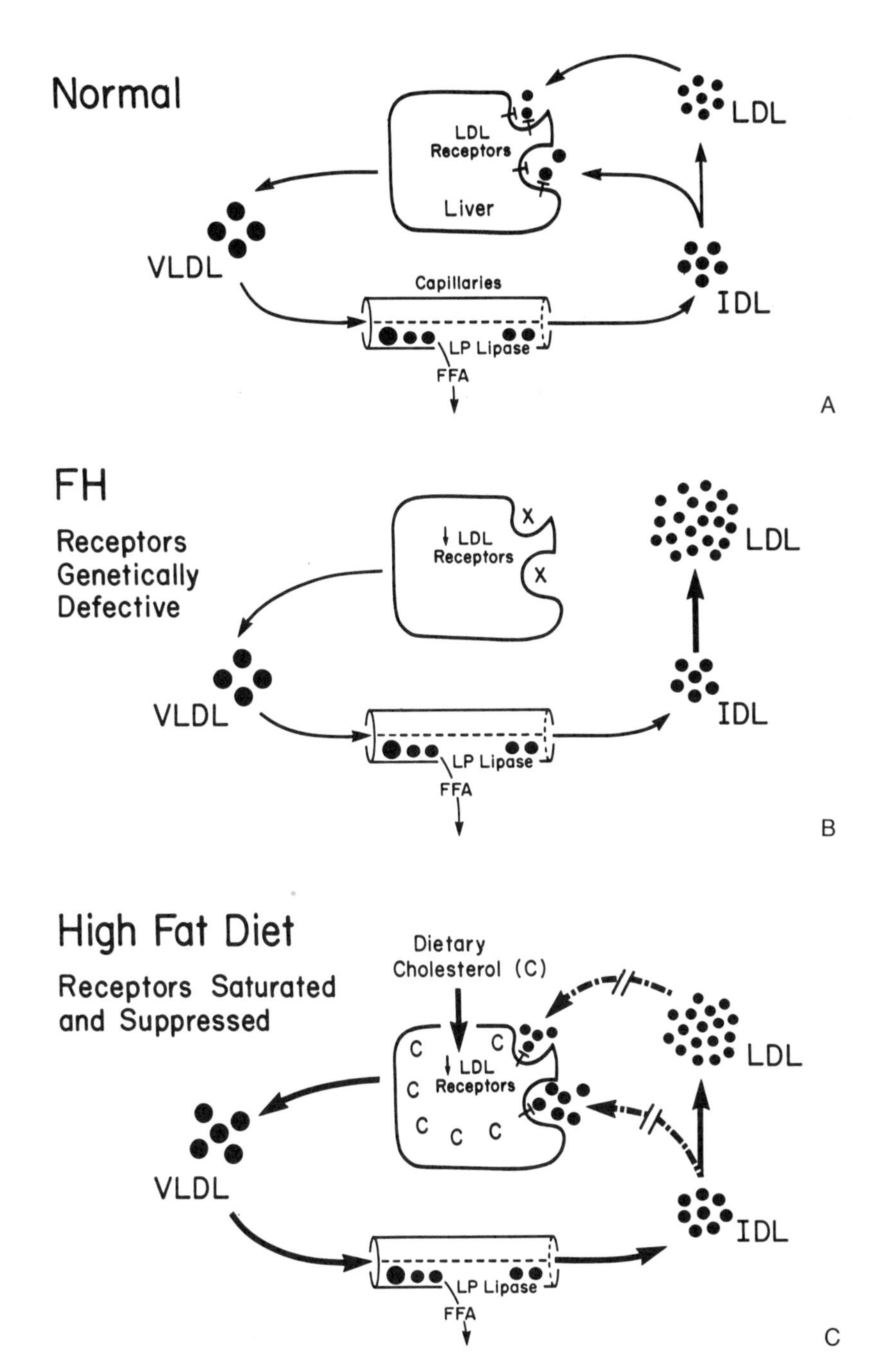

FIG. 1. Schematic model of the mechanism by which low-density lipoprotein (LDL) receptors in the liver control both the production and catabolism of plasma LDL in **(A)** normal human subjects, **(B)** in individuals with familial hypercholesterolemia (FH), and **(C)** in non-FH individuals who ingest diets high in cholesterol and saturated fat. FFA, free fatty acids, IDL, intermediate-density lipoprotein; VLDL, very low density lipoprotein. Modified from Brown and Goldstein (ref. 1).

THE LDL RECEPTOR: PROTEIN, mRNA, GENE

The human LDL receptor is a single-chain transmembrane glycoprotein of 839 amino acids (Fig. 2) (3). It specifically binds lipoproteins that contain apo B-100 (such as LDL) and apo E (such as IDL). The ligand-binding domain comprises the NH_2-terminal 292 amino acids and is composed of a cysteine-rich sequence of 40 amino acids that is repeated seven times with minor variations. The cytoplasmic domain, composed of 50 amino acids at the COOH-terminal end of the protein, serves to direct the receptor to coated pits where the bound LDL is rapidly internalized. Between these two ends of the protein there is a 400-amino acid region that is homologous to the precursor for epidermal growth factor, a 58-amino acid region that con-

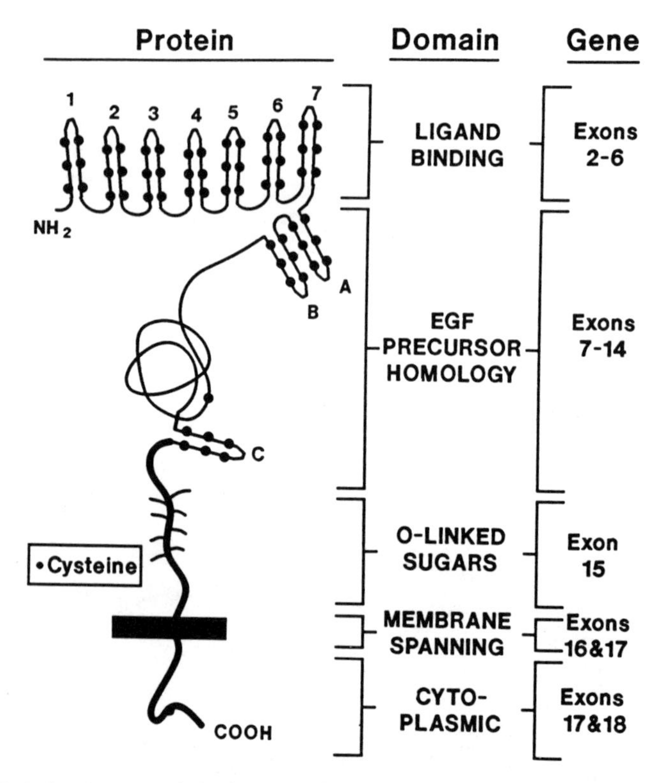

FIG. 2. Domain structure of the human LDL receptor protein and its relation to the exon organization of the gene. The domains of the 839-amino acid mature protein are shown at the left and the corresponding exons encoding the protein domains at the right. Exon 1 (not shown) encodes the 21-amino acid signal sequence, which is cleaved from the mature protein during synthesis in the endoplasmic reticulum. EGF, epidermal growth factor.

tains multiple O-linked carbohydrate chains, and a 22-amino acid membrane-spanning region (3).

The human LDL receptor is encoded by a 5.3-kilobase (kb) mRNA. About half of the mRNA constitutes a long 3′ untranslated region that contains three members of the Alu family of middle repetitive DNAs (4). The human LDL receptor gene, located on the distal short arm of chromosome 19 (p13.1-p13.3) (5), spans 45 kb, and is divided into 18 exons and 17 introns (6). Many of the exons share an evolutionary history with exons of other genes, an observation consistent with the notion that the LDL receptor gene was assembled by exon shuffling. Figure 2 shows how the exons correlate with the functional domains of the protein (3,6).

MUTATIONS IN THE LDL RECEPTOR GENE

LDL receptors can be reduced as a result of two mechanisms: genetic (Fig. 1B) and acquired (Fig. 1C). Genetic reductions occur in familial hypercholesterolemia (FH), a common autosomal dominant disorder that is caused by multiple allelic mutations in the LDL receptor gene (2).

When LDL receptors are diminished in FH, IDL particles are not cleared rapidly by the liver and they are converted into LDL in increased amounts (Fig. 1B). Thus, a decrease in hepatic LDL receptors increases the production of LDL. At the same time, the rate of removal of LDL from blood is diminished. The combination of increased production and diminished removal leads to a disproportionate elevation in plasma LDL levels when hepatic LDL receptors are decreased (2).

Approximately 1 in every 500 persons in most populations of the world inherits a single copy of a mutant LDL receptor gene and thus has the heterozygous form of FH (2). The cells of these individuals produce about half the normal number of LDL receptors. As a result, LDL is removed from the circulation at half the normal rate; the lipoprotein accumulates in blood to levels 2-fold above normal; and heart attacks occur typically in the fourth and fifth decades. Heterozygous FH causes about 5% of all heart attacks in people under age 60 (2).

Rarely, two FH heterozygotes marry and produce a child who inherits two mutant genes at the LDL receptor locus. These children are referred to as FH homozygotes, although they are often not true homozygotes in that they may inherit different mutant LDL receptor genes from each of their parents. FH homozygotes have a much more severe clinical syndrome than do the heterozygotes. Their cholesterol levels are 6- to 10-fold above normal and they usually suffer heart attacks in early childhood (2).

Over the past 17 years, fibroblast cultures from 149 FH homozygotes from 134 unrelated families have been analyzed in our laboratory. These cell

TABLE 1. *Numerology of LDL receptor mutations among 134 FH homozygotes in the Dallas collection*

61 True homozygotes =	61 Possible alleles
73 Compound heterozygotes =	146 Possible alleles
	207 Total possible alleles
	−24 Redundant alleles[a]
	183 Probable alleles (upper limit)

[a]Identical mutant alleles observed in more than one FH homozygote.

strains are designated as the Dallas collection. Approximately 45% of the FH homozygotes in the Dallas collection are true homozygotes as determined by haplotype analysis (7). The remaining 55% are compound heterozygotes with two different mutant alleles. The 134 unrelated FH homozygotes in the Dallas collection could have as many as 183 mutant alleles as revealed by the calculations shown in Table 1. It is likely that the true number of alleles is even higher than we estimate since we have seen only three examples in which the same mutation was encountered in FH homozygotes from unrelated families (7). In certain ethnic populations specific LDL receptor mutations have achieved a high frequency via a founder effect. Examples include the French Canadians, Afrikaners, Christian Lebanese, and Finns (discussed below). In these populations the number of different mutant alleles is much less than in outbred populations.

Twenty-four of the 183 mutant alleles in the Dallas collection (13%) have major structural rearrangements detectable by Southern blotting (7). This group includes 3 insertions and 21 deletions. The 87% of mutations that have no overt structural rearrangement are predominantly point mutations or small inframe deletions. To date, we have not identified any mutations that primarily disrupt RNA splicing, but this most likely reflects a selection bias in that we have concentrated primarily on mutations that produce receptors with altered structures, and we have not studied many null alleles. Figure 3 shows the location of 33 mutations in the LDL receptor gene that have been sequenced at the DNA level in Dallas. Each FH family analyzed to date (with the few exceptions mentioned above) has had a different mutation.

Most of the deletions and insertions in the LDL receptor gene have arisen because of recombination between ubiquitously located repetitive Alu-type elements. A detailed discussion of Alu-mediated recombination in the generation of LDL receptor mutations has been presented elsewhere (7). In brief, of the 28 large deletions in the LDL receptor genes reported in the literature, eight have been sequenced, and seven of these involve an Alu repeat at one or both mutation endpoints (7).

At a functional level, LDL receptor mutations can be divided into five classes based on their phenotypic effects on the protein (Fig. 4). *Class 1 mutations* fail to produce immunoprecipitable protein (null alleles). They in-

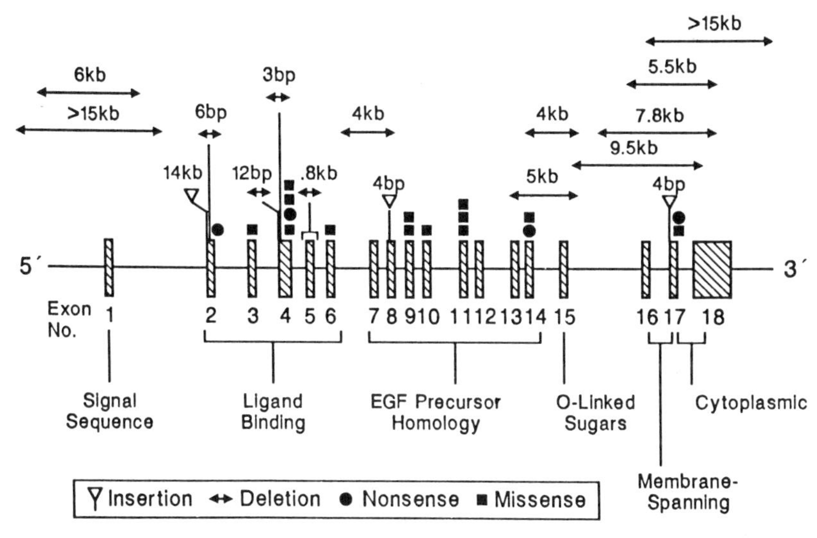

FIG. 3. Mutations in the LDL receptor gene. Exons, shown as hatched boxes, are separated by introns, which are drawn to approximate scale. Mutations that have been sequenced in Dallas are shown above the gene. EGF, epidermal growth factor.

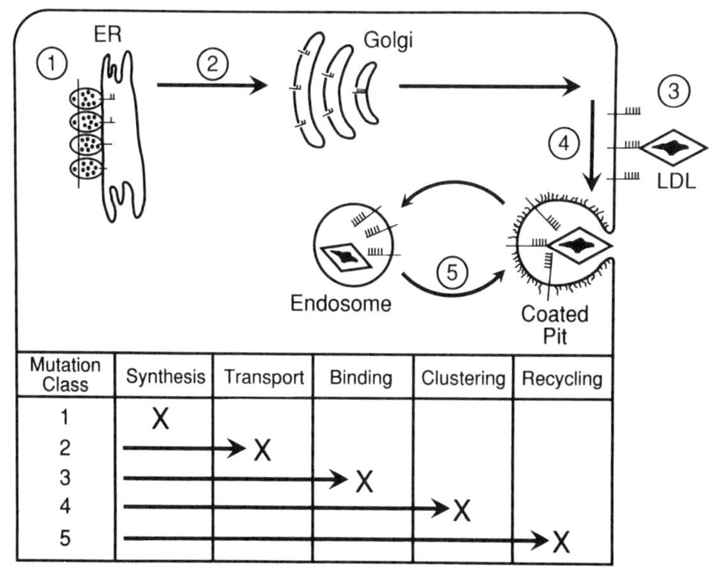

FIG. 4. Five classes of mutations at the low-density lipoprotein (LDL) receptor locus. These mutations disrupt the receptor's synthesis in the endoplasmic reticulum (ER), transport to the Golgi, binding of apolipoprotein ligands, clustering in coated pits, and recycling in endosomes. Each class is heterogeneous at the DNA level.

volve large deletions or nonsense mutations that can occur anywhere in the gene. *Class 2 mutations* encode proteins that do not fold properly after synthesis and are blocked, either partially or completely, in transport between the endoplasmic reticulum and the Golgi complex (transport-defective alleles). They arise from missense mutations or small inframe deletions that occur either in the ligand-binding domain or in the epidermal growth factor (EGF) precursor homology domain. *Class 3 mutations* encode proteins that are synthesized and transported to the cell surface, but fail to bind LDL normally (binding-defective alleles). They arise from missense mutations or inframe deletions involving one of the seven exons encoding the ligand-binding domain. *Class 4 mutations* encode proteins that move to the cell surface and bind LDL normally, but are unable to cluster in clathrin-coated pits and thus do not internalize LDL (internalization-defective alleles). They arise from missense or nonsense mutations in the cytoplasmic domain. *Class 5 mutations* encode receptors that bind and internalize ligand in coated pits, but fail to discharge the ligand in the endosome and fail to recycle to the cell surface (recycling-defective alleles). They arise from inframe deletions that remove one or more exons encoding the EGF precursor homology domain.

LDL RECEPTOR MUTATIONS IN INBRED POPULATIONS

In most populations of the world, the frequency of heterozygous FH is ~ 0.2% (2). However, in certain populations the frequency of FH is much higher, owing to one or more mutations that have reached high frequency because of a founder effect. The highest frequency of FH in the world is in the Afrikaner population of South Africa where the prevalence of FH is 5-fold higher than in the European population from which it originated (8). The present-day Afrikaners comprising a population of ~ 3 million are descended from ~ 2,000 original settlers, mostly from Holland, Germany, and France, who emigrated to the Cape of South Africa in the 17th and 18th centuries. In the 19th century they moved into the interior of the country, the Transvaal, where they remained largely isolated from the surrounding populations. Among 12 unrelated FH homozygotes of Afrikaner descent, 96% of the mutant alleles consisted of two missense mutations (9). The overall frequencies of these two mutations in Afrikaners are presently being determined, but it is likely that they alone explain the high frequency of FH in this population.

The frequency of FH heterozygotes in the French Canadian population has been estimated to be 1 in 270 based on the number of FH homozygotes identified in Quebec province in 1981, and an even higher frequency (1 in 154) was found in the northeastern region of the province (7,10). The 5.3 million modern French Canadians are descended from about 8,000 French settlers who emigrated to Quebec Province between 1608 and 1763 (11), founding an agrarian population that has remained physically and socially isolated. Five mutations account for a total of 76% of the mutant LDL receptor alleles in French Canada (12,13). The most common of these five mutations, a large deletion at the 5′ end of the LDL receptor gene, was found in 59% of 130 French Canadian FH heterozygotes from Montreal (12).

Historically, the Christian Lebanese population has played an important role in delineating the genetics of FH. In 1964 Khachadurian (14) described a high frequency of FH homozygotes in this population. He used this population to delineate, for the first time, the clinical and genetic differences between heterozygotes and homozygotes. The Dallas collection of 134 FH homozygotes contains six with a Christian Lebanese or Syrian ancestry. All were found to be homozygous for a nonsense mutation in exon 14 (ref. 15 and unpublished observations), which has not yet been identified in any other population. The exact frequency of this mutation in Lebanon has not been determined, but the high incidence of homozygosity suggests that it is common.

The general European population, like the North American population, has a plethora of different LDL receptor mutations that cause FH. One exception is the Finnish population, where a single mutation involving a large deletion at the 3′ end of the gene is found in at least one-third of unrelated Finnish individuals with heterozygous FH (16). The prevalence of FH in the Finnish population has not been determined, but the high frequency of a single mutation suggests that the occurrence of FH in Finland may be higher than in the general European population.

REGULATORY DEFECTS IN THE LDL RECEPTOR

The vast majority of individuals with elevated LDL cholesterol levels are not FH heterozygotes: they develop hypercholesterolemia even though they possess two normal LDL receptor genes. Strong circumstantial evidence suggests that many of these cases of hypercholesterolemia are caused by acquired deficiencies attributable to faulty feedback regulation of LDL receptor genes (1). Feedback suppression of LDL receptors was originally demonstrated in tissue culture cells, and it has now been demonstrated in livers of animals from several species (17,18). We attribute this repression to a defense mechanism that protects cells against overaccumulation of intracellular cholesterol. When cholesterol builds up within the cell, the sterol represses transcription of the gene for the LDL receptor, and this reduces the number of LDL receptors. By this mechanism liver cells reduce their uptake of LDL, thereby protecting themselves against an overaccumulation of intracellular cholesterol, but at the expense of an elevated level of LDL in blood.

Sterol-mediated suppression of the LDL receptor gene is mediated by a *cis*-acting DNA sequence of approximately 10 base pairs in length that is located in the promoter portion of the 5′ flanking region of the gene (19). To date, no mutations in individuals with FH have been mapped to this transcriptional regulatory element presumably because of the restricted length of this segment in contrast to the much longer DNA segment that encodes the receptor protein (10 base pairs vs. 45,000 base pairs).

Nearly everyone in Western societies eats a relatively high fat diet, yet some suppress their LDL receptors more than others. One source of variability may lie in polymorphisms in the genes encoding the regulatory proteins that control receptor production by interacting with the *cis*-acting sterol regulatory element. Some regulatory proteins may bind sterols with higher affinity than others, or they may bind to the DNA element more tightly. Once the putative *trans*-acting regulatory proteins responsible for sterol-mediated repression are identified, it will be of great interest to clone the genes and to study their genetic variation in individuals who respond differently to a similar high fat diet.

SUMMARY: ATHEROSCLEROSIS AS A MULTIFACTORIAL TRAIT

The progress in understanding the LDL receptor may now provide the foundation that will support a concerted attack on other genes and the environmental factors that contribute to atherosclerosis in the general population. Individual human beings vary greatly in their propensity to develop atherosclerosis. One reason for this variability is the variability in plasma LDL cholesterol levels. On average, individuals with higher LDL levels tend to have heart attacks earlier than individuals with lower LDL levels. Some of the reasons for the variability in plasma LDL levels have been discussed above. But LDL receptors are not the only determinants of plasma LDL levels. Other proteins control the rate of synthesis of very low-density lipoprotein (VLDL), the rate of conversion of VLDL to IDL, and the rate of conversion of IDL to LDL. Variations in any of these processes will indirectly affect LDL levels. In addition, mutations in the genes for apo B-100 and apo E, the two ligands for the LDL receptor, can directly affect LDL levels by altering the binding of lipoproteins to the LDL receptor.

Even if we understood all of the reasons for variability in LDL levels, we would not understand the variable sensitivity of individuals to atherosclerosis. Among individuals with identical plasma LDL levels, susceptibility to atherosclerosis varies widely. We know something about a few of the factors that mediate this vulnerability. Identified "risk factors" such as smoking and hypertension play a role. Cigarette smoke may damage oxidatively the LDL particle as it circulates through the lung, and it may also damage the vascular endothelium. Hypertension increases the stress upon cells in the artery wall. An increased influx and retention of cholesterol-laden lipoproteins occurs at sites of increased stress.

In addition to these widely recognized risk factors, there is growing evidence for the involvement of another lipoprotein, designated Lp(a) (20).

Lp(a) is an unusual lipoprotein consisting of one molecule of a kringle-containing protein, apo(a), attached covalently in disulfide linkage to the apo B-100 of one LDL particle. Lp(a) is found in relatively large amounts in approximately 20% of Caucasian individuals. The variability in Lp(a) levels arises because of variations in the gene encoding apo(a) in the population. The presence of Lp(a) in blood has been associated with an increase in the incidence of myocardial infarctions. This effect of Lp(a) is enhanced if an individual also has an elevated level of LDL (21).

In the future it is hoped that we will learn more about the systemic and local factors that determine the resistance of blood vessels to atherosclerotic deposition. Such information will probably emerge from a better knowledge of the cells in the atherosclerotic plaque, including endothelial cells, smooth muscle cells, and macrophages. By learning the nature of the gene products that control the behavior of these cells we hope to learn why some individuals are resistant to atherosclerosis.

REFERENCES

1. Brown MS, and Goldstein JL. A receptor-mediated pathway for cholesterol homeostasis. *Science* 1986;232:34–47.
2. Goldstein JL, and Brown MS. Familial hypercholesterolemia. In *The Metabolic Basis of Inherited Disease*. Chapter 48. 6th Edition, C.R. Scriver, A.L. Beaudet, W.S. Sly, and D. Valle, eds. New York: McGraw-Hill Publishing Company, 1989;1215–1250.
3. Goldstein JL, Brown MS, Anderson RGW, Russell DW, and Schneider WJ. Receptor-mediated endocytosis: Concepts emerging from the LDL receptor system. *Annu Rev Cell Biol* 1985;1:1–39.
4. Yamamoto T, Davis CG, Brown MS, et al. The human LDL receptor: A cysteine-rich protein with multiple Alu sequences in its mRNA. *Cell* 1984;39:27–38.
5. Lindgren V, Luskey KL, Russell DW, and Francke U. Human genes involved in cholesterol metabolism: Chromosomal mapping of the loci for the low density lipoprotein receptor and 3-hydroxy-3-methylglutaryl-coenzyme A reductase with cDNA probes. *Proc Natl Acad Sci USA* 1985;82:8567–8571.
6. Südhof TC, Goldstein JL, Brown MS, and Russell DW. The LDL receptor gene: A mosaic of exons shared with different proteins. *Science* 1985;228:815–822.
7. Hobbs HH, Russell DW, Brown MS, and Goldstein JL. The LDL receptor locus in familial hypercholesterolemia: Mutational analysis of a membrane protein. *Annu Rev Genet* 1990;24:133–170.
8. Seftel HC, Baker SG, Sandler MP, et al. A host of hypercholesterolaemic homozygotes in South Africa. *Br Med J* 1980;281:633–636.
9. Lange Y, and Matthies HJG. Transfer of cholesterol from its site of synthesis to the plasma membrane. *J Biol Chem* 1984;259:14624–14630.
10. Moorjani S, Roy M, Gagne C, et al. Homozygous familial hypercholesterolemia among French Canadians in Quebec Province. *Arteriosclerosis* 1989;9:211–216.
11. Laberge C. Prospectus for genetic studies in the French Canadians, with preliminary data on blood groups and consanguinity. *Bull Johns Hopkins Hosp* 1966;118:52–68.
12. Hobbs HH, Brown MS, Russell DW, Davignon J, and Goldstein JL. Deletion in the gene for the LDL receptor in majority of French Canadians with familial hypercholesterolemia. *N Engl J Med* 1987;317:734–737.
13. Leitersdorf E, Tobin EJ, Davignon J, and Hobbs HH. Common low-density lipoprotein receptor mutations in the French Canadian population. *J Clin Invest* 1990;85:1014–1023.
14. Khachadurian AK. The inheritance of essential familial hypercholesterolemia. *Am J Med* 1964;37:402–407.

15. Lehrman MA, Schneider WJ, Brown MS, et al. The Lebanese allele at the LDL receptor locus: Nonsense mutation produces truncated receptor that is retained in endoplasmic reticulum. *J Biol Chem* 1987;262:401–410.
16. Aalto-Setala K, Helve E, Kovanen PT, and Kontula K. Finnish type of low density lipoprotein receptor gene mutation (FH-Helsinki) deletes exons encoding the carboxy-terminal part of the receptor and creates an internalization-defective phenotype. *J Clin Invest* 1989;84:499–505.
17. Brown MS, and Goldstein JL. Lipoprotein receptors in the liver: Control signals for plasma cholesterol traffic. *J Clin Invest* 1983;72:743–747.
18. Spady DK, and Dietschy JM. Dietary saturated triacylglycerols suppress hepatic low density lipoprotein receptor in the hamster. *Proc Natl Acad Sci USA* 1985; 82:4526–4530.
19. Goldstein JL, and Brown MS. Regulation of the mevalonate pathway. *Nature* 1990;343:425–430.
20. Utermann G. The mysteries of lipoprotein(a). *Science* 1989;246:904–910.
21. Seed M, Hoppichler F, Reaveley D, et al. Relation of serum lipoprotein(a) concentration and apolipoprotein(a) phenotype to coronary heart disease in patients with familial hypercholesterolemia. *N Engl J Med* 1990;322:1494–1499.

Gene Replacement

Etiology of Human Disease at the DNA Level,
edited by Jan Lindsten and Ulf Pettersson.

12

Gene Transfer and Gene Therapy

Principles, Prospects, and Perspective

Richard C. Mulligan

*Whitehead Institute for Biomedical Research,
Cambridge, Massachusetts, and Department of Biology, Massachusetts Institute of Technology, Cambridge, Massachusetts 02142*

Over a very short time, recombinant DNA technology has revolutionized the practice of experimental biology and dramatically impacted upon our current understanding of biological processes at the molecular level. Through the ability to identify, isolate, and manipulate specific regions of the mammalian chromosome and to reintroduce those genetic sequences back into animal cells, it has been possible to begin to understand the role of specific cells and gene products in critical decisions in cellular differentiation, and to understand the basic mechanisms of gene regulation themselves. In addition to stimulating dramatic advances in the basic sciences, recombinant DNA technology has also already made its mark in the area of medicine in several important ways. First, the ability to engineer the expression of large amounts of a polypeptide encoded by a specific DNA sequence has made possible the availability of a variety of biologically active gene products (e.g., enzymes, cytokines, growth factors) hitherto too scarcely found in the human body to permit characterization and to predict their usefulness clinically. A variety of such "genetically engineered" proteins of clinical import have already been identified and are being effectively administered to patients. A second important contribution has involved the use of recombinant DNA technology to define the role of specific gene products in the pathogenesis of various human diseases. Perhaps the most impressive advances in this area have come from the DNA mapping studies that have led to the identification of genes and gene products responsible for several common inherited diseases, such as cystic fibrosis and muscular dystrophy (see Chapter 1). In addition to the implications of these studies for the prenatal diagnosis of diseases, it is almost certain that further studies of the function or dysfunction of "disease" genes will directly lead to improved clinical management.

In spite of the advances in medicine outlined above, "gene therapy," perhaps the most obvious and certainly the most direct means of applying recombinant DNA techniques to the practice of medicine, is only beginning to be put in practice. Broadly defined, gene therapy refers to the use of mammalian gene transfer techniques to introduce genetic sequences into specific cells of a patient in order to achieve some clinical benefit. Most often, gene therapy is considered in the context of the treatment of inherited diseases resulting from well-characterized single gene defects, primarily because the relevant gene to be transferred is known, as is the relevant target cell. However, it is increasingly clear that gene transfer might be effectively applied to the treatment of a wide range of acquired diseases as well, if the relevant genes and target tissue were identified, and the procedures for gene transfer were appropriate.

In this rather interpretative review, the notion of utilizing gene transfer as a form of therapy for inherited and acquired disease will be considered, primarily from the standpoint of technical feasibility and breadth of application. To set the framework for the rest of the review, the various ways in which gene transfer might be used to treat disease will first be outlined in a general way. Next the characterization of different existing techniques for mammalian gene transfer will be described. Since retroviral-mediated gene transfer has emerged as the most likely technique to be employed in gene therapy protocols in the near future, the principles, characteristics, and safety of this gene transfer system will be discussed in most detail. The remainder of the review will mainly be devoted to a discussion of the feasibility of gene therapies for specific inherited and acquired diseases, involving when possible existing preclinical data supporting the efficacy and safety of the proposed technique. Finally, some general comments regarding the future direction of gene therapy research and the impact of this new technology on medicine will be made.

STRATEGIES FOR GENE THERAPY

In the same way that the diversity of human disease has necessitated the development of a wide range of different conventional therapies, the successful treatment of disease using gene therapy will require a variety of novel strategies, each tailor-made for a particular clinical situation. Nevertheless, in order to begin to understand the basic concepts and potential breadth of application of gene therapy, it is useful to categorize in a general way the types of genetic interventions that might be useful in the treatment of disease, and then consider scenarios for their implementation.

One important way to classify genetic interventions is with respect to their intended longevity. In the classic notion of gene therapy, the goal of treatment is a lifelong supply of "genetically corrected" cells. For this requirement to be met, not only must the transferred gene persist and function, but the cells into which the gene has been transferred must persist and function as well. This latter point is quite an important issue when the target cell has a naturally short half-life. For other applications, however, it may be important either for the transduced cell to survive for only short periods of time, or for expression of a transferred gene to be short-lived or modulated in a specific way. While techniques are becoming available for eliminating transduced cells at defined time points (1), we are not yet able to control vector gene expression effectively in a simple way. Nevertheless, it may be possible to obtain short-lived gene expression, through the use of transient gene expression techniques (see below).

A second informative means of classifying genetic interventions is with respect to the nature of intended benefit. For a number of inherited diseases, the intent of intervention is to "correct" a genetic defect that manifests itself only in a particular cell type. Successful treatment would therefore require transfer of the relevant gene specifically into that target cell or its progenitor. Such an intervention would likely be necessary for the treatment of diseases such as β-thalassemia, sickle cell anemia, cystic fibrosis, and muscular dystrophy. In contrast, in the case of a number of diseases, a genetic defect may affect the expression of a gene product that is not critical for the survival and function of the cell in which it is synthesized, but rather critical for the proper function of other cells, organs, or biological processes. Accordingly, in this case, the target cell for gene transfer may not need to be the same cell that normally synthesizes the gene product. Such an indirect intervention may be possible for the treatment of clotting disorders, hormone deficiencies, and any other disorders that result from the lack of a suitable circulating concentration of a protein product. This type of approach may also be applicable to disorders resulting from insufficient detoxification of specific harmful products or to acquired diseases in which the systemic administration of a protein product may be beneficial [see discussion of human immunodeficiency virus (HIV)]. A final mode of intervention that should be mentioned involves attempts to inhibit the expression of a gene product, rather than to express a product. Such an approach may or may not necessitate expression of a protein product.

For implementation of either of these genetic interventions, two general scenarios for gene therapy need to be considered. In the first scenario, termed the ex vivo approach, cells from a particular location of the body are removed from the patient, cultured in vitro for a time sufficient for efficient gene transfer, and subsequently returned to the patient. Technically, this approach demands that (i) the relevant cells are accessible to withdrawal from the patient; (ii) the cells can be cultured in vitro, without sacrificing their ultimate function in vivo; and (iii) the cells can be returned to the patient in such a way that they persist and function. The other scenario, the in vivo approach, involves the direct transfer of genes to cells naturally residing in the body. Accordingly, there are no restraints upon the ability to culture the cells in vitro or to transplant the cells effectively. However, for this approach to succeed (i) there need be (in most cases) a means to target transduction of the desired cell type selectively; (ii) the cells must be receptive to gene transfer (often, cell replication is a prerequisite for integration); and (iii), sufficient numbers of cells must be transducible.

While the applicability of either scenario to a particular disease depends on a variety of issues, the biological characteristics of the desired target cell for gene transfer are perhaps most relevant. In Table 1, a partial list of potentially important target cells is presented. While virtually all of the cell types are accessible to withdrawal from the body, they vary in their capacity for in vitro culture and their subsequent capacity to be transplanted. Bone-marrow-derived stem cells, for example, are generally quiescent in vivo (2); while conditions have been developed for the long-term culture of some hematopoietic cell types derived from the stem cell (3), no culture system that provides for the proliferation of stem cells in the absence of differentiation currently exists. Practically, this means that bone marrow stem cells can only be cultured in vitro for short periods of time if they are to retain their ability to reconstitute bone marrow transplant recipients. Pancreatic β cells are also very difficult to culture, although the cells can be transplanted in specialized circumstances (4). While hepatocytes can be cultured for some time in vitro (5), only a low level of cell division can be obtained. In addition, even short-term culture of hepatocytes can dramatically affect their expression of differentiated gene products (6) (although this effect of in vitro culture might be reversible). No effective method yet exists for hepatocyte transplantation, and therefore the effect of in vitro culture on transplantability is uncertain. Lymphocytes can be cultured for long periods of time in vitro, but only under specialized culture conditions do the lymphocytes retain their proper immune function after transplantation in vivo (7). Keratinocytes and

TABLE 1. *Potential target cells for gene transfer*

Cell type	Ease of culture	Ease of transplantation
Hematopoietic stem cells	+	+ + +
Lymphocytes	+ +	+
Keratinocytes	+ + +	+ + +
Fibroblasts	+ + +	?
Myoblasts	+ +	+ + +
Endothelial cells	+ +	+ +
Islet cells	+	?
Hepatocytes	+	?
Glial cells	+ +	?

endothelial cells, on the other hand, can be readily cultured in vitro, and can subsequently be transplanted (8,9). Fibroblasts can also be cultured in vitro, yet their long-term survival after transplantation is unclear (10,11).

In light of the above, it is clear that both the in vitro culture of cells and their transplantability can be critical issues bearing on the feasibility of an ex vivo gene therapy. Indeed, these two issues have become the focal point of much of current gene therapy research. Although the in vivo approach towards gene therapy alleviates these problems, other issues emerge, primarily related to the accessibility and susceptibility of the target cells to gene transfer. With few exceptions, the efficiency of gene transfer obtained with any of the available systems for gene transfer is insufficient to generate a suitable number of transduced cells by direct in vivo gene transfer. Moreover, the techniques for targeting gene transfer to specific cells are only in the earliest stages of development.

METHODS FOR GENE TRANSFER

In light of the critical importance in defining the function of cloned mammalian genes and their products in the context of a mammalian cell, there has been intense interest over the past 10–15 years in the development of methods for efficiently introducing DNA into mammalian cells. As a consequence of these efforts, a variety of different gene transfer methods, predominantly physical or chemical in nature, now exist to meet the varied interests of investigators (Table 2). While the choice of a specific gene transfer system obviously depends on the specific application of interest, the goal of most gene transfer experiments is to deliver DNA sequences efficiently into mammalian cells in such a way that the sequences are appropriately

TABLE 2. *Commonly used methods for mammalian gene transfer*

	Application	
Method	Transient expression	Stable transformation
$CaPO_4$ precipitation	+	+
DEAE dextran	+	–
Electroporation	+	+
Protoplast fusion	+	+
Lipofection	+	+
Microinjection (cell)	+	+
Microinjection (intact muscle)	+	?
Viral vectors (see Table 3)	+	+
Ligand/DNA conjugates	+	?

+, method is commonly used. –, method not commonly used, ?, application not clear.

recognized by the transcriptional machinery of the cell. With one or two exceptions, this goal dictates that the genetic material reach the nucleus of the recipient cells in an intact and functional form. Interestingly, employing many of the techniques outlined in Table 2, it is possible to achieve the transfer of DNA into a reasonably high proportion of treated cells (10–50%), at short times (24–72 hours) after treatment (12). When it has been examined, the DNA so introduced exists in the nucleus in a free, intact form, not associated with chromosomal DNA sequences (13). These studies demonstrate that residence within chromosomal DNA sequences is not an absolute prerequisite for gene expression. Since for many applications, a rapid assay for gene expression is desired, there has been tremendous interest in exploiting this feature of gene transfer, termed "transient expression."

Often, however, it is desirable to generate stable cell lines expressing a particular gene product. Unfortunately, while DNA can be very efficiently transferred to cells over the short term, the proportion of cells that retain transferred sequences in a stable and heritable fashion is usually extremely low (0.1–0.01%), regardless of the gene transfer system employed. While the reason for this finding is unclear, it would appear that persistence of the DNA (usually through its integration into chromosomal DNA sequences) is a rate-limiting step in gene transfer. Two exceptions to this rule appear to be the direct microinjection of DNA into cells, which results in both high (100%) transient expression and moderate (10%) stable expression (14), and retroviral-mediated transfer, which can be extraordinarily efficient (100%) (see below). Because of the low efficiency of stable DNA transfer obtained with the majority of gene transfer methods, it is usually necessary to employ methods to identify and isolate the rare transformants within a population of cells. If the transferred gene confers upon the cell a dominant phenotype (e.g., an oncogene or membrane protein), it is sometimes possible to isolate transformants directly on the basis of the transferred gene's expression. Typically, however, it is necessary to employ other means to detect transformants. In most cases, genes encoding products (15,16) that permit the selective survival of cells expressing the products are either physically linked to the DNA sequence of interest or simply mixed with the DNA prior to transfection (even physically unlinked molecules become efficiently ligated after the molecules enter the cell) (17). The use of these genes, termed "dominant selectable markers," usually necessitates the use of specialized media, toxic to normal untransformed cells, and, more importantly, requires the culture of transfected cells for a time sufficient to effect the death of nontransfected cells and sufficient for the clonal outgrowth of successfully transduced cells (usually 10–14 days).

Depending on the goals of the gene transfer experiment, one of two general methods for obtaining the expression of a desired gene product is usually employed. In cases where the proper regulated expression of a gene is critical, both in terms of absolute levels of expression and the tissue specificity of expression, it is customary to transfer into cells either a cloned segment of chromosomal DNA that encodes all of the necessary information for proper expression, or a genetically engineered "minigene," a truncated version of the chromosomal gene in which the minimal essential genetic elements for proper expression are juxtaposed in an appropriate form. While the use of minigenes is particularly important when the intact chromosomal DNA sequences are too large to be efficiently transferred into cells, the approach is not always straightforward, since often, important signals for expression of a gene are distributed along large stretches of DNA in a complex arrangement. The transfer of chromosomal DNA sequences or minigenes into cells is usually accomplished through the use of a vector. In the parlance of gene transfer, a vector is broadly defined as any genetic element that facilitates the transfer of genetic material into cells and/or the expression of genes within cells (Fig. 1). In some cases, vectors also possess additional features, which influence the mode of persistence of the transferred genetic material (e.g., autonomous replication vs. stable integration), or, particularly in the case of viral vectors, make possible the conversion of the DNA vector to the appropriate biological agent ultimately used for gene transfer (see below). For chromosomal DNA transfer, a vector that encodes a dominant selectable marker that provides a means of identifying stable integrants is often linked to the sequences, or simply cotransfected with them. In other situations, chromosomal DNA sequences are incorporated into viral vectors in order to facilitate entry of the sequences into cells. To

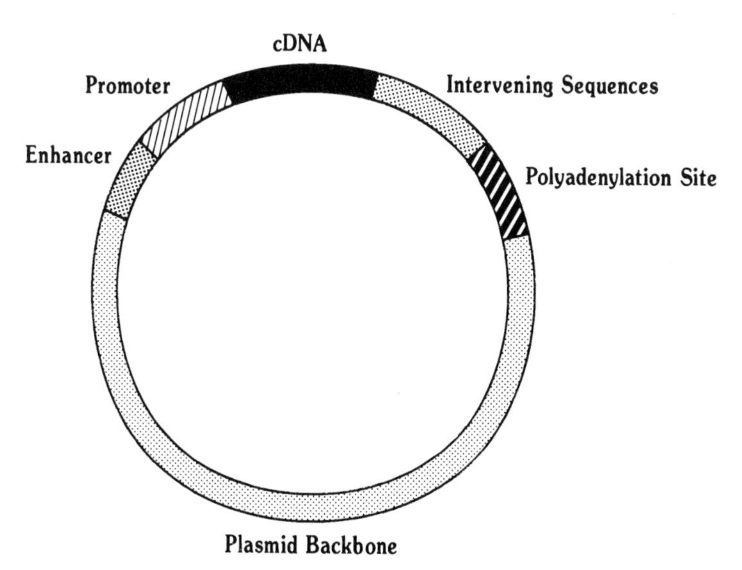

FIG. 1. A typical mammalian expression vector.

date, only moderate success has been achieved through the transfer of intact chromosomal sequences or minigenes. Often, variable levels of expression are obtained, which are dependent on the particular site of integration and number of integrated copies. However, since a number of transferred genes do show normal levels of expression, regardless of their site of integration (18,19), it may simply be that further genetic elements need to be included for genes that are not expressed properly after gene transfer.

In many situations, the absolute level of expression and tissue specificity of expression is not crucial. More importantly, often it is desirable to express a gene in a cell that normally does not express the gene. For these applications, mammalian expression vectors are usually employed. These vectors provide for the expression of inserted protein coding sequences, usually complementary DNA (cDNA) sequences by providing all of the genetic signals (e.g., promoters, enhancers, splicing signals, signals for the polyadenylation of RNA transcripts) necessary for the transcription of inserted sequences (Fig. 1). Sometimes the vectors contain selectable markers as well. Through the use of very well-characterized genetic elements derived from specific cellular or viral genes, the expression of a majority of inserted DNA sequences has been possible using these types of vectors.

In stably transformed cells, the transferred DNA most often is found integrated into chromosomal DNA sequences, at random locations in the genome. When more than one copy of the transferred sequences is maintained, the copies usually reside in the same chromosomal location, arranged in a head-to-tail manner (17). Although far less frequent, it is possible to obtain homologous recombinational events in which transferred sequences that are homologous to a specific chromosomal sequence become integrated at that specific location (20,21). Since these "legitimate" recombinational events typically represent only a small fraction of total integrations (1/50–1/1,000), a selective scheme to identify the homologous events in the background of excess nonhomologous events is usually employed (21). The reasons for the relatively high level of nonhomologous recombinational events in mammalian cells are not clear.

The use of homologous recombination in gene transfer studies has attracted a lot of attention recently, primarily because it represents a powerful way to introduce genetic modifications into specific loci in the mammalian genome. Most current interest has focused on the use of homologous recombination to disrupt genes in murine embryonal stem (ES) cells, and subsequently to use the modified ES cells to generate genetic alterations in the mouse germ line (22). However, the use of homologous recombination to "correct" the function of defective genes has also been reported (23). While such an application indicates the possibility of using homologous recombination techniques in the context of gene therapy, a number of critical technical obstacles must be overcome before the techniques could be widely applied. First, as pointed out above, the overall efficiency of generating stable integrants in mammalian cells is quite low, and targeted (homologous) integration events represent only a small fraction of cell integration events. In addition, the absolute frequency of targeted events can vary tremendously, depending on the locus and/or the characteristics of the targeting vector. Accordingly, at the current time, application of homologous recombination techniques to gene therapy would in general necessitate the ability to culture target cells in vitro sufficiently long to permit the isolation and expression of suitable numbers of "corrected cells." As noted above, few potential target cells can be cultured in vitro for such a time without losing important characteristics for function in vivo. Hopefully, methods will become available for the culture of these cell types that preserve more of their normal biological characteristics. There is also great current interest in developing methods for increasing both the absolute frequency of homologous recombination events and the ratio of targeted to nontargeted events. A second issue regarding homologous recombination that must be resolved before techniques could be widely applied in gene therapy is the applicability of the methods to correcting mutations resulting from large deletions or chromosomal rearrangements. Currently, most studies with homologous recombination have involved either the targeted insertion of relatively small lengths of DNA or the targeted correction of point mutations or small deletions.

In spite of the apparent obstacles, the use of homologous recombination techniques for gene therapy is an important goal of current research, for the technique offers several important advantages over methods that yield nonhomologous integration events. Perhaps the most important of these advantages is that expression of the gene of interest is accomplished through "repair" of the normal endogenous gene, rather than by integration and expression of an engineered gene in random chromosomal locations. Since the expression of a majority of engineered genes after gene transfer is sensitive to chromosomal position, activation of an endogenous gene locus is undoubtedly more likely to result in appropriate expression and regulation of the gene. In addition, repair of a defective gene in situ by homologous recombination eliminates the possible deleterious consequences of a random chromosomal insertion.

RETROVIRAL VECTORS

While the physical and chemical methods for gene transfer listed in Table 2 have met many of the needs of investigators, the utility of these methods in the context of gene therapy is limited. First, primary cultures of cells are often exquisitely sensitive to in vitro culture conditions, and do not survive many of the gene transfer treatments. More importantly, as pointed out earlier, many of the cell types that would be used in gene therapy scenarios cannot be cultured in vitro sufficiently long to permit selection of transduced cells. Finally, few of the methods are applicable to direct in vivo gene transfer because of their efficiency of transduction.

Because of the need for more efficient gene transfer systems, considerable attention has been given to the use of animal viruses as agents for gene transfer, since viruses are in general able to enter cells efficiently both in vitro and in vivo and interact with the cell's genetic machinery in a number of different ways. As in the case of the other methods for gene transfer described above, virus vectors have been developed that are useful both for transient gene expression studies and for stable transformation (Table 3). Viruses such as SV40, polyoma, adenovirus, and vaccinia undergo a life cycle that includes extensive replication of the viral genome and death of the host cell, and therefore are particularly useful for transient expression studies. In contrast, retroviruses, herpesviruses, and adeno-associated viruses can interact with cells without causing cell death, and therefore have been considered for use in stable transduction studies (24).

For a number of reasons, retroviral vectors appear to be the viral vector system most likely to be employed in initial gene therapy protocols. Unlike many viruses, which enter cells, replicate themselves, and kill the host cell, retroviruses accomplish the production of progeny virus through a rather unique life cycle, which in most cases does not lead to death of the infected cell (25) (Fig. 2). Retrovirus particles are composed of two copies of an RNA genome encapsidated into a rather complex virus particle structure containing both viral and cellular components. The viruses enter cells primarily on the basis of interactions between a viral protein, termed the envelope protein, present on the surface of the virus particle, and cellular receptors for the virus on recipient cells. Depending on the retrovirus, entry of the genome into cells may result from direct fusion of the viral and cell membranes or through a receptor-mediated endocytosis-like pathway. Uncoating of the virus in the cytoplasm reveals a core-like virus structure consisting of a number of viral proteins bound to the dimer RNA genome. Replication of the viral genome then occurs in several stages. First, the RNA genome is converted to several double-stranded DNA forms through a reverse transcrip-

TABLE 3. *Viral vector systems*

Virus	Transient expression	Stable transformation
SV40	+	−
Polyoma	+	−
Adenovirus	+	±
Retrovirus	−	+
Herpesvirus	±	+
Vaccinia	+	−

+, commonly used. −, not commonly used. ±, only useful under specialized circumstances.

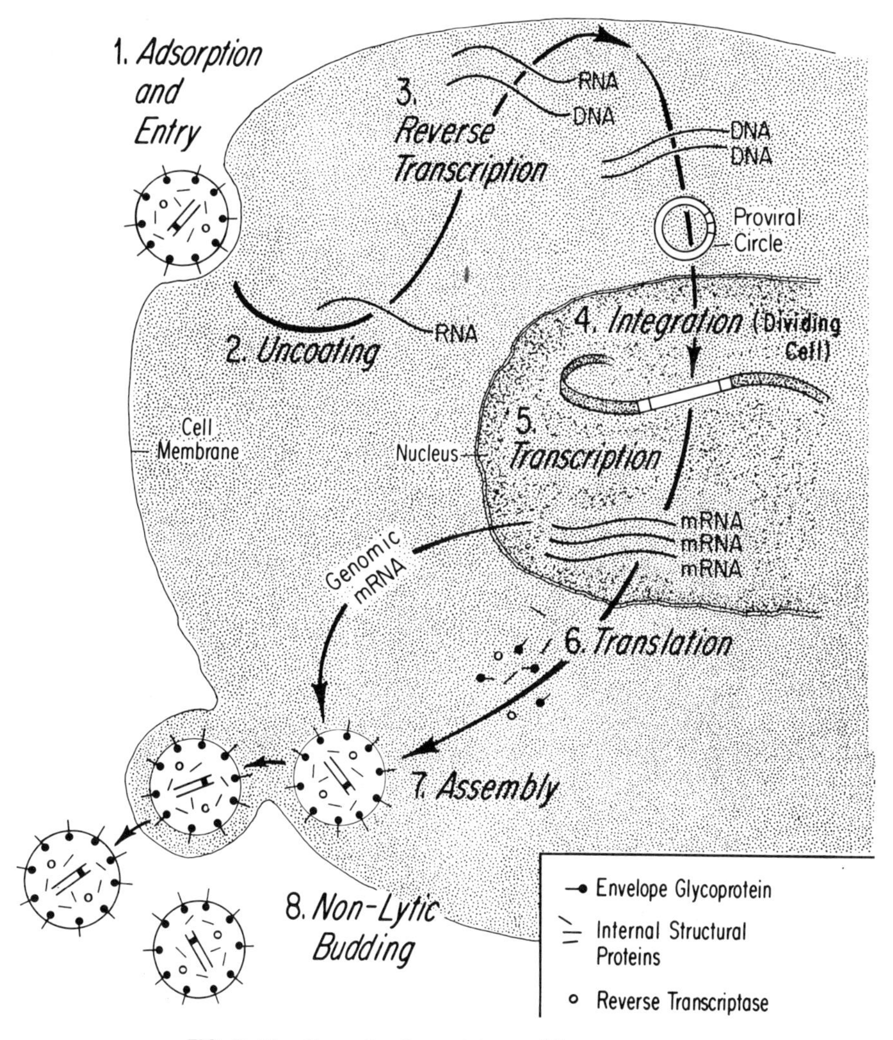

FIG. 2. The life cycle of a prototype wild type retrovirus.

tion process, involving primer tRNA molecules and reverse transcriptase present in the core-like structure. Next, an appropriate double-stranded form of the genome moves to the nucleus of the infected cells and is integrated into cell chromosomal DNA sequences through the action of an integrase, another viral-encoded protein also present in the uncoated core-like structure. Upon integration of the viral sequences, the genome, now represented as DNA (termed a *provirus*), becomes a template for the transcription of both the full-length genomic RNA destined for inclusion into a new virus particle, and RNAs encoding the various gene products necessary to assemble the viral particle. Finally, in a process not well understood, the virus particle is assembled, presumably near the inner surface of the infected cell's membrane, and viable particles are released from the cell via a budding process that does not result in cell lysis.

An important feature of the replication process is that the encapsidated RNA in virus particles need not encode any of the viral proteins, since the reverse transcription and integration of the viral genome depends only on the activity of previously synthesized viral products that are carried into the infected cell in the virus particle. The viral RNA does, however, need to encode a number of cis-acting features for replication, including the encapsidation or packaging site (26) (see below) and sequences necessary for reverse transcription and integration of the viral genome. Practically, this means that recombinant viral RNAs carrying foreign genetic sequences could in principle be encapsidated and replicated, as long as the necessary cis-acting signals in the viral RNA are preserved. Another important feature of the replication process quite relevant to the scenarios for gene therapy outlined above is that cell replication is necessary for proviral integration to occur.

Retroviruses thus naturally accomplish with great efficiency two of the most important steps in the gene transfer process: entry into cells and integration of genetic material in a stable and heritable fashion. While the host range of retroviruses does vary, depending upon the species from which the virus is isolated, collectively the viruses are able to infect most mammalian cells both in vitro and in vivo. To date, although vectors have been developed from many different types of retroviruses, most progress has been made in the development of avian and murine retrovirus vectors. Since murine viruses will most likely be used in the first clinical gene therapy protocols, they will be considered in detail here.

Murine retroviruses are classified on the basis of the cells they are able to infect, or, more precisely, on the basis of the receptor that the viral env protein interacts with. The two most important types of murine retroviruses for vector purposes are the ecotropic viruses, which infect only rodent cells, and amphotropic viruses, which infect a wide range of mammalian cell types, including human. Unfortunately, little is known about the distribution of ecotropic or amphotropic receptors on different cell types of a particular species, and therefore it is not known, for example, whether amphotropic virus will successfully enter all cells of the human body.

The strategy for adapting retroviruses as vectors has been based on the fact that while RNA is the genetic material of the virus, our ability to "cut and paste" DNA using recombinant DNA techniques is far more advanced than the existing technology for manipulating RNA. Accordingly, the generation of recombinant retrovirus genomes is accomplished at the proviral DNA level, and subsequently, the chimeric proviral DNA sequences are used as a template for generation of the appropriate recombinant RNA. The general strategy for generating recombinant vectors is shown in Fig. 3. One component of the vector system is a plasmid construct that encodes the minimal proviral sequences necessary to generate a viral RNA possessing the signals for "packaging," reverse transcription, and integration. This necessitates preservation of portions of the long terminal repeat (LTR) required for transcription of the proviral genome (promoter and enhancer of the LTR and polyadenylation signal), the tRNA provirus binding site, the ψ sequence, and a region of the proviral genome directly upstream of the 3′ LTR. Insertion of foreign DNA is made into the proviral transcriptional unit in place of the viral protein coding portions of the genome. Transfer of the resulting chimeric transcriptional unit into mammalian cells that recognize the viral transcriptional signals leads to the generation of a chimeric RNA that is recognized by the virus/cell machinery responsible for the encapsidation of viral RNA into particles, should that machinery be present in the cell. Since the recombinant genome does not itself encode any viral proteins, transfer of a chimeric transcriptional unit into cells lacking the packaging machinery leads only to the generation of a chimeric RNA similar to any other messenger RNA in the cell.

The second key component of the retrovirus vector system is a specialized

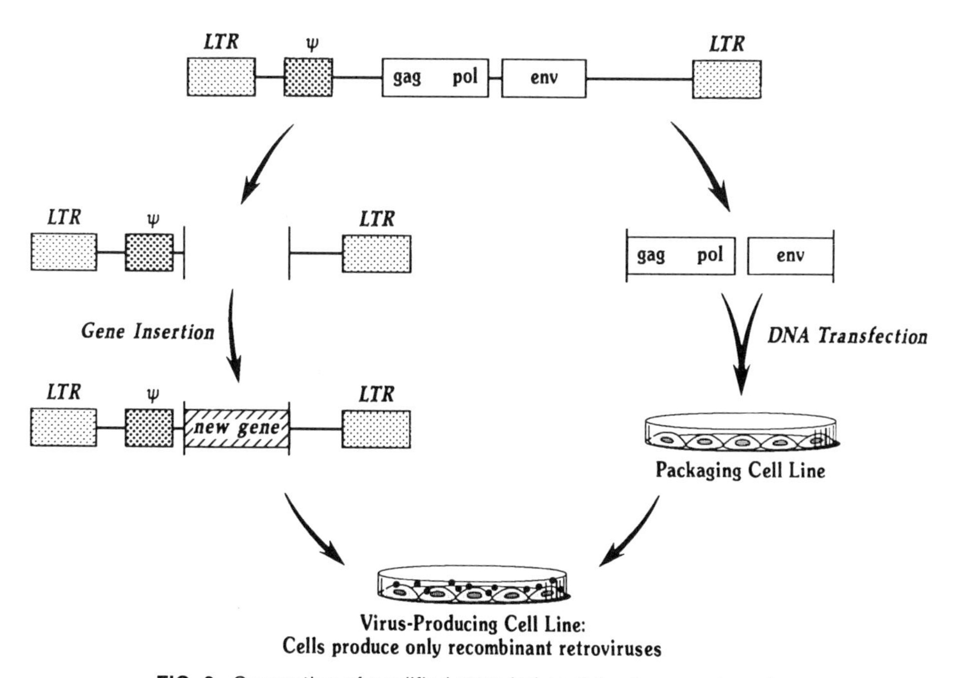

FIG. 3. Generation of modified retroviral particles for gene transfer.

cell line that provides the viral proteins necessary for encapsidation of the chimeric RNA described above. Conceptually, the simplest version of such a cell line is simply a cell line infected with replication-competent virus. Such a cell line obviously synthesizes all of the viral proteins necessary for encapsidation, since the viral genome, in part responsible for the synthesis of some of those proteins, is itself encapsidated into infectious particles, which are subsequently released from the cells. Transfer of a chimeric transcriptional unit into such a cell line results in the synthesis and encapsidation of the chimeric RNA, and ultimately the release of the recombinant virus particles along with the replication-competent particles newly made by the infected cells (27,28).

While such a system for generating recombinant virus is suitable for a number of studies, in most cases, it is desirable to generate recombinant virus free of replication-competent virus, so that the target cell for gene transfer incorporates only the recombinant viral genome of interest. The expression of viral proteins in the target cell can alter the phenotype of the cells in an undesirable way and lead to the spread of both replication-competent virus and recombinant virus to other host cells. To achieve the production of stocks of recombinant retroviruses free of replication-competent viruses, specialized cell lines, termed helper-free packaging cells, have been generated that synthesize the necessary viral proteins for the generation of virus particles, yet are incapable of encapsidating any of the RNAs encoding the viral proteins (6,29–34). In the "first generation" of such lines (26,29,30), expression of the viral proteins was accomplished by transfer into cells (by classic DNA transfection techniques) of a defective proviral genome, which encoded all of the viral proteins, yet lacked a signal necessary for the RNA synthesized from that proviral DNA to be encapsidated itself. Two versions of packaging cells were made in this way to provide for the generation of recombinant virus possessing either ecotropic or amphotropic host range. While these cell lines have proved quite useful for a variety of investigators, virus generated from those cells is often contaminated by low levels of virus particles containing ψ^- genomes and, less frequently, by replication-competent virus. This is because even in the absence of the ψ sequence, a low but detectable level of encapsidation of the resulting RNA can take place (35). Transmission of that defective genome to recipient cells results in integration and expression of a ψ genome. Recombinant events occurring during viral replication that involve a ψ^- genome and a ψ^+ recombinant genome can also lead to a fully replication-competent genome being transferred to recipient cells.

To improve upon the packaging cell lines, several approaches have been taken. First, to reduce the likelihood that RNAs encoding viral functions could be packaged, and to reduce the frequency of recombinational events between recombinant RNAs and the packaging RNAs, expression constructs for the viral functions have been generated that contain only the minimum necessary viral protein-coding sequences. Second, to reduce the capacity of encapsidated packaging RNAs to be transferred to cells, the 3′ LTR sequences needed for efficient reverse transcription and integration have been replaced with nonretroviral signals encoding the polyadenylation of transcripts. Lastly, to decrease the likelihood that recombinational events could generate replication-competent genomes, the viral functions needed for encapsidation have been expressed individually through the use of expression constructs introduced into different regions of chromosomal DNA. Packaging cell lines generated using the above strategies (31–34) (particularly cell lines in which the individual viral functions are expressed separately) have been examined in great detail and with great sensitivity for their ability to generate replication-competent virus or viruses capable of transferring any viral functions, and have been found to be unable to give rise to either contaminating virus species.

As is the case with conventional mammalian vectors, retrovirus vectors are designed for the purpose of either facilitating the transfer of chromosomal DNA sequences (or marker sequences) into cells or to facilitate both gene transfer and expression of specific protein-coding sequences. A very general summary of retrovirus vector design is shown in Fig. 4. By far the simplest and most well-characterized type of retrovirus vector is one in which viral transcriptional signals in the 5′ LTR are used both to generate the genomic RNA destined for packaging, and the mRNA encoding the gene product of interest. In the earliest versions of this type of vector, protein-coding sequences were inserted in place of the gag coding sequences, in such

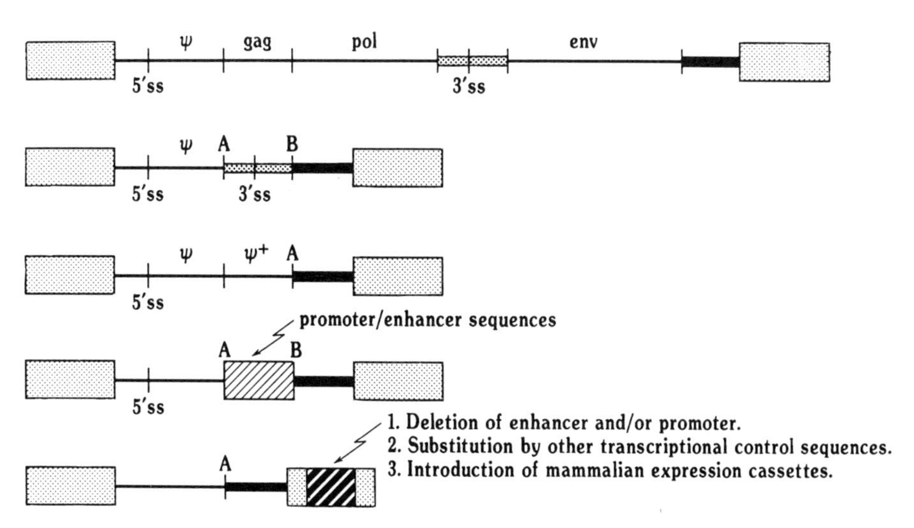

FIG. 4. Summary of retrovirus vector design.

a way that the initiation codon for translation of the inserted gene was positioned upstream of the initiation codon for gag translation or in place of the env-coding region. In the case of insertions in the gag region, the full-length viral RNA would be translated to express the foreign protein, while in the case of the env insertions, expression of the insert would be achieved by a chimeric, spliced, subgenomic RNA analogous to the normal spliced env RNA. Often, both the gag and env regions were replaced with foreign sequences, typically by the insertion of protein-coding sequence of interest in the gag position, and selectable marker sequences (to monitor transduction) into the env position (36). Since insertions of some sequences at the gag position interfered with the generation of the spliced mRNA encoding the selectable marker, modified versions of these vectors were generated that contained an intact separate transcriptional unit for expression of a selectable marker (or gene of interest) located downstream of the gag insertion site (37,38). A further modification of this type of vector involved the inclusion of additional viral sequences in the gag region, a modification shown to increase the titer of the resulting viruses (39). The magnitude of the increase in titer obtained with this modification appears to depend dramatically on other structural features of the vectors (40,41). Since in these latter vectors (termed gag+) the introduction of protein coding sequences downstream of the start of the gag-coding region may not result in efficient translation of the desired gene product, new gag+ vectors have been constructed in such a way that the protein-coding sequence of interest is positioned in place of the env gene, and thereby expressed by a transcript analogous to the normal subgenomic spliced env RNA (Robbins P, Guild B, Mulligan RC, unpublished data).

In some cases it is advantageous to control the expression of an inserted gene in a specific way, rather than obtain the constitutive expression afforded by LTR-driven transcription. Accordingly, a whole series of vectors have been designed in which transcriptional elements (promoters, enhancers, etc.) are positioned within the proviral transcriptional unit upstream of the inserted gene. In most cases, the "new" transcription unit is oriented in the same direction as the proviral transcriptional unit. Alternatively, it is possible to place the new transcriptional unit in the opposite orientation. These latter constructs are more problematic, however, presumably due to the greater likelihood that the sequences oriented in this way will interfere with generation of the proviral transcript. Similarly, intact chromosomal DNA sequences encoding a gene of interest can also be introduced within a proviral transcriptional unit. However, in this application, the sequences are usually oriented opposite to the direction of proviral transcription (42), since chromosomal DNA sequences oriented in the same direction can interfere with proviral transcription (if the gene's own polyadenylation site is used in preference to the polyadenylation site in the 3′ LTR) and can suffer deletions during viral passage (since intervening sequences present in the gene will likely be processed during the generation of full-length viral transcripts).

Because viral transcription can potentially affect the expression of transcriptional units located within the provirus transcriptional unit, independently of orientation, vectors have been developed that undergo a loss of viral sequences necessary for efficient viral transcription upon integration into chromosomal DNA (42–44). This is accomplished by carefully removing from the 3′ LTR of the vector enhancer sequences or both enhancer and promoter sequences, without removing sequences essential for integration. Since the enhancer and promoter regions (termed U3) found in each LTR upon proviral integration are actually copied from the region of viral RNA copied from the 3′ LTR alone, full-length viral transcripts generated from vectors with 3′ LTR deletions are able to undergo normal reverse transcription, yet give rise to integrated proviruses in which both LTRs now lack the critical transcriptional control sequences. This feature of retrovirus replication can be exploited generally to generate integrated proviruses containing any sequence in both LTRs (45,46).

The amount and concentration (often termed titer) of recombinant retroviruses that can be produced using available vectors and packaging cells can be the critical parameter for assessing the feasibility of transferring genes into a particular cell type, and, accordingly, an understanding of the factors that influence viral titer is of considerable interest. The two most important determinants of titer appear to be the absolute quantity of packaging proteins synthesized in the packaging cell lines, and the absolute quantity of recombinant RNA produced. A useful comparison can be made between the amount of viral RNA and protein made in cells infected with replication-competent virus, and the amount of vector RNA and packaging protein in packaging cells stably transfected with a retroviral vector construct. In general, titers of murine retroviruses range from 10^6 to 10^7 infection units/ml culture fluid. In contrast, the titers of recombinant retroviruses reported in the literature generally range between 10^5 and 10^6 units/ml or below. Several properties of the packaging cell lines currently in use suggest that the level of viral protein synthesized often approaches but does not generally exceed the level of protein synthesized in replication-competent virus-infected cells. Accordingly, while it is actually not clear what component of the virus expression system (RNA or protein) limits replication-competent virus production, it appears that in most cases, the amount of recombinant RNA produced in the packaging cells is the most important factor in the determining titer of recombinant viruses. Indeed, techniques have been developed for the identification of the best clone of recombinant retrovirus-producing cells that involve assays of culture supernatant for RNA content (Huszar D, unpublished data). In addition to the amount of RNA produced in cells, other characteristics of the RNA, some known and some yet to be determined, appear to influence viral titer. Certainly, the length of RNA influences its capacity to be packaged. In addition, the overall structure of the RNA is important, since the 5′ and 3′ ends of the viral RNA are necessary for replication. For instance, the insertion of sequences into a proviral transcriptional unit that interfere with full-length viral RNA formation, or even reduce the stability of the viral transcript, can appreciably affect virus production. Unfortunately, little is known regarding the relationship of secondary or tertiary RNA structure to encapsidation efficiency or the efficiency of reverse transcription. However, the dramatic variations in titer reported by different groups using quite similar viral constructs suggest that a number of characteristics of RNA structure important for efficient virus production have yet to be understood.

Safety Issues Relating to Retrovirus Vectors

In concluding the discussion on retroviral vectors, it is important to note that while these vectors offer some clear advantages over other existing methods of gene transfer for the practice of gene therapy, the system is not ideal, for a number of reasons. First, and perhaps most importantly, retroviral-mediated gene transfer leads to integration of the desired genetic sequences into random locations in the recipient cell chromosomal DNA. Since it is clear that at least under some circumstances, such random insertions can disrupt the activity of specific cellular genes and influence the expression of adjacent cellular sequences (25), retroviral-mediated gene transfer has the potential for altering transduced cells in an undesirable way. Of greatest concern is the possibility that specific insertion of proviral DNA may activate or deregulate the expression of growth control genes, and thereby contribute to the evolution of a malignant cell. The paradigm for such a scenario clearly exists, since a number of replication-competent retroviruses appear to induce neoplasia through the insertion of proviral sequences adjacent to well-known protooncogenes (25). Although such insertions are not thought to be themselves sufficient for the development of a fully malignant phenotype, they likely play an important role in the rather complex process of oncogenesis. Unfortunately, it has been very difficult to quantitate experimentally the risks of retroviral-mediated gene transfer in this regard, either in vitro or in vivo. To date, no reliable assay in vitro for oncogenic events of this type has been developed, perhaps because the genetic changes simply cannot be scored in vitro. There have been animal studies performed in order to assess the risk of retrovirus-mediated gene transfer in vivo, which have involved either the injection of replication-competent virus or recombinant viruses into the circulation or the transplantation of cells infected in vitro (47). While no overt signs of disease have ever been detected in these types of studies, it is unclear that the studies were performed in a meaningful way, since a most important parameter of such safety studies, the total number of proviral insertions obtained per animal, was never determined in a rigorous way. Overall, on the basis of current information, it is probably prudent to conclude that there may indeed be some risk of oncogenesis due to retroviral-mediated gene transfer in spite of the difficulty in assessing the probability of such an event. While admittedly difficult, such a risk should be factored into the overall risk/benefit analysis of any proposed gene therapy scenario. It is also important to keep in mind that the risk of oncogenesis due to the insertion of transferred genetic sequences is not unique to retroviral-mediated gene transfer; any gene transfer method that results in nonhomologous integration events poses the same risk.

Other concerns regarding retroviral-mediated gene transfer include the possibility that the recombinant virus used in gene therapy protocols might be contaminated with replication-competent retrovirus or the possibility that recombinant proviral genomes present in the cells of patients might be mobilized to spread to other cellular locations via pseudotype formation and/or recombinational events initiated by infection of the transduced cell by naturally occurring human viruses. Fortunately, with the development of the latest generation of packaging cell lines and new sensitive techniques for detecting the presence of helper virus sequences, the first concern is no longer an issue. The second scenario is also extremely unlikely, in light of the inability to detect the encapsidation and transfer of murine retrovirus genomes by any human virus.

Finally, regardless of the gene transfer system employed in a gene therapy protocol, one important safety issue that is often overlooked is the potentially deleterious effect of the in vitro culture of primary cell types upon their subsequent persistence and function after transplantation. As pointed out earlier, many cells lose important properties as a consequence of their culture in vitro, and often the changes are irreversible. While the death of some portion of transplanted cells due to their previous culture history may not present any problem in many gene therapy scenarios, the consequences of the acquisition of abnormal growth characteristics through the in vitro culture of cells need to be considered in detail.

SELECTION OF DISEASES AS CANDIDATES FOR GENE THERAPY

Historically, two inherited diseases, β-thalassemia and Lesch-Nyhan syndrome, were heralded by many as the most likely first targets for gene therapy, in large part due to the wealth of information available at the time regarding the molecular biology underlying the diseases. More recently, however, as more diseases have been defined in similar molecular terms, and as the technologies necessary for achieving successful gene therapy have been more clearly defined, new criteria for the selection of a disease as a candidate for gene therapy have emerged, and, not surprisingly, new "most likely" candidates for gene therapy have emerged as well.

The most obvious prerequisites for the development of a gene therapy protocol for a particular disease include (i) knowledge and availability of the gene to be transferred; (ii) identification of the relevant target cell; (iii) development of gene transfer approaches, either ex vivo or in vivo, suitable for transducing sufficient numbers of target cells; and (iv) development of gene transfer vectors capable of appropriate gene expression in the target cells. While for inherited diseases, identification and acquisition of the relevant gene is becoming increasingly less of a problem, due to the powerful impact of DNA mapping technologies, for many acquired diseases, determination of the relevant gene to be transformed remains a formidable challenge. Determination of the relevant target cells for gene transfer, on the other hand, can even be an important issue with regard to inherited diseases, since for both technical and biological reasons, the optimal target cell for gene transfer may not always be the end cell that manifests pathology. In a number of diseases, for example, the progenitor cells that give rise to affected cells may be a more appropriate cell for gene transfer than the affected cell itself.

From a purely technical standpoint, the major issues impacting upon the suitability of a particular disease as a candidate for gene therapy relate to the accessibility of the cells to gene transfer, and specifically in the cases of ex vivo approaches, the ability of the relevant target cells to be manipulated in vitro, and subsequently successfully transplanted. As will become clear from the discussion below, while many studies have demonstrated the ability to complement specific mutations in vitro by transfer of the relevant gene, the extension of these in vitro studies to more relevant in vivo contexts has been exceedingly difficult.

GENE TRANSFER STUDIES IN HEMATOPOIETIC CELLS

For a number of reasons, a major focus of gene therapy research to date has been the development of methods for introducing genes into hematopoietic cells, and the use of transplantation techniques to reintroduce the genetically modified cells into recipients. One obvious reason for this intense interest is that many of the most well-characterized diseases affect cells of the blood. The genetic lesions responsible for quite a number of these diseases have been identified, and much is known about the characteristics of the affected cells. From a technical standpoint, cells of the blood are uniquely accessible and manipulatable in vitro, at least for short periods of time. More importantly, several unique features of the development of blood cells (48) make possible the transplantation of hematopoietic cells in such a way that they persist and function in vivo. In this regard, the single most important feature of hematopoiesis in experimental animals and humans is that all of the different hematopoietic cell types found in the circulation are generated over the course of an animal's lifetime by a relatively small population of cells found in the bone marrow, termed stem cells. This process is dictated in part by the short life span of most hematopoietic cell types and in part by the need to replenish elements of the blood destroyed by various environmental insults. Operationally, stem cells are defined as having the ability both to extensively self-renew themselves and to give rise to the entire spectrum of hematopoietic cell types. Experimentally, the existence of such cells has been best demonstrated through bone marrow transplantation (BMT) studies in humans and other species. BMT involves a two-step process in which the recipient is first treated by irradiation or chemotherapy in order to eliminate endogenous hematopoiesis, and then transplanted with a relatively small number of cells either previously harvested from the bone marrow of the recipient or obtained from a tissue-matched donor. When successful, it can be shown that the introduced cells are able to reconstitute all elements of the blood, for the lifetime of the recipient.

While BMT is commonly used clinically in the treatment of cancer and bone marrow aplasias, it has also been used successfully in the treatment of a number of inherited diseases affecting hematopoietic cells (see Chapter 18). In such situations, allogeneic rather than autologous BMT is performed, since the goal of the treatment is to reconstitute the recipient with nonmutant hematopoietic cells. The success of this type of treatment is important in the context of gene therapy, since it indicates that, in the case of a number of inherited diseases, hematopoietic cells are the sole source of pathology, since the replacement of only hematopoietic cells results in effective treatment of the disease. While these results have suggested the obvious possibility of using virus-transduced autologous cells in a stem cell/BMT scenario for the treatment of inherited diseases involving hematopoietic cells, they have also raised questions about the need for gene therapies for diseases affecting hematopoietic cells, since providing the patients with cells harboring a perfectly normal gene is clearly preferable to transferring cells that have been genetically engineered. Unfortunately, at the current time, the applicability of allogeneic bone marrow transplantation to inherited diseases is limited by the difficulty in identifying acceptable sources of donor marrow for transplantation. While significant advances have already been made in understanding the lethal immunological reactions that often accompany allogeneic BMT (49), the procedure can currently be applied to only a subset of patients in need of treatment. Hopefully, advances will be made in the future that dramatically extend the applicability of this approach.

Because of the limitations of allogeneic BMT, there remains a need at the current time for effective new treatments for inherited diseases affecting hematopoietic cells. In light of their characteristics, hematopoietic stem cells may represent ideal target cells for genetic modifications aimed at treating these diseases. First, the cells are very accessible to removal from the body and manipulation in vitro at least for short times. Second, the techniques for transplantation of the cells are well described and already employed in clinical practice. Third, because of the tremendous expansion of cell numbers during the early phases after BMT, only a relatively small number of cells need be transduced in vitro. Last, the method of BMT results in a continuous source of all hematopoietic cell types, thus eliminating the need for the continual retransplantation of the cell type of interest.

Most research aimed at determining the feasibility of a genetic therapy for inherited diseases involving hematopoietic cells has focused on the development of protocols for efficiently transducing reconstituting stem cells and the development of suitable vectors that would provide stable, appropriate expression of the gene of interest. Because of the difficulties and expense involved in BMT studies in large animals, most effort has been directed towards studies in the mouse, since the techniques of BMT are well established, and the supportive care necessary for survival posttransplantation is minimal. To date, retroviral-mediated gene transfer has been the gene transfer system of choice, primarily due to the need to transduce a cell population (stem cells) that comprises only a small proportion (.01%) of cells isolated from the bone marrow. Since proviral integration necessitates replication of the target cells, yet stem cells are largely a quiescent cell population, much effort has been put into the development of infection protocols that employ the use of different growth factors thought to induce the replication of stem cells (50–53). The greatest stumbling block in the effort to develop transduction protocols for stem cells has been the confusion among investigators as to the most appropriate assay for the transduction of reconstituting stem cells. In early studies, most investigators examined the transducibility of bone marrow cells capable of forming either multilineage colonies in vitro [colony-forming units (CFU-C) assays] or of giving rise to multilineage colonies in the spleen of lethally irradiated recipients injected previously with transduced cells (CFU-S or spleen colony assays). These assays are reasonably rapid, and clearly score for progenitor cells capable of multilineage differentiation. However, the relationship of CFU-C- or CFU-S-forming cells to reconstituting cells has been, and remains uncertain. Indeed, having worked out conditions for transducing stem cells and having defined optimal vectors for gene expression based on the assays, a number of studies involving the analysis of mice reconstituted for long periods of time with transduced cells concluded that only low levels of vector-mediated expression were obtainable. Accordingly, much attention was shifted to the development of new expression vectors, since the data suggested that while many of the conventional vectors worked well in vitro and in short-term reconstitution assays (e.g., CFU assay), they were not capable of long-term expression in vivo.

It now appears that, paradoxically, while retroviral vector design may in some cases be critical for obtaining both short- or long-term expression in vivo (see below), the most important parameter for long-term expression may simply be the efficiency of transduction of long-term reconstituting cells. In other words, the inability to detect gene expression after long-term reconstitution in many cases likely reflected the poor efficiency of transduction of cells with long-term reconstituting capacity, rather than problems with vector expression. In our own studies, we have shown that viral titer is a critical parameter bearing on the efficient transduction of long-term reconstituting cells and that, unfortunately, the titers necessary for efficient gene transfer are not commonly achieved with most recombinant retroviral genomes (54). Our studies also indicate that, at least in some cases, very conventional vectors in which the viral transcriptional sequences are used to promote expression of the inserted sequences (see Fig. 4) can function very well in vivo (54). A most likely explanation for the ability to detect efficient gene transfer and expression both in vitro and at short times after transplantation, but not after long periods of reconstitution, is that populations of progenitor cells distinct from the long-term reconstituting stem cells exist that do contribute to the early process of engraftment and are easily infected by recombinant retroviruses, yet do not persist over long periods of time. Evidence for the existence of such a class of cells has recently been obtained in stem cell purification experiments.

In spite of the problems encountered, it is now clear from the work of a variety of investigators that murine cells capable of the complete and permanent reconstitution of lethally irradiated recipients can be efficiently transduced using retroviral vectors (50–59). In addition, as will be described in more detail below, significant progress has been made towards the goal of achieving the appropriate expression of specific "disease" genes in murine hematopoietic cells. Nevertheless, one area of investigation that needs to be pursued before clinical protocols involving gene transfer into hematopoietic cells can be contemplated is gene transfer/BMT experiments involving larger animals.

Not surprisingly, analogous experiments involving hematopoietic cells from larger animals have not kept pace, because of the lack of suitable in vitro assays for reconstituting stem cells. Nevertheless, several promising avenues of investigation are now being pursued. Because of the need for a large animal model for gene transfer/BMT studies, several investigators are pursuing (or have pursued) BMT studies in both dogs and monkeys. To date, only a few animals have been reconstituted with cells putatively transduced in vitro. In studies by Stead et al. in the dog (60), and Kantoff et al. in the monkey (61), in which bone marrow cells were exposed to virus in vitro and subsequently transplanted, no evidence was presented to indicate that the long-term reconstituting stem cells were transduced, although transduced progenitors could be detected in vitro, prior to transplantation, and, in the case of the monkey experiments, a very small number of transduced cells could be detected in vivo, albeit only transiently. Since the success of these types of experiments may depend critically upon the proliferative states of the bone-marrow-derived stem cells, there is great current interest in examining the response of human and monkey cells to the wide range of available hematopoietic growth factors. The recent identification and characterization of a factor that acts in conjunction with other growth factors to stimulate the proliferation of stem cells is particularly exciting in this regard (62).

Another extremely important area of research involves the long-term culture of human bone marrow cells. Using techniques analogous to those developed for murine cells, a number of groups have described long-term marrow culture systems that maintain significant levels of human progenitor cells over time (63,64). In general, the systems depend upon both soluble growth factors provided to the cultures and adherent stromal elements. An exciting recent development is the generation of cloned murine stromal cell lines capable of maintaining in vitro progenitor activity, CFU-S activity, and long-term reconstituting activity for at least short periods of time (65). Although there have been reports of the use of long-term cultured marrow cells for BMT in humans (66), it has not yet been clearly shown that such a system can be safely and effectively used for human BMT. It would appear that this issue should be vigorously pursued, since for the development of successful protocols for transducing human hematopoietic stem cells, an in vitro system is critical.

Still another interesting yet highly speculative area of investigation involves the use of immunocompromised mice as hosts for human hematopoietic cells (67–69). Several groups have already reported the persistence of human progenitors in mice after the transfer of different cell types. While the results do indicate that multipotent progenitors can be maintained, and while it is likely that the systems can be improved significantly, it will be very hard, if not impossible, to provide convincing evidence that a stem cell capable of reconstituting patients can be maintained.

Finally, there is also great current interest in developing the methods for purifying hematopoietic stem cells. In the mouse, a number of groups have now reported protocols that result in a very significant enrichment of long-term reconstituting activity (70–73). The ability to culture such cells for long periods in vitro, perhaps by employing the use of recently identified growth factors, is another important long-term goal of gene therapy research, since such an approach would potentially make possible the use of homologous recombination techniques to correct specific genetic lesions in an ideal way.

GENE THERAPY FOR SPECIFIC DISEASES INVOLVING HEMATOPOIETIC CELLS: PRECLINICAL DATA

ADA Deficiency

Mutations in the adenosine deaminase (ADA) gene in humans lead to a fatal immunodeficiency disease in which both B- and T-lymphocyte functions are severely impaired (74). A majority of ADA-deficient patients present clinically at a very early age with recurrent infections and the failure to thrive and usually succumb to infections by the age of 2, if untreated. While it is widely held that the abnormal deoxynucleoside pools that develop both intracellularly and in the circulation as a consequence of ADA deficiency are responsible for the profound lymphocyte dysfunction observed, the biochemical basis for the selectivity of this toxicity is still not clearly established. Up until several years ago, treatment for ADA deficiency consisted of the usual supportive care used for immunocompromised patients (e.g., antibiotic therapy and isolation techniques), and either repeated red blood cell transfusions aimed at correcting the deoxynucleotide pool imbalance by a form of enzyme replacement, or allogeneic bone marrow transplantation. Bone marrow transplantation can be curative, and has been the treatment of choice, if an HLA-identical donor is available. More recently, HLA-haploidentical BMT has been performed on ADA-deficient patients with encouraging results as well (75,76) (also see Chapter 19). Red cell transfusions have proved to affect deoxynucleotide pools significantly and in some cases improve immune function, but overall the results obtained with those treated have been disappointing (77).

In the past several years, a new form of enzyme replacement therapy for ADA deficiency has been investigated involving the infusion of bovine ADA enzyme that has been covalently attached to monomethoxypolyethylene glycol (PEG). The modified enzyme, termed PEG-ADA, possesses a greatly increased half-life in the circulation and appears to be much less immunogenic than the enzyme itself (78). Many of the existing ADA-deficient patients in the world have now been treated with the modified enzyme, with extremely encouraging results (79 and Herschfield MS, personal communication). In addition to normalizing the deoxynucleoside pools in most patients, the immune status of these patients is dramatically improved, as evidenced by both in vitro analyses and clinical examination.

Before the advent of PEG-ADA therapy, the poor prognosis for ADA-deficient patients, the biological characteristics of ADA deficiency, and the evidence that allogeneic bone marrow transplantation was an effective treatment all suggested that ADA deficiency might be an ideal target for gene therapy. Most considered a scenario in which bone marrow cells would be removed from the patient, transduced in vitro with an ADA-expressing retrovirus, and subsequently returned to the patient in a procedure very analogous to conventional autologous BMT. In theory, if gene transfer into the reconstituting stem cell could be achieved, and if expression of human enzyme could be achieved in the entire spectrum of hematopoietic cell types, the treatment would be as effective as successful HLA-identical transplantation. Other scenarios involving the transplantation of transduced T cells have also been proposed and now form the basis for the first clinical trials involving gene therapy (see below). Currently, however, the successful preliminary results obtained with PEG-ADA rather dramatically shift the risk/benefit ratio of a gene therapy for ADA deficiency and raise important questions about the suitability of ADA deficiency as a first candidate for gene therapy. Nevertheless, the preclinical data obtained to date regarding ADA gene transfer and expression represent an important advance in the development of methods generally applicable to diseases affecting hematopoietic cells and are therefore worthy of discussion here.

In the case of the stem cell infection/BMT scenario proposed for ADA deficiency, preclinical studies were designed to demonstrate that hematopoietic stem cells could indeed be efficiently transduced with a functional human ADA gene, and that the transferred gene would function in all of the differentiated descendants of the stem cells, since the earliest stage of lymphocyte development at which a deficiency of ADA is deleterious is not known. As pointed out above, with only a few exceptions, these issues have been addressed only in the mouse. Most early studies with retrovirus constructs encoding human ADA employed vectors designed to express both the ADA cDNA and a specific selectable marker. In some cases, the vectors employed the viral LTR for promotion of both ADA and marker expression, while others employed both the LTR and a well-characterized nonretroviral promoter element positioned within the provirus transcriptional unit to provide for expression of the two genes. In short-term assays for transduced progenitor cells, such as the CFU-C or CFU-S assay, variable results regarding expression of human ADA sequences were obtained. Some, but not all of the retrovirus vectors that employed the viral LTR for expression of the sequences yielded detectable expression of ADA, and only a subset of internal promoter vectors demonstrated expression. For example, while Belmont et al. were able to demonstrate expression in both CFU-C and CFU-S assays using LTR-based vectors (56,80,81), McIvor et al. could not detect any ADA expression with viruses employing the same LTR for expression (82). Similarly, while Williams et al. (83) were unable to detect ADA expression in CFU-S using vectors employing the SV40 early promoter, Lim et al. reported success with vectors employing the SV40 sequences in progenitor colony assays (84). In retrospect, it appears likely that the variability in results obtained in the above studies owed more to the variations in vector design than to the inherent activity of the different transcriptional elements in primitive hematopoietic cells. From more recent experience with retrovirus vectors, it is clear that comparisons among transcriptional elements need to be performed in vector constructs possessing otherwise identical structures, since many variables (e.g., sequence context, presence of splicing signals, inclusion of expression cassettes for selectable markers, mRNA stability) can profoundly influence expression from a construct.

More recent studies aimed at assessing the activity of different ADA-containing constructs at long times after bone marrow reconstitution have also provided no clear message as to the transcriptional efficiency of different promoter elements. A number of groups have now demonstrated long-term expression of human ADA in reconstituted murine BMT recipients. In the case of Wilson et al. (58) and Moore et al. (60), LTR-based vectors were shown to promote ADA expression effectively, while the studies of Lim et al. (85) and Wilson et al. (54) demonstrated that internal promoters (e.g., human pgk and chicken β actin) could function as well. However, because of the variable titers of the vectors and differences in transduction protocols, it has been quite difficult to compare the performance of the different vectors. One issue commonly raised has been whether the Mo MuLV enhancer sequences are efficiently utilized in transduced cells derived from infected stem cells. Bowtell et al., for example, have reported that the MuLV enhancer functioned poorly in this context, and that vectors containing the myeloproliferative sarcoma virus (MPSV) enhancer sequences (sequences derived from the LTR of the murine MPSV) in place of the MuLV sequences were capable of significantly better expression (86). Moore et al., on the other hand, have recently reported significant levels of ADA expression in long-term reconstituted recipients using a standard MuLV-based vector. Similarly, the LTR-based vector employed by Wilson et al. (54) that was shown to function well in hematopoietic cells has recently been shown to contain only MuLV sequences, and not MPSV enhancer sequences as previously reported. However, since this latter vector also contains a mutation that may influence transcriptional activity (87), the specific role of the MuLV enhancer remains unclear in those studies. ADA-expressing vectors containing other hybrid LTRs have also been shown to perform well in hematopoietic cells (59), but again, no effective comparison to other vectors has been possible.

As pointed out earlier, the above studies and others have collectively demonstrated that the reconstituting stem cell in the mouse can be efficiently transduced, and that expression of inserted genes can be obtained in the mature cells derived from transduced stem cells. Unfortunately, in the case of human ADA expression, it is still uncertain from the murine studies whether adequate levels of human ADA would be expressed using the existing vectors. While the existence of immunocompetent patients possessing less than 10% of normal ADA activity in blood cells (74) suggests that the level of expression necessary to ameliorate the symptoms of ADA deficiency may not be high, the lack of information regarding the potential importance of expressing high levels of the enzyme in early lymphoid progenitor cells makes it quite difficult to make an informed assessment of the feasibility of gene therapy for ADA deficiency based on the murine gene transfer experiments. Analogous studies performed in the dog and monkey have not led to any evidence of transduction of a reconstituting stem cell, and therefore, shed no light on issues of vector-mediated ADA expression in clinically relevant cell types.

A second scenario for the treatment of ADA deficiency by gene therapy that is already being explored in the clinic involves attempts to bolster the immune response of ADA-deficient patients through the removal of mature T lymphocytes, transduction of the cells in vitro by ADA-expressing retroviruses, expansion of the cells in vitro to large numbers, and reintroduction of the expanded transduced cells into the patients (88). In this scenario, it has been postulated that clinical benefit would result from several features of the protocol. First, the introduction of large numbers of T cells expressing ADA would provide a novel source of ADA enzyme in the circulation, and might enhance lymphocyte function much as PEG-ADA improves immune function. Second, and more importantly, if sufficient numbers of transduced lymphocytes remained in the circulation after transplantation and if the collection of cells represented a repertoire of T-cell receptor specificities sufficient to combat the wide range of immunological insults faced during one's lifetime, the procedure might lead more directly to improved immune function.

In light of the current implementation of this type of protocol in the clinic, the paucity of preclinical data in support of the feasibility of the protocol is disappointing. Investigators have demonstrated the ability to remove T cells from ADA-deficient patients, to transduce them efficiently with ADA-expressing retroviruses, and to amplify the cells to large numbers in vitro. The resulting cells show reasonable levels of human ADA activity and improved resistance to high deoxynucleoside concentrations in tissue culture media. However, in the few studies performed, it has not been possible to demonstrate preservation of the T-cell repertoire after in vitro culture. In fact, using a very crude measure of repertoire involving characterization of the T-cell surface phenotype (e.g., $CD4^+$ or $CD8^+$ lymphocytes), the expansion of T cells has been shown to alter the $CD4^+/CD8^+$ ratios of cells very significantly; this ratio is an obviously important parameter of normal immune function (88). A second key issue that has not been resolved is the ability of such amplified, transduced T lymphocytes to persist in vivo after subsequent transplantation. Fathman and coworkers have shown that while it is possible to transplant murine lymphocytes effectively, success depends very critically upon the previous culture history of the cells (7). In particular, many commonly used methods for culturing T cells in vitro render the cells exquisitely dependent on the lymphokine interleukin-2 (IL-2) for their survival, and therefore not suitable for survival in vivo. The only data obtained in human clinical trials bearing on these issues derives from the recent tumor-infiltrating lymphocyte (TIL)-gene transfer experiments conducted by Rosenberg and coworkers (89) (see below). In those studies, amplified T lymphocytes were indeed exquisitely dependent on IL-2, as evidenced by the marked loss of transplanted lymphocytes in the presence of large doses of IL-2, and the complete loss of the cells after withdrawal of IL-2. In light of these findings, it is surprising that culture studies in vitro and subsequent transplantation studies in the mouse or monkey were not an integral part of the preclinical studies in support of the above gene therapy scenario. The major data presented in support of the protocol involved studies in which lymphocytes from normal patients or patients with ADA deficiency were either mock-infected or infected with human ADA-expressing retroviruses and subsequently introduced into immunodeficient mice (Bordignon C, unpublished data). The investigators found that while ADA-deficient lymphocytes were unable to engraft the mice, either normal or vector-transduced ADA deficient lymphocytes were capable of engraftment. While the selective transplantability of genetically corrected T lymphocytes was taken to imply a "selective advantage" of the cells over non-transduced cells, it is difficult to understand the relevance of this result in the context of the gene therapy protocol. First, the capacity of amplified cells to be transplanted was not in fact investigated in a meaningful way, since the cells were not cultured under the same conditions (nor to the same extent) as was proposed for the clinical protocol. Second, the experiments provide little or no evidence for the engraftment of a significant repertoire of functional lymphocytes. While the transplanted cells that were vector-infected showed a response in vitro to tetanus toxin, the patient from which these cells were taken had been exposed to tetanus toxin and had showed a response. Neither normal or vector-treated transplanted cells were able to mount a response in vivo without prior exposure to antigen. Finally, the ability of vector-transduced cells to engraft does not imply that there is a selective advantage in vivo for ADA-expressing lymphocytes. It is quite possible, in fact likely, that the exposure of ADA-deficient cells to the specific in vitro culture conditions employed compromised their capacity to be transplanted. In this regard, it is not even clear from the experiments reported that normal cells, ADA-deficient cells, and transduced cells were treated in the same way in vitro.

A final disappointment is that the clinical protocol for gene therapy for ADA deficiency which was approved and is now in progress offers little opportunity to assess the potential efficacy of the treatment. The patients involved in the study will have been and will continue to be treated with PEG-ADA throughout the study. Since most of these patients are being effectively treated using PEG-ADA, assessments of the consequences of lymphocyte transplantation will amount to detection and quantitation of transduced cells in the patient's circulation over the course of the experiment.

β-thalassemia

β-thalassemias represent a class of diseases in which the synthesis of the β-globin polypeptide is impaired (90). In general, reduction in β-globin synthesis leads to a relative excess of the α-globin chain, which is unable to form the stable functional hemoglobin tetramer structure. Precipitation of the insoluble α chains results in the formation of inclusion bodies in early erythroblast precursors, which in turn dramatically affects the ability of the erythroid precursors to give rise to normal mature erythrocytes. In the most severe cases of β-thalassemia, a profound anemia results, and is accompanied by massive proliferation of erythroid precursors in the bone marrow and extramedullary sites, a compensating response to the low oxygen-carrying capacity of the blood. Although a number of factors influence the severity of β-thalassemias, in the most severe cases the disease is characterized by a hemolytic anemia that leads to hepatosplenomegaly and congestive heart failure, as well as a variety of developmental defects. Current treatment for severe β-thalassemia consists primarily of repeated transfusions, aimed at decreasing the anemia and suppressing the massive ineffective erythropoiesis (91). A critical and lethal side effect of transfusion therapy is iron overload. Because of this complication, transfusion therapy is now accompanied by periodic iron chelation therapy, using desferrioxamine. While such therapy has dramatically improved over the last 10 years, the long-term survival and quality of life of patients treated by transfusion and chelation therapy are uncertain. More recently, a number of centers have reported encouraging results with allogeneic bone marrow transplantation (91). Since BMT can in principle provide a life-long cure for the disease, it is becoming increasingly performed, particularly when a histocompatible donor exists.

In light of the need for more effective treatments of β-thalassemia and the success obtained in treating the disease with BMT, a gene therapy protocol much like the stem cell gene transfer/BMT protocol proposed for treatment of ADA deficiency would appear appropriate. However, in contrast to the case of ADA deficiency, in order to achieve clinical benefit from a gene therapy for β-thalassemia, there would likely be far more stringent requirements for gene expression. For instance, since the β-globin gene is only expressed in cells of the erythroid lineage, it is clear that expression of an inserted human β globin in erythroid cells would be critical. In light of the pathological consequences of a relative excess of β-globin chain synthesis observed in α-thalassemia, however, ectopic expression of the transferred β-globin gene in nonerythroid tissues would likely be undesirable. More importantly, the need to provide a balanced synthesis of α and β globin in erythroid cells suggests that the absolute amount of β globin expressed may be critical. If too little β globin is expressed, no benefit will likely accrue, yet if greater than normal levels of expression occur, an α-thalassemia-like syndrome could in principle result.

Unfortunately, gene transfer experiments involving the human β-globin gene have shown that obtaining the normal regulated expression of transferred human β-globin genes is not straightforward. In studies involving either erythroid cells in culture or transgenic animals, it has been clearly shown that while cloned DNA sequences encompassing both the structural gene and directly adjacent 5′ and 3′ associated sequences are expressed in an erythroid-specific fashion, the absolute expression levels are variable and generally well below normal levels. Moreover, the expression of the transferred genes does not appear to depend linearally on the number of copies of the gene that are transferred (e.g., copy-number-independent), and is critically dependent upon the chromosomal site of integration (e.g., position-dependent). Over the past few years, however, several investigators have identified and characterized additional DNA sequences located quite far (approximately 20–50 kb) from the structural gene that are necessary and sufficient for copy-number-dependent, position-independent expression of the human β-globin gene (18,92–94). These sequences were originally identified because they represented a region of chromatin hypersensitive to DNAse digestion, and therefore potentially a region involved in the activation of β-globin gene expression. While the regions originally described were relatively large (10 kb), studies in erythroid cells and with transgenic mice have delineated a much smaller region of noncontiguous DNA segments (less than 1 kb) that possess full transcriptional activity (95).

Not surprisingly, preclinical studies aimed at determining the feasibility of a gene therapy for β-thalassemia have focused primarily on the issue of β-globin gene expression. The greatest technical challenge appears to be the generation of high-titer retroviruses that encode both the β-globin structural gene and the DNAase hypersensitive sequences, faithfully transmit the recombinant genome to cells, and result in the tissue-specific, copy-number-dependent, position-independent expression of the human β-globin gene. Before the requirement for the DNAse hypersensitive sites were known, a number of laboratories reported on the construction and characterization of recombinant retroviral genomes encoding the β-globin chromosomal gene itself (42,53,96–100). In those studies, the gene was introduced into vectors in a transcriptional orientation opposite to that of the provirus, so that the polyadenylation site of the β-globin gene would not interfere with proviral transcription and so that the intervening sequences of the β-globin gene would not be lost during replication and transfer of the recombinant genome to cells. In our own studies, we found that recombinant genomes containing the β-globin gene could indeed be generated and faithfully transmitted to cells, although the viral titers obtained were low (42,96). Hematopoietic stem cell/BMT experiments with these viral constructs, although quite difficult because of the low viral titers, did indicate the ability to transfer the recombinant genome to stem cells, and erythroid-restricted expression of the transferred gene was demonstrated (96). However, the absolute level of β-globin expression was quite low, due both to the inefficient transduction of stem cells and the low level of expression obtained per proviral copy (approximately 1–2% of normal mouse β-globin levels). In more recent studies by Bender et al. (100) and Bodine et al. (53), the level of β-globin expression obtained per proviral copy remained at the 1–2% level, although the transmissibility of the β-globin viruses was improved.

Current efforts are being directed towards the generation of recombinant genomes that include the DNAse-hypersensitive sequences. Towards that end, one important line of investigation has been the further delineation of the minimal DNAse-hypersensitive sequences sufficient for regulated β-globin expression mentioned above, since the original sequences as reported were too large to be incorporated into a retroviral genome. While few data have been published to date regarding efforts to include the sequences in vector constructs (101), it appears that a general problem confronting investigators is that the sequences can both adversely affect the transmissibility of the resulting recombinant genome and lead to the transfer of rearranged genomes. The retention of selectable markers in such constructs and the subsequent use of selective schemes to identify transduced cells may compound these problems. Overall, the results to date suggest that the precise site of insertion and/or orientation of the sequences may be critical. In our own studies, we find that both the hypersensitive sequences and the β-globin structural gene itself can independently reduce transmissibility of the resulting recombinant genomes and that the effects of the two sequences together in one vector cannot always be predicted on the basis of results obtained with constructs that incorporate each element alone (Sadelain M, Mulligan RC, unpublished data). In general, this finding implies that a wide variety of different constructs including both sequences need to be tested empirically.

Ultimately, a key technical goal will be the generation of a virus encoding both the structural gene and the transcriptional control sequences that transmits at high efficiency, since high-titer virus will be critical for efficient transduction of bone-marrow-derived hematopoietic stem cells. Obviously, however, it is also critical that the β-globin gene and associated sequences express in the context of an integrated proviral genome in a manner equivalent to the constructs shown previously to express the β-globin gene appropriately in transgenic mice. In this regard, specific features of retrovirus vectors that influence viral transcriptional activity may prove quite important.

OTHER GENE THERAPY APPLICATIONS INVOLVING CELL TRANSPLANTATION STRATEGIES

As pointed out earlier, while in some cases a target cell for gene therapy is dictated by the disease itself, in many cases the only requirements for a target cell are that it can be efficiently transduced, and transplanted in such a way that the cells persist, function, and communicate with the circulation.

Surprisingly, while such a requirement might not appear to represent a major technical problem, problems with cell transplantation probably represent the single greatest obstacle to the development of a large number of potential gene therapies.

Over the years, a variety of cell transplantation strategies have been proposed. At one end of the spectrum is the barrier device technology, which involves the use of a container to house the cells of interest, and the direct connection of ports of the container to the circulation. In such a system, cells are contained physically through the use of an appropriate membrane and in some cases contained immunologically through the use of membranes that inhibit the passage of immunoglobulins. For cells to survive within the container, they must obtain necessary nutrients from the blood perfused through the container, and often must be attached to an appropriate matrix material. The cells also must possess growth characteristics that allow them either to remain quiescent over long periods of time or to replicate and die in a manner that yields a constant cell number. A slightly more invasive cell transplantation technique involves the encapsulation of cells and their subsequent implantation, usually subcutaneously. In this strategy, the pore size of the encapsulating material is key, both for exchange of nutrients, diffusion of the gene product of interest, and protection of the cells from immunological attack. In this technology, no direct contact with the circulation is made. Finally, at the other end of the spectrum are all transplantation strategies in which the cells are directly integrated into a tissue or organ of the recipient. In such cases, contact with the circulation is dictated by the site of transplantation itself, and the persistence of the cells is dictated by their natural properties in the environment in which they are placed. Again, the cells may persist by virtue of their quiescence, or through their ability to renew themselves constantly over time.

While progress has been made with each of the approaches outlined above, it is fair to say that at the current time, it has not been possible with any of the technologies to demonstrate successful long-term transplantation, successful sustained expression of a transferred gene, and efficient delivery of a gene product into the circulation. In the case of barrier devices, most attention has been directed towards the transplantation of islet cells capable of expressing insulin in a regulated fashion. Several groups have now reported the sustained control of glucose homeostasis in recipients rendered diabetic by experimental means (4,102,103). However, problems with the devices appear to include thrombotic events that interfere with perfusion of the cells or release of insulin into the blood and physical breakage of the membranes. No studies have yet been reported with transduced islet cells. Since the islet cell represents a highly unique cell type, it is unclear whether the conditions used for the retention of islet cells in the devices will be applicable to other cell types. Future research is necessary to define both the optional cell type for such a device and the appropriate matrix material for holding the cells.

Encapsulation technologies have also been used in order to transplant islet cells (104). However, in those studies, fibrous growth around the implanted cells due to foreign body reactions appeared to be a major problem. More recent work by Aberscher and coworkers (103), who employed a new encapsulation material and a variety of different cell types, appears promising. However, transduced cells have not yet been used with the technology.

Experiments aimed at transplanting primary cells directly into an existing tissue have to date represented the major direction of cell transplantation research by investigators interested in gene therapy applications. Among the first cells to be transduced and transplanted with the aim of providing a constant supply of a gene product were keratinocytes. Keratinocytes can be cultured in vitro in such a way that the cells can be detached from the culture dish as an intact epithelial sheet (106). More importantly, studies by Green and coworkers have convincingly demonstrated that such an intact epithelial sheet could be efficiently transplanted to burn patients, and the cultured cells could form a long-lasting functional epidermis (8). It was also shown that a significant portion of the surface area of a patient's skin could be effectively replaced by cultured epithelium, thus indicating that a large absolute number of cells could be transplanted. In light of those clinical results, and the accessibility of such transplants, the possibility of using such a transplant system for the purposes of continuous delivery of a polypeptide was explored by Morgan et al. (107). Human cultured keratinocytes were shown to be capable of being transduced by amphotropic retroviruses, and capable of expressing and secreting human growth hormone. Studies involving the transplantation of human keratinocytes expressing human growth hormone to nude mice, however, indicated that while expression of growth hormone could be detected in extracts prepared from cells excised from the transplanted mice, the hormone could not be detected in the blood of such recipients. Several other gene products expressed in keratinocytes in the same way were also not found to be detectable in the blood of transplant recipients (J. Morgan, unpub. results). These results raised important issues regarding the accessibility of the transplanted cells to the bloodstream, and suggested that diffusion of a gene product from the epidermis to the capillary beds within the dermis may be generally inefficient. Recent results, however, from Taichman and coworkers suggest that this may not always be the case. Fenjves et al. (108) have recently shown that Apo E, an endogenous gene product expressed in human keratinocytes that is secreted in vitro, is quite efficiently delivered into the bloodstream of nude mice transplanted with human keratinocytes. While it is possible that a specific feature of Apo E (for example its capacity to transport lipid) may account for its ability to be efficiently introduced into the bloodstream, the finding suggests that the possibility of using keratinocytes for the continuous delivery of polypeptides into the circulation should be reevaluated.

The skin fibroblast is another cell type that has been studied as a potential transplantable cell for the delivery of gene products into the circulation.

While the transplantation of fibroblasts has not been studied much clinically, and little is known regarding the self-renewal capacity and half-life of the cells, the availability of large quantities of fibroblasts and the ease with which they can be cultured makes them attractive candidates for gene therapy/cell transplantation studies. Studies to date involving either the subcutaneous transplantation of fibroblasts on collagen matrix or the intraperitoneal injection of fibroblasts on collagen-coated beads have indicated at least short-term persistence of the cells and their capacity to release a genetically engineered product into the circulation (10,11). Unfortunately, to date there have been no reports of the persistent expression of a transduced product delivered via a fibroblast implant. This may suggest either that expression of the transduced gene is influenced by the time after implantation (perhaps due to immunological reactions), that the accessibility of fibroblasts to the circulation may change with time, or, at least in some cases, that the implant itself does not survive for long periods of time.

Other cells that have been examined experimentally as potential candidates for transplantation include hepatocytes, endothelial cells, and myoblasts. In the case of hepatocytes, much effort has been devoted towards both direct injection of the cells into the spleen, portal circulation, or peritoneum, or the injection of cells attached to beads into various locations (109–111). Most of this work has been performed in mice and rats. More recent approaches to the transplantation of hepatocytes have involved the use of biodegradable polymers as substrates for cell attachment and the use of angiogenic factors to induce neovascularization at the site of implantation (112,113). At the current time, it is too early to determine whether such an approach will lead to long-term persistence of the cells, since the fate of the neovasculature is uncertain. Nevertheless, the approach is a very interesting one. Myoblasts have also recently been used in transplantation studies with encouraging results (114). Endothelial cell transplantation strategies have been considered primarily for the study and treatment of cardiovascular disease (9,115–117), and will be described below.

While most of the above cell transplantation strategies would, if successful, make possible the continuous delivery of polypeptides to the circulation of the recipient, one important characteristic of the approaches that may ultimately determine their relative practical utility is their degree of invasiveness. From this standpoint, barrier devices are particularly interesting, since, in principle, the transduced cells would be prevented from direct contact with the cells and tissues of the recipient, and therefore would not present a risk if transduction of the cells altered their phenotype in an undesirable way. On the other hand, the requirement of directly anastamosing the device to the circulatory system necessitates an invasive procedure. All strategies that involve the direct injection of cells into tissues and organs pose a significantly greater risk from a safety standpoint because of the lack of containment of the cells, and the inability to control the absolute number of cells that contribute to the expression of a desired gene product over time.

However, in the case of a number of these strategies, particularly those involving keratinocytes and subcutaneous implants, the accessibility of the grafts to manipulation and/or removal reduces to some extent the risk of the approaches.

Although not an approach involving cell transplantation, a final approach to the continuous delivery of a gene product to the circulation that should be considered is the use of gene transfer to transduce cells directly in vivo. As pointed out earlier, successful direct in vivo gene transfer will depend on a number of factors and in only a few cases has progress been made in this area. One promising area of investigation, however, involves the direct injection of mammalian expression vector constructs into muscle (118). Another strategy involves the use of specific ligand/receptor interactions to promote the targeting of ligand/plasmid DNA complexes to specific cell types (119). With both techniques, it has been possible to demonstrate significant amounts of gene expression, at least for short periods of time. While the mode of persistence of the transferred DNA sequences is not yet fully understood, it appears that in both cases, the DNA is found in an unintegrated form. Whether these approaches will result in sustained long-term expression of sequences is unclear at the present time, as are the consequences of this particular mode of persistence of the transferred DNA upon cell viability and function.

In conclusion, many approaches towards the continuous delivery of polypeptides to the circulation of recipients have been proposed, and significant progress has been made on a number of fronts. However, none of the techniques has yet met the test of time. It is very likely that, ultimately, the successful development of methods for implanting cells and delivering gene products to the circulation will depend on the efforts of investigators in widely divergent disciplines (e.g., materials science, cell adhesion, angiogenesis). The formation of multidisciplined research teams to solve these problems, therefore, is critical.

GENE THERAPY FOR ACQUIRED DISEASE

In contrast to the case with most inherited diseases, a critical issue in considering any acquired disease as a candidate for gene therapy is the choice of the gene to be transferred. In some acquired diseases, gene products or cell types critical to the pathological process may be known. If so, strategies might be devised to interfere with the expression or activity of such a gene product, or to ablate the offending cell type. Such strategies may be particularly applicable in the case of infectious and autoimmune diseases. Often, however, little is known regarding the role, if any, of specific gene products in the observed pathology. Accordingly, before such diseases can be approached by gene therapy, information regarding the molecular biology underlying the disease must be obtained. As will be pointed out below, gene transfer is potentially a very powerful tool for obtaining such information.

Overall, most acquired diseases fit into the category of diseases for which additional information regarding the molecular cause of pathology must be obtained before an effective therapy can be devised. Nevertheless, it has been possible to consider plausible scenarios for the treatment of a number of major diseases including cancer, acquired immunodeficiency syndrome (AIDS), and diseases of the cardiovascular system. Collectively, the scenarios outlined below for the treatment of these diseases are useful in considering the breadth of application of gene therapy.

Cancer

In spite of the extraordinary progress made in the understanding of cancer at the molecular level, to date surprisingly little of this information has had impact upon the clinical management of cancer. Hopefully, this will soon change. In the meantime, there remains a great need for more effective therapies for a wide variety of metastatic cancers. Towards that end, much recent attention has been given to the concept of the immunotherapy of cancer (120). In general terms, such an approach involves attempts to activate various arms of the immune response to kill cancer cells more effectively. Interestingly, the extent to which immune surveillance normally plays a role in the elimination of tumor cells is surprisingly unclear. For instance, while patients with various immunodeficiency syndromes often show an increased risk of cancer, the spectrum of cancers that they show increased susceptibility to is remarkably narrow (121). Regardless of the normal role of the immune system in fighting cancer, however, our rapidly increasing knowledge of the role of specific gene products, particularly cytokines, in the immune response suggests that it may well be possible to manipulate the immune response empirically in many effective ways.

Most studies to date in the area of cancer immunotherapy have focused on attempts to enhance the quantity and specific cytotoxicity of lymphocytes reactive with tumor cells. Such studies have resulted in the development of two clinical protocols for the treatment of metastatic cancer. In the case of lymphocyte-activated killer (LAK) cell therapy (122), peripheral blood lymphocytes are removed from the patient, expanded to large numbers in high doses of IL-2, and returned to the patient. The patient is then given large doses of IL-2 systemically. While some controversy exists regarding the cell types that are amplified in vitro, evidence suggests that both natural killer (NK)-like and cytotoxic T lymphocytes can be amplified and that the activity of the amplified cells in vivo is likely mediated through antigen-independent cytotoxic activity of the cells (123). The LAK cell protocol has been shown to be effective in a small proportion of patients with otherwise untreatable metastatic cancer, although the benefit of LAK cell therapy versus systemic IL-2 therapy alone is uncertain (124). A second related protocol, TIL therapy (125), which has also proved successful particularly in the treatment of advanced metastatic melanoma, involves the in vitro culture and amplification of lymphocytes present in a resected tumor mass, and again, subsequent return of the cells to the patient, along with high doses of IL-2. Amplified TIL cells appear to have the properties of antigen-specific cytotoxic lymphocytes (126). Although TIL cells cultured in vitro can be shown to possess specific cytotoxicity against the tumor from which the cells were isolated, the clinical efficacy of TIL therapy has not been shown to correlate well with the extent of specific cytotoxicity found in vitro (127). Therefore, the mechanism by which TIL therapy can be effective is not yet clear. Nevertheless, it has been proposed that TIL therapy results in the homing of tumor-reactive antigen-specific T lymphocytes to the sites of metastatic foci.

Several years ago, Rosenberg and coworkers proposed a gene transfer experiment designed to gain further information regarding the factors that influence the outcome of TIL therapy (128). Specifically, the investigators sought to use retroviral-mediated gene transfer to mark TIL genetically, so that the fate of the cells could be monitored after their introduction into the patient. Although previous imaging studies using conventionally marked cells had demonstrated persistence of the cells over a time period of several weeks, and a high local concentration of cells at sites of tumor (129), the imaging techniques could not be extended to later time points.

The gene transfer experiment was approved by the appropriate review groups and initiated in 1989. Results of the study of five patients were recently reported (130). Overall, the data indicated that soon after infusion of the cells, the absolute number of TIL that could be detected dropped significantly, yet could be detected in most cases throughout the 21-day course of IL-2 therapy. Thereafter, in most cases, transduced cells were undetectable in the circulation, even using very sensitive assays. In addition, marked cells could be detected at the site of tumor in a number of instances, at various times after infusion. Unfortunately, few of the important issues regarding TIL therapy could be resolved in the study. In particular, while the TIL could in some cases be detected at the tumor site, none of the data presented demonstrated specific homing of TIL to tumor. More importantly, because of the obvious limitations upon obtaining tumor and normal tissue material, it was not possible to understand whether the efficacy of the procedure correlated with detection of TIL at the tumor site. The major conclusions that could be drawn for the data concerned the half-life of TIL in vivo, and the safety of the procedure.

In order to improve TIL therapy, Rosenberg and coworkers have proposed augmentation of the expression of cytokines in TIL through gene transfer techniques (131). In the near future, a clinical protocol will be initiated in which TILs transduced with a retroviral genome encoding human tumor necrosis factor will be used. Subsequent protocols will likely be prepared in which other cytokines shown to possess antitumor activity will be expressed in TIL. While the concept of using genetically engineered TIL in cancer therapy is promising, the ultimate success of the approach will likely depend upon the supposition that TILs do indeed track specifically to tumor, and that antigen-specific cytotoxicity does indeed account for the antitumor effects of the treatment. As pointed out above, it is not clear at the current time that these suppositions are correct. If, for example, TIL therapy is effective in some cases for other reasons (e.g., the infusion of large numbers of TIL leads to optimal levels and ratios of specific cytokines in the circulation or lymphoid compartments), genetic modification of TIL may have no effect or be deleterious.

Another scenario for the immunotherapy of cancer involves attempts to enhance the immune response directed against a tumor through genetic modification of tumor cells that would directly affect the immunogenicity of the cells themselves. This approach is based primarily on the emerging data suggesting that a variety of specific proteins, particularly cytokines and cell adhesion molecules, are critical for initiating and perpetuating an immune response. Particularly in the case of T lymphocytes, it is now quite clear that specific cytokines play a major role in activation of the cells, cytotoxicity, and the generation of memory (132). Several groups have now shown in mice that weak or nonimmunogenic tumors (tumors that are normally not rejected upon transplantation) are efficiently rejected when the tumor cells are engineered to express various cytokines (133–136). Moreover, preliminary results indicate that, in some cases, animals "immunized" with such genetically modified cells are able to mount an effective antitumor response to a subsequent challenge with unmodified tumor cells (134,136). These results raise the very general question of why some tumors are nonimmunogenic, and whether, in fact, the inability of the immune response to deal effectively with certain cancers results from specific deficits in cytokine production. From a practical standpoint, the data suggest that it may be possible to "vaccinate" patients against their own tumor through a protocol involving removal of a portion of the tumor, transduction of the cells in vitro with the appropriate expression constructs, and reimplantation of the tumor cells in the patient. In such a scenario, it would obviously be critical that the genetically modified cells elicit an effective lasting immune response aginst other metastic foci in the patient. Accordingly, preclinical studies are necessary to define which genetic modifications are best able to induce protective responses against nontransduced cells. In this regard, it will be most important to demonstrate the regression of an existing nontransduced tumor mass through vaccination(s) with transduced tumor cells. It will also be important to devise ways to obtain effective vaccination using inactivated cells. Finally, it will be critical to demonstrate the need for genetic modification of the tumor cells, since vaccination with untransduced cells has already been examined (137). Experiments of this type are ongoing in our laboratories and others.

Two other scenarios for the treatment of cancer by therapy are worthy of consideration. The first approach involves the transfer of drug resistance to specific cells of the body particularly sensitive to cancer chemotherapy (e.g., bone marrow and intestinal epithelium). In this approach, the goal would be to develop procedures that would allow the treatment of specific cancers with greater doses of chemotherapy than are now currently possible. Gene products able to confer resistance to a number of chemotherapeutic agents used in the clinic have been identified, and in some cases, the relevant gene has been identified and isolated. Several of these genes have been introduced into retrovirus vectors and have been shown to confer the expected drug resistance (30,138–140). While in principle this approach to improving conventional chemotherapy is plausible, it remains highly speculative. First, it is unclear that increasing the dose of chemotherapy for specific cancers would be efficacious. Second, and perhaps most importantly, at the current time the hematopoietic system is the only sensitive cell compartment in which it will likely be feasible to confer resistance. Unless methods are developed for conferring resistance upon other sensitive tissues, such as intestinal epithelium and cells of the heart, it is unclear that doses of the most useful chemotherapeutic agents could in fact be elevated.

Finally, it has been proposed that introduction of recessive oncogenes into tumor cells that lack those genes could form the basis for a cancer treatment (141). This suggestion is based on the finding that in a number of human tumors, both alleles of a number of such genes appear to be specifically inactivated (142). Moreover, introduction of a functional recessive oncogene construct into tumor cells unable to express that oncogene appears to alter the growth characteristics and tumorgenicity of the cells (141). Unfortunately, the loss of endogenous recessive oncogene expression is not thought to be the only genetic or epigenetic alteration that occurs in tumor cells, and therefore, it is unclear whether restoration of recessive oncogene expression will be sufficient to revert the transformed phenotype. In addition, from a practical standpoint, it is extremely unlikely that a sufficient proportion of a tumor burden could be genetically modified to be useful, since the cells would generally be inaccessible and the methods for targeted in vivo gene transfer have not been developed.

Although this last scenario is an unlikely one, it is likely that knowledge of growth control at the molecular level, particularly with regard to the role of recessive oncogenes, will lead to improved cancer therapy. The recent identification of sequence-specific DNA binding proteins that interact with a variety of growth control genes (Robbins F, Horowitz J, Mulligan RC, unpublished) provides only one example of such new targets for intervention in cancer therapy.

Acquired Immunodeficiency Syndrome

Development of an effective treatment for AIDS represents perhaps the most important scientific challenge of this century, and accordingly, all potential strategies, however speculative, need to be considered. Genetically based approaches that have been proposed can be divided into several categories. A first category includes all approaches towards the continuous delivery of a product that might inhibit any stage of the HIV life cycle. Technically, this approach would involve the use of either cell implants or direct in vivo gene transfer to provide a continuous source of the product of interest. Clearly, soluble CD4 molecules (143) might be delivered in this way, although at the current time, it is unclear whether sufficient quantities of the proteins could be produced by the available techniques. It is also possible that specific antibodies or peptides that might inhibit various stages of the life cycle might be delivered in the same fashion, although again it is unclear that sufficient quantities could be produced. A second category of intervention involves intracellular immunization schemes (144), whereby the target cells for HIV infection are rendered resistant to infection or unable to produce progeny virus after infection. Technically, this approach would probably necessitate a hematopoietic stem cell/BMT protocol similar to that discussed earlier for inherited diseases, in conjunction with some form of azothymidine (AZT) therapy. Although it is possible that gene transfer could be restricted to T cells, this will not likely be useful, in light of the other known cellular reservoirs of HIV and the inherent problem with T-cell culture and transplantation described earlier. Broder and colleagues have recently reviewed the stages of the HIV life cycle in some detail (143), and identified a variety of potential interventions that might form the basis for an intracellular immunization scheme. In general these include the inhibition of viral gene expression through the use of (1) dominant mutant forms of HIV proteins involved in the biogenesis of viral RNA or viral particles, (2) antisense constructs, and (3) ribozymes. For any of these intracellular immunization schemes to succeed, it is critical not only that the viral replication be efficiently inhibited, but also that expression of the relevant protein or RNA not have undesirable side effects on any of the transduced cells.

A final category of genetic therapy that has attracted little attention involves cell-based vaccination schemes analogous to those outlined earlier for the treatment of cancer. Different normal cell types that are capable of being transplanted could be engineered to express specific HIV proteins in conjunction with specific cytokines and subsequently transplanted, in order to augment the immune response against HIV-infected cells. While the reasons for the inability of the immune system to clear virus efficiently in the early stages of HIV infection are not clear, it is possible that enhanced cytokine production in situ may be beneficial.

Cardiovascular Disease

In considering gene therapy for acquired diseases of the cardiovascular system, the choice of genes to transfer is particularly problematic, in light of the complexity of biological events that likely contribute to disease processes such as atherosclerosis. Clearly, before gene therapy can have a major input upon the treatment of cardiovascular disease, much more must be learned regarding the molecular biology of the cardiovascular system. Since the vasculature most often represents the ultimate site of cardiovascular pathology, it is particularly important to characterize further normal and abnormal vascular function. This will require a much greater understanding of the cell types, cell-cell interactions, and gene products involved.

In spite of the complexity of the problem, it is likely that gene transfer will ultimately prove to be a powerful therapeutic tool for treatment of cardiovascular disease, most likely in a scenario involving the transfer of transduced cells or genes themselves to local regions of vessel injury or pathology. In advance of therapeutic applications, however, gene transfer will also likely play a central role in defining gene products with therapeutic potential. In this latter regard, gene transfer can be useful in several ways. First, retroviral-mediated gene transfer can be used to mark cells genetically, and therefore track their fate (146). Recombinant genes encoding *E. coli* β-galactosidase (Lac Z), for example, allow infected cells to be identified in situ, by virtue of expression of the bacterial enzyme. Because retroviruses require replication of the target cell for proviral integration, such marking studies also provide data on the previous proliferative history of the transduced cell. Accordingly, retroviral-mediated gene transfer could be used, in principle, to determine both the origin and type of cells involved in processes such as atherosclerosis, and the proliferative responses of specific cell types to vascular injury.

Perhaps more importantly, gene transfer provides a means of ectopically expressing gene products implicated in vascular disease processes (e.g., adhesion molecules, cytokines, growth factors/receptor, enzymes) as well as other gene products capable of inhibiting the specific functions of gene products contributing to disease. Such an approach is a powerful means of assessing the overall role of individual gene products in complex biological events and for assessing the feasibility of blocking those events through the expression of other gene products. Finally, gene transfer offers a means of restricting such ectopic expression to specific cell types or to specific anatomical locations.

In an effort to apply the above principles to the study and treatment of cardiovascular disease, our laboratory and others have begun to develop the necessary reagents and techniques for accessing cells of the vascular wall. Most important has been the development of methods for transducing endothelial cells in vitro and transplanting those cells back onto denuded arterial segments in vivo, and methods for the direct in vivo transduction of endothelial cells on vessel surfaces (116,147 and Conte M, Birinyi LK, Mulligan RC, in preparation). Much work remains, however, to characterize the properties of such "resurfaced" or "transduced" vessels, and to determine their suitability for assessing the effects of the ectopic expression of specific gene products. These studies are necessary prerequisites for identifying candidate genes useful for therapy. In more practical ongoing studies, the potential use of genetically modified endothelial cells seeded on synthetic vascular grafts as a means of improving the performance of small-caliber vascular protheses is being explored (115). Genetic interventions being considered include the expression of thrombolytic agents such as tPA, peptides which inhibit platelet adhesion and aggregation, and inhibitors of smooth muscle cell proliferation.

CONCLUSIONS

While it is certainly true that gene transfer has "come of age," gene therapy has not yet. Nevertheless, many of the necessary elements are in place, and it is likely that within 10 years, gene transfer will have made a significant impact upon the practice of medicine. Currently, we are passing through the "honeymoon" phase of gene therapy research, in which the bounds of the technology are being defined primarily through in vitro studies involving a variety of genes and cell types. One unfortunate consequence of the great media attention paid to this phase of research is that the technical significance of many of the studies to date has not been kept in perspective relative to the accomplishments ultimately necessary to devise an effective clinical protocol, in even among some investigators in the field. For example, while the clinical protocols involving gene therapy that were recently approved represent extremely important advances in the public understanding and acceptance of this new technology, the preclinical studies performed in support of the protocols did not represent many critical technical advances. Furthermore, the implementation of the protocols in the clinic is unlikely to move the field ahead significantly, from a technical standpoint.

The next stage of gene therapy research, which involves the examination of issues of efficacy and safety in in vivo model systems, is clearly a difficult one, particularly in light of the need for expertise in a wide variety of previously unconnected scientific and medical disciplines. Nevertheless, it is a most critical phase that will define in a more useful way the practical breadth of the technology. In this phase, it will be important to explore further classical scenarios for gene therapy, but also to explore ways in which gene transfer might be used to improve conventional medical procedures. For example, as discussed earlier, it has been suggested that it would be preferable to use allogeneic bone marrow transplantation rather than gene therapy to treat inherited diseases involving hematopoietic cells, since there would be no issues regarding gene expression levels or the safety of gene insertion. However, the applicability of this approach is currently limited, due to the problems associated with allogenic bone marrow transplantation. Similar arguments could also be made for the use of allogenic cells in the treatment of other disorders (e.g., disorders of the liver, neurological disorders, and others). Gene transfer needs to be considered as a potential means to improve such procedures. A further extension of this concept is that, in addition to its potential as a therapeutic tool, gene transfer may prove extremely useful for defining those molecular characteristics of inherited and acquired diseases important for the development of conventional medical therapies.

ACKNOWLEDGMENTS

I would like to acknowledge the members of my laboratory for their critical contributions to the formulation of many of the issues discussed in this review, and to thank Dr. Sam Broder for convincing me to write such a review.

REFERENCES

1. Heyman RA, Borrelli E, Lesley J, et al. *Proc Natl Acad Sci USA* 1989;86:2698.
2. Becker AJ, McCulloch EA, Siminovitch L, Till JE. *Blood* 1965;26:296.
3. Dexter TM, Spooner F, Simmons P, Allen TD. In: Wright DG, and Greenberger JS, eds. *Long term bone marrow cultures*. New York: Alan R. Liss, Kroc Foundation Series, 1984;18:57.
4. Proceedings of the First International Congress on Pancreatic and Islet Transplantation. *Diabetes,* 1989;38(Suppl 1):1.
5. Enat R, Jefferson DM, Ruz-Opazo N, Gratmaitan Z, Leinwald LA, Reid LM. *Proc Natl Acad Sci USA* 1984;81:1411.
6. Clayton DF, Darnell JE. *Mol Cell Biol* 1983;2:1552.
7. Fathman CG, Fitch FW, Denis AK, Witte ON. In: Paul W, ed. *Fundamental immunology,* 2nd ed. New York: Raven Press, 1989.
8. Gallico GG, O'Connor NE, Compton CC, Kehinde O, Green H. *N Engl J Med* 1984;311:448.
9. Stanley JC, Lindenauer SA, Graham LM, et al. Vascular grafts. In: *Vascular surgery,* 2nd ed. New York: Gruned & Stratton, 1986;365.
10. St Louis D, Verma IM. *Proc Natl Acad Sci USA* 1988;85:3150.
11. Palmer TD, Thompson AR, Miller AD. *Blood* 1989;73:438.
12. Banerji J, Rusconi S, Schaffner W. *Cell* 1981;27:299.
13. Scangos G, Ruddle FH. *Gene* 1981;14:1.
14. Capecchi MR. *Cell* 1980;22:479.
15. Mulligan RC, Berg P. *Proc Natl Acad Sci USA* 1981;78:2072.
16. Southern PJ, Berg P. *J Mol Appl Genet* 1982;1:327.
17. Wigler M, et al. *Cell* 1979;16:777.
18. Grosveld F, van Assendelft GB, Greaves DR, Kollis G. *Cell* 1987;51:975.
19. Greaves DR, et al. *Cell* 1989;56:979.

20. Smithies O, Gregg RG, Boggs SS, Korlewski MA, Kucherlapati RS. *Nature* 1985; 317:230.
21. Mansour SL, Thomas KR, Capecchi MR. *Nature* 1988;336:348.
22. Bradley A, Evans MJ, Kaufman MH, Robertson E. *Nature* 1984;309:255.
23. Thompson S, Clarke AR, Pow AM, Hooper ML, Melton DW. *Cell* 1989;56:313.
24. Gluzman Y, ed. *Eukaryotic viral vectors*. Cold Spring Harbor, New York: Cold Spring Harbor Laboratory, 1982.
25. Weiss R. *RNA tumor viruses*. Cold Spring Harbor, New York: Cold Spring Harbor Laboratory, 1984.
26. Mann R, Mulligan RC, Baltimore D. *Cell* 1983;33:153.
27. Wei CM, et al. *J Virol* 1981;39:935.
28. Mulligan RC. In: Gluzman Y, ed. *Eukaryotic viral vectors*. Cold Spring Harbor, New York: Cold Spring Harbor Laboratory, 1982:133.
29. Cone R, Mulligan RC. *Proc Natl Acad Sci USA* 1984;81:6349.
30. Miller AD, Law MR, Verma IM. *Mol Cell Biol* 1985;5:431.
31. Miller AD, Buttimore C. *Mol Cell Biol* 1986;6:2895.
32. Bosselman RA, Hsu RY, Bruszewski J, Hu F, Martin F, Nicholson M. *Mol Cell Biol* 1987;7:1797.
33. Danos O, Mulligan RC. *Proc Natl Acad Sci USA* 1988;85:6460.
34. Markowitz D, Goff S, Bank A. *J Virol* 1988;62:1120.
35. Mann R, Baltimore D. *J Virol* 1985;54:401.
36. Cepko CL, Roberts BE, Mulligan RC. *Cell* 1984;37:1053.
37. Korman AJ, et al. *Proc Natl Acad Sci USA* 1987;84:2150.
38. Wagner EF, Vanek M, Vennstrom B. *EMBO J* 1985;4:663.
39. Armentano D, et al. *J Virol* 1987;61:1647.
40. Guild BC, Finer M, Housman DE, Mulligan RC. *J Virol* 1988;62:3795.
41. Adam MA, Miller AD. *J Virol* 1988;62:3802.
42. Cone RD, Weber-Benarous A, Baorto D, Mulligan RC. *Mol Cell Biol* 1987;7:887.
43. Yu SF, Von Ruben T, Kantoff PW, et al. *Proc Natl Acad Sci USA* 1986;83:3194.
44. Hawley RG, et al. *Proc Natl Acad Sci USA* 1987;84:2406.
45. Stuhlmann H, Jaaenisch R, Mulligan RC. *Mol Cell Biol* 1989;9:100.
46. Hantzopoulos PA, Sullenger BA, Ungers G, Gilboa E. *Proc Natl Acad Sci USA* 1989;86:3519.
47. Cornetta K, et al. *Hum Gene Ther* 1990;1:15.
48. Metcalf D. *The molecular control of blood cells*. Cambridge, MA: Harvard University Press, 1988.
49. Martin PJ, et al. *Blood* 1990;76:1464.
50. Williams DA, Lemischka IR, Nathan DG, Mulligan RC. *Nature* 1984;310:476.
51. Keller G, Paige C, Gilboa E, Wagner EF. *Nature* 1985;318:149.
52. Lemischka IR, Raulet DH, Mulligan RC. *Cell* 1986;45:531.
53. Bodine DLM, Karlsson S, Nienhuis AW. *Proc Natl Acad Sci USA* 1989;86:8897.
54. Wilson JM, Danos O, Grossman M, Raulet DH, Mulligan RC. *Proc Natl Acad Sci USA* 1990;87:439.
55. Dick JE, Magli MC, Huszar D, Phillips RA, Bernstein A. *Cell* 1985;42:71.
56. Moore KA, et al. *Blood* 1990;75:2085.
57. Lim B, Apperley JF, Orkin SH, Williams DA. *Proc Natl Acad Sci USA* 1989;86:8892.
58. Osborne W, Hock RA, Kaleko M, Miller AD. *Hum Gene Ther* 1990;1:31.
59. van Beuschem VW, Kukler A, Einerhard MPW, et al. *J Exp Med* 1990;172:729.
60. Stead RB, Kwok WW, Storb R, Miller AD. *Blood* 1988;71:742.
61. Kantoff PW, et al. *J Exp Med* 1987;166:219.
62. Witte ON. *Cell* 1990;63:5.
63. Sutherland HF, et al. *Blood* 1989;74:1563.
64. Andrews RG, Singer JW, Bernstein ID. *Blood* 1990;172:355.
65. Williams DA, et al. *Mol Cell Biol* 1988;8:3864.
66. Chang, et al. *Lancet* 1986;1:294.
67. McCune JM, et al. *Science* 1988;241:1632.
68. Mosier DE, et al. *Nature* 1988;335:256.

69. Namikawa R, et al. *J Exp Med* 1990;172:1055.
70. Visser JMW, Bauman JGJ, Mulder AK, Eliason JF, de Leeve AM. *J Exp Med* 1984;59:1576.
71. Spangrude GJ, Heimfeld S, Weissman IL. *Science* 1988;241:58.
72. Ploemacher RE, Brons RHL. *Exp Hematol* 1989;17:263.
73. Jones RJ, Wagner JE, Celano P, Zicha MS, Sharkis SJ. *Nature* 1990;347:181.
74. Kredich NM, Hershfield MS. In: Scriver CR, Beudet AL, Sly WS, Valle D, eds. *The metabolic basis of inherited disease,* 6th ed. New York: McGraw-Hill, 1989;40.
75. Reisner Y, Kapoor N, Kirkpatrick D, et al. *Blood* 1983;61:341.
76. Markert ML, Herschfield MS, Schiff RI, Buckley RH. *J Clin Immunol* 1987;7:389.
77. Polmer SH. In: *Enzyme defects and immune disfunction,* Ciba Foundation Symposium 68. New York: Excerpta Medica, 1979:213.
78. Danos S, Abuchonski A, Park YK, Davis FF. *Clin Exp Immunol* 1981;46:649.
79. Levy Y, Hershfield MS, Fernandez-Mejia C, et al. *J Pediatr* 1988;113:312.
80. Belmont, et al. *Nature* 1986;322:385.
81. Belmont JW, et al. *Mol Cell Biol* 1988;8:5116.
82. McIvor, et al. *Mol Cell Biol* 1987;7:838.
83. Williams DA, Orkin SH, Mulligan RC. *Proc Natl Acad Sci USA* 1986;83:2566.
84. Lim B, Williams DA, Orkin SH. *Mol Cell Biol* 1987;7:3459.
85. Lim B, Apperley JF, Orkin SH, Williams DA. *Proc Natl Acad Sci USA* 1989;86:8892.
86. Bowtell DD, Johnson GR, Kelso A, Cory S. *Mol Biol Med* 1987;4:229.
87. Barklis E, Mulligan RC, Jaenisch R. *Cell* 1986;47:391.
88. NIH Clinical Protocol, approved Spring, 1990.
89. Rosenberg SA, et al. *N Engl J Med* 1990:570.
90. Weatherall DJ, Clegg JB, Higgs DR, Wood WG. In: Scriver CR, Beudet AL, Sly WS, Valle D, eds. *The metabolic basis of inherited disease,* 6th ed. New York: McGraw-Hill, 1989, p. 2281.
91. Fosburg MT, Nathan DG. *Blood* 1990;76:435.
92. Tuan D, et al. *Proc Natl Acad Sci USA* 1985;82:6384.
93. Forrester, et al. *Nucleic Acids Res* 1987;15:10159.
94. Ryan TM, Behringer RR, Martin NC, Townes TM, Palmiter RD, Brinster RC. *Genes Dev* 1989;3:314.
95. Collis P, Antoniou M, Grosveld F. *EMBO J* 1990;9:233.
96. Dzierzak EA, Papayannoupoulou T, Mulligan RC. *Nature* 1988;331:35.
97. Bender MA, Miller AD, Gelinas RE. *Mol Cell Biol* 1988;8:1725.
98. Miller AD, Bender MA, Harris EAS, Kaleko M, Gelinas RE. *J Virol* 1988;62:4337.
99. Karlsson S, Bodine D, Perry L, Papayannopoulou T, Nienhuis AW. *Proc Natl Acad Sci USA* 1988;85:6062.
100. Bender MA, Gelinas RE, Miller AD. *Mol Cell Biol* 9:1426.
101. Novak U, Harns EAS, Forrester W, Groudine M, Gelinas R. *Proc Natl Acad Sci USA* 1990;87:3386.
102. Araki Y, Solomon BA, Basile RM, Chick WL. *Diabetes* 1985;34:854.
103. Maki T, et al. *Transplantation* 1990, in press.
104. Lim F, Sun AM. *Science* 1980;210:908.
105. Abescher P, et al. *Brain Res* 1988;448:364.
106. Rheinwald JG, Green H. *Cell* 1975;6:331.
107. Morgan JR, Barrandon Y, Green H, Mulligan RC. *Science* 1987;237:1476.
108. Fenjves ES, Gordon DA, Pershing LK, Williams DL, Taichman LB. *Proc Natl Acad Sci USA* 1989;86:8803.
109. Demetriou AA, Whiting JF, Feldman D, et al. *Science* 1986;233:1190.
110. Parker-Ponder K, et al. *Proc Natl Acad Sci USA* 1990, in press.
111. Wilson JM, et al. *Proc Natl Acad Sci USA* 1990, in press.
112. Vacanti J, et al. *J Pediatr Surg* 1987;23:3.
113. Thompson JA, Hardenschild CC, Anderson KD, DiPietro JM, Anderson WF, Maciag T. *Proc Natl Acad Sci USA* 1989;86:7928.
114. Patridge TA, et al. *Nature* 1989;337:176.
115. Wilson JM, Birinyi LK, Salomon RN, Libby P, Callow AD, Mulligan RC. *Science* 1989;244:1344.

116. Nabel EG, Plantz G, Boyce RM, Stanley JC, Nabel GJ. *Science* 1989;244:1324.
117. Dichek DA, Neville RF, Zweibel JA, Freeman SM, Leon MB, Anderson WF. *Circulation* 1989;80:1347.
118. Wolff JA, Malone RW, Williams P, et al. *Science* 1990;247:1465.
119. Wu GY, Wu CH. *J Biol Chem* 1988;263:14621.
120. Rosenberg SA, Terry WD. *Adv Cancer Res* 1977;25:323.
121. Ioachim HL. The opportunistic tumors of immune deficiency. In: Van de Woude GF, Klein G, eds. *Advances in cancer research*. San Diego: Academic Press, 1990;54:301.
122. Rosenberg SA, et al. *N Engl J Med* 1985;313:1485.
123. Phillips JH, Lanier LL. *J Exp Med* 1986:814.
124. Rosenberg SA, et al. *J Natl Cancer Inst* 1990:82, in press.
125. Rosenberg SA, Spiess P, Lafreniere R. *Science* 1986;233:1318.
126. Barth RJ, Bock SN, Mule JJ, Rosenberg SA. *J Immunol* 1990;141:1531.
127. Topalian, et al. *J Clin Oncol* 1988;6:839.
128. *Hum Gene Ther* 1990;1:73.
129. Fischer B, Packard BS, Reed EJ, et al. *J Clin Oncol* 1989;7:250.
130. Rosenberg SA, et al. *N Engl J Med* 1990:570.
131. NIH Clinical Protocol, approved Spring, 1990.
132. Miyajima A, et al. *FASEB J* 1988;2:2462.
133. Tepper R, Pattengale PK, Leder P. *Cell* 1989;57:503.
134. Fearon ER, et al. *Cell* 1990;60:397.
135. Watanabe, et al. *Proc Natl Acad Sci USA* 1989;86:9456.
136. Gansbacher B, Zier K, Daniels B, Cronin K, Bannerji R, Gilboa E. *J Exp Med* 1990;172:1217.
137. Hoover HC, et al. *Cancer Res* 1984;44:1671.
138. Williams DA, et al. *J Exp Med* 1987;166:210.
139. Guild BC, et al. *Proc Natl Acad Sci USA* 1988;85:1595.
140. Pastan I, et al. *Proc Natl Acad Sci USA* 1988;85:4486.
141. Huang HJ, et al. *Science* 1988;242:1563.
142. Klein G. *Science* 1989;238:1339.
143. Smith DH, et al. *Science* 1987;238:1704.
144. Baltimore D. *Nature* 1988;335:395.
145. Mitsuya H, Yarchoan R, Broder S. *Science* 1990;249:1533.
146. Price J, Turner D, Cepko C. *Proc Natl Acad Sci USA* 1987;84:156.
147. Nabel EG, Plautz G, Nabel GJ. *Science* 1990;249:1285.

Etiology of Human Disease at the DNA Level,
edited by Jan Lindsten and Ulf Pettersson.

13

Mutating Genes in the Germ Line of Mice

Rudolf Jaenisch

Whitehead Institute for Biomedical Research, Massachusetts Institute of Technology, Cambridge, Massachusetts 02142

The introduction of genes into the germ line of mammals and the successful expression of the inserted gene in the organism have allowed genetic manipulation of animals on an unprecedented scale. Transgenic animals have been instrumental in providing new insights into mechanisms of development and developmental gene regulation, into the action of oncogenes and into the intricate cell interactions within the immune system. Furthermore the transgenic technology offers exciting possibilities for genetically dissecting the mammalian genome and for generating precise animal models for human genetic diseases. It is this last application that will be the focus of this report.

The first animals carrying experimentally introduced foreign genes were derived in 1974 by microinjection of simian virus 40 (SV40) DNA into the blastocyst cavity (1,2). The presence of the injected DNA in a number of somatic tissues derived from mice that had been injected as embryos was demonstrated by DNA reassociation kinetics. However, integration of the viral DNA into the germ line was not demonstrated in these early experiments. Germ line transmission of foreign DNA was detected soon thereafter in subsequent studies when mouse embryos were exposed to infectious Moloney leukemia retrovirus (M-MuLV), which resulted in the generation of the first transgenic mouse strain (3). In 1981 a number of groups reported the generation of transgenic mouse strains that were derived from zygotes injected with recombinant DNA (4–8). Pronuclear injection of cloned DNA has become the most popular technique to derive animals that express a given transgene in a tissue-specific manner (for review see 9,10).

The introduction of foreign DNA sequences into the mammalian germ line is highly mutagenic and can be used to generate mice carrying a wide variety of mutations. Strategies that have been used successfully for the generation of mutations in mice include: (a) insertional mutagenesis, (b) generation of dominant negative mutations and (c) targeted mutations by homologous recombination (for reviews see 10,11). Insertional mutagenesis causes mutational changes by disrupting the function of an endogenous gene. The inserted foreign sequences serve as a molecular tag and allow the cloning of the mutated gene. Most insertional mutations in transgenic mice are recessive and have been induced by infection of embryos or embryonic stem (ES) cells with retroviruses or by microinjection of recombinant DNA into the pronucleus (12–19). The frequency of inducing insertional mutations is about 5–10%. Because the foreign sequences can insert at many different chromosomal locations, the target gene of insertional mutagenesis cannot be predicted. The approach may, however, lead to the isolation of previously unknown genes. In contrast, dominant negative mutations can result from expression of a mutant gene whose product interferes with the function of the endogenous gene. Such mutations may result in a well-defined phenotype.

Mouse strains carrying a precisely engineered mutant gene can be generated by homologous recombination. This approach was pioneered in the laboratories of Smithies and Capecchi (20,21) and involves the introduction of a suitably constructed vector into a target gene of an embryonic stem (ES) cell (22). Mutant ES cells are microinjected into a host blastocyst to generate germ line chimeras that will transmit the mutant gene to the next generation. A number of targeted genes have been transmitted through the germ line and include the *HGPRT* gene (23,24), the β2-microglobulin gene (25), the *c-abl* gene (26) and the *IGF-2* gene (27).

TABLE 1. *Summary of mutations induced by various transgenic techniques*

Method of mutagenesis	Mutant phenotype	Mutated gene	Ref.
A. Retrovirus-induced insertional mutation			
Mov13	Homozygous: lethal, midgestation	*Col1a1*	16
	Heterozygous: mild form of osteogenesis imperfecta	*Col1a1*	32
Mov34	Lethal, postimplantation	(Highly conserved genes)	17,33
Mpv17	Adult, kidney failure	(Highly conserved genes)	34
Mpv20	Lethal, 16-cell stage	?	
B. Dominant negative mutation			
Expression of mutated gene (Gly→Cys mutation in α1(I) collagen helix)	Lethal form of osteogenesis imperfecta	*Col1a1*	36
C. Targeted mutation			
Homologous recombination in embryonic stem cells	Defect in T-cell development: Deficiency of CD8^{+} killer cells	β2-microglobulin	25,47

In my laboratory all three strategies have been used to generate mutant mouse strains. The following article briefly describes the derivation and phenotype of these mutations. Our results are summarized in Table 1.

INSERTIONAL MUTAGENESIS

Of 70 transgenic mouse strains carrying a single proviral insert, 4 were found to carry a recessive lethal mutation when bred to homozygosity (Mov13, Mov34, Mpv17, Mpv20; see Table 1). Host sequences flanking the provirus were isolated in each case and shown to specify a transcript whose expression was disrupted by the respective provirus. While the identity of only one of the four mutated genes is known to date (Mov13), sequence analysis suggests that the provirus disrupted a protein-encoding gene in two other mutant strains (Mov34, Mpv17). No sequence information about the nature of the gene mutated in Mpv20 mice has been obtained as yet.

Mov13

This strain was derived by microinjection of Moloney leukemia virus into postimplantation mouse embryos (13). The virus insertion in this strain caused an embryonic lethal mutation with death of homozygous embryos at day 13 of gestation. The mutated gene was cloned and identified as coding for the α1 chain of type I collagen (16). The proviral insertion induced changes in the methylation pattern and chromatin conformation of the collagen gene and was shown to interfere with transcriptional initiation, causing a block in type I collagen synthesis in homozygous embryos. This resulted in death at midgestation after rupture of major blood vessels (28). The Mov13 strain has been useful in studying the role of collagen in development (29) and for the molecular analysis of structural mutations in the α1(I) collagen gene (30). A detailed study of collagen expression in different lineages showed that the mutant gene was normally expressed in ondontoblasts resulting in normal tooth development (31). This surprising result indicated that the provirus acted as a tissue-specific mutagen causing a complete block of collagen transcription in most but permitting gene expression in some lineages.

Given the nature of the mutation carried in the Mov13 strain, it was of interest to evaluate whether this strain would represent a model for osteogenesis imperfecta type I (OI I). This disease is a mild, dominantly inherited connective tissue disorder characterized by bone fragility, progressive hearing loss, dental defects and other symptoms. A recent detailed study demonstrated that Mov13 mutant mice heterozygous for the null allele strikingly resemble human patients with OI I (32). The Mov13 strain may, therefore, provide an opportunity to investigate the effect of reduced amount of collagen I on the structure and integrity of the extracellular matrix and may represent a system in which therapeutic strategies to strengthen connective tissue can be developed.

Mov34

This mutation was found to arise from the integration of an M-MuLV-derived provirus in an as yet unidentified gene. Embryos homozygous for the Mov34 integration were found to develop normally to the blastocyst stage and die after implantation into the uterus (17). The proviral integration was shown to disrupt a transcription unit located on chromosome 8 (33). The Mov34 gene transcript is highly conserved in evolution and the *Drosophila* homolog appears to encode a protein with 62% amino acid identity to the predicted murine protein. Recently, antibodies to a peptide predicted from the amino acid sequence of the Mov34 protein were shown to detect a protein of approximately 39 Kd (Vassalli and Jaenisch, unpublished).

Mpv17

This mutant mouse strain was generated by exposing mouse embryos to a recombinant retrovirus (MPSV) (34). Adult mice homozygous for the Mpv17 integration developed nephrotic syndrome and chronic renal failure. Histologically, affected kidneys showed a progressive sclerosis of the glomeruli with deposition of hyaline material in the glomeruli and in renal tubules. Similar glomerular lesions are seen in patients with progressive deterioration of renal function due to various renal disorders. Sequences flanking the proviral integration were cloned and shown to be highly conserved during evolution. The flanking probe detected a 1.7-kb RNA that is ubiquitously expressed during embryogenesis and in the adult with high levels in kidney, brain and heart. This RNA was not detected in any tissue of homozygous animals, suggesting that the provirus interferes with expression of stable Mpv17 RNA. Sequence analysis of the cDNA suggested that the gene codes for a peptide of 176 amino acids with a hydrophobic region that may reveal membrane association of the putative Mpv17 protein. The Mpv17 mutant is a potentially useful experimental system for studying mechanisms leading to renal disorders in man that are not well understood on the genetic or molecular level.

DOMINANT NEGATIVE MUTATIONS

Another strategy for generating mutants with a precisely predetermined phenotype is to alter a cloned gene by site-directed mutagenesis so that it encodes a mutant product capable of inhibiting the function of the wild-type gene. Such mutations have been termed "antimorphs" or, more recently, "dominant negative mutations" (35). In the case of multimeric proteins, such mutations may cause the formation of nonfunctional multimers. The main advantage of this strategy is that it requires only expression of the mutant gene product and not the inactivation of the endogenous wild-type gene in order to realize the mutant phenotype in a cell. To test the feasibility of this approach in the animal, a point mutation analogous to mutations seen in patients with osteogenesis imperfecta II was introduced into the murine pro α1(I) collagen gene in vitro. Substitution of a single glycine residue in the pro α1(I) collagen gene was shown recently to be associated with this dominant perinatal lethal disease in humans. When introduced into transgenic mice, expression of as little as 10% mutant RNA of the total pro α1(I) collagen NA caused a dominant perinatal lethal phenotype that resembled the human condition (36). This kind of approach is likely to be useful for the genetic analysis of many other proteins that form multimeric structures, such as proteins of the cytoskeleton, and could provide defined animal models for human diseases in the absence of mutations in the endogenous gene of interest.

HOMOLOGOUS RECOMBINATION IN EMBRYONIC STEM CELLS

We (25) and others (37) have used the technique of homologous recombination to disrupt the β2-microglobulin (β2-m) gene. β2-m is a polypeptide of relative molecular mass 12,000 (12K), which is activated in mouse embryos by the two-cell stage, and associates with the heavy chain of the polymorphic major histocompatibility complex (MHC) class I proteins encoded by the H2-K, H2-L/D and Qa/Tla loci (38). The role of MHC class I proteins in the presentation of antigens to the immune system and in the development of the T-cell repertoire is well documented (39,40), but it has been suggested that MHC molecules have other nonimmunological functions, for example, as differentiation antigens (41,42), or in the function of hormone receptors (43) or as olfactory cues influencing mating behavior (44). Furthermore, it has been demonstrated that β2-m associates with the Fc receptor in neonatal gut cells (45) and may serve as a chemotactic protein in the fetal thymus (46). The most direct means of unravelling the many biological functions of β2-m was to create a mutant mouse with a defective β2-m gene.

A cloned fragment of the β2-m gene containing a neomycin-resistant gene inserted into the second exon was introduced into ES cells. Drug-resistant colonies were screened for cells carrying a mutant β2-m gene disrupted by homologous recombination. The targeting frequency of the gene was high, yielding 1 targeted cell in 25 to 100 G418r colonies (25,37).

Mice homozygous for the disrupted β2-m gene developed normally and were fertile in the absence of detectable β2-m protein and expressed little if any functional MHC class I antigen on the cell surface (47,48). These results therefore question the widely held view that class I molecules have an essential role in the embryonic development of vertebrates. The mutant mice did reveal, however, a defined defect in T-cell development and T-cell function. They showed a normal distribution of $CD4^{+}8^{+}$ and $CD4^{+}8^{-}$ T cells, but had no mature $CD4^{-}8^{+}$ T cells and were defective in $CD4^{-}8^{+}$ cell-mediated cytotoxicity. These results strongly support earlier evidence (39,40) that MHC class I molecules are crucial for positive selection of cytotoxic T cells. The β2-m-deficient mutant mice represent a model system that should allow precise evaluation of the role of MHC class I antigens in complex biological interactions such as tumorigenesis and the response to viral and other infectious agents and in organ transplantation.

CONCLUSION

The last few years have witnessed an extraordinary increase in the use of transgenic animals. Methods of manipulating embryos and transferring genes have been refined and now constitute standard procedures used for a variety of purposes. Each of the three methods for generating transgenic animals has distinct advantages for some and disadvantages for other applications. Pronuclear injection of recombinant DNA is the method of choice for obtaining expression of a foreign gene in almost any specific tissue. Retroviruses or retroviral vectors are superior when genetic tagging of chromosomal loci, for example, for insertional mutagenesis, or of cells for lineage studies are desired. Finally, the most recently developed method of generating transgenic animals from ES cells allows in principle the derivation of mice with any genetic or phenotypic characteristics for which in vitro screening or selection methods are available. The prospect for generating recessive or dominant mutations in preselected genes not only will permit the derivation of precise animal models for human hereditary diseases but also will mark the beginning of a systematic genetic dissection of developmental processes that will radically change the future of experimental mammalian genetics.

REFERENCES

1. Jaenisch R, Mintz B. Simian virus 40 DNA sequences in DNA of healthy adult mice derived from preimplantation blastocysts injected with viral DNA. *Proc Natl Acad Sci USA* 1974;1250–1254.
2. Jaenisch R. Infection of mouse blastocysts with SV40 DNA; normal development of the infected embryos and persistence of SV40-specific DNA sequences in the adult animals. *Cold Spring Harbor Symp Quant Biol* 1975;39:375–380.
3. Jaenisch R. Germ line integration and Mendelian transmission of the exogenous Moloney leukemia virus. *Proc Natl Acad Sci USA* 1976;73:1260–1264.

4. Costantini F, Lacy E. Introduction of a rabbit β-globin gene into the mouse germ line. *Nature* 1981;294:92–94.
5. Brinster R et al. Somatic expression of herpes thymidine kinase in mice following injection of a fusion gene into eggs. *Cell* 1981;27:223–231.
6. Gordon JW, Ruddle FH. Integration and stable germ line transmission of genes injected into mouse pronuclei. *Science* 1981;214:1244–1246.
7. Harbers K, Jähner D, Jaenisch R. Microinjection of cloned retroviral genomes into mouse zygotes: integration and expression in the animal. *Nature* 1981;293:540–542.
8. Wagner EF, Stewart T, Mintz B. The human β-globin gene and a functional viral thymidine kinase gene in developing mice. *Proc Natl Acad Sci USA* 1981;78:5016–5020.
9. Gridley T, Soriano P, Jaenisch R. Insertional mutagenesis in mice. *Trends Genet* 1987;3:162–166.
10. Jaenisch R. Transgenic animals. *Nature* 1988;240:919–922.
11. Capecchi MR. The new mouse genetics: altering the genome by gene targeting. *Trends Genet* 1989;5:70–76.
12. Covarrubias L, Nishida Y, Mintz B. Early postimplantation embryo lethality due to DNA rearrangements in a transgenic mouse strain. *Proc Natl Acad Sci USA* 1986;83:6020–6024.
13. Jaenisch R, Harbers K, Schnieke A, Löhler J, Chumakov I, Jähner D, Grotkopp D, Hoffmann E. Germ line integration of Moloney murine leukemia virus at the Mov 13 locus leads to recessive mutation and early embryonic death. *Cell* 1983;32:209–216.
14. Kuehn MR, Bradley A, Robertson EJ, Evans MJ. A potential animal model for Lesch Nyhan syndrome through introduction of HPRT mutation into mice. *Nature* 1987;326: 295–298.
15. Mahon KA, Overbeek PA, Westphal H. Prenatal lethality in a transgenic mouse line is the result of a chromosomal translocation. *Proc Natl Acad Sci USA* 1988;85:1165–1168.
16. Schnieke A, Harbers K, Jaenisch R. Embryonic lethal mutation in mice induced by retrovirus insertion into the α 1(I) collagen gene. *Nature* 1983;304:315–320.
17. Soriano P, Gridley T, Jaenisch R. Retroviruses and insertional mutagenesis in mice: proviral integration at the Mov 34 locus leads to early embryonic death. *Genes Dev* 1987;1:366–375.
18. Wilkie TM, Palmiter RD. Analysis of the integrant in Myk-103 transgenic mice in which males fail to transmit the integrant. *Mol Cell Biol* 1987;7:1646–1655.
19. Woychik RP, Stewart TA, Davis LG, Eustachio PD, Leder P. An inherited limb deformity created by insertional mutagenesis in a transgenic mouse. *Nature* 1985;318:36–40.
20. Thomas KR, Capecchi MR. Site-directed mutagenesis by gene targeting in mouse embryo-derived stem cells. *Cell* 1987;51:503–512.
21. Doetschman T, Gregg RG, Maeda N, Hooper ML, Melton DW, Thompson S, Smithies O. Targeted correction of a mutant HPRT gene in mouse embryonic stem cells. *Nature* 1987;330:576–578.
22. Bradley A, Evans M, Kaufman MH, Robertson E. Formation of germ-line chimaeras from embryo-derived teratocarcinoma cell lines. *Nature* 1984;309:255–256.
23. Thompson S, Clarke AR, Pow AM, Hooper ML, Melton DW. Germ line transmission and expression of a corrected HPRT gene produced by gene targeting in embryonic stem cells. *Cell* 1989;56:313–321.
24. Koller BH, Hagemann LJ, Doetschman T, et al. Germ-line transmission of a planned alteration made in a hypoxanthine phosphoribosyltransferase gene by homologous recombination in embryonic stem cells. *Proc Natl Acad Sci USA* 1989;86:8927–8931.
25. Zijlstra M, Li E, Sajjadi F, Subramani S, Jaenisch R. Germ-line transmission of a disrupted β 2-microglobulin gene produced by homologous recombination in embryonic stem cells. *Nature* 1989;342:435–438.
26. Schwartzberg PL, Goff SP, Robertson EJ. Germ-line transmission of a c-*abl* mutation produced by targeted gene disruption in ES cells. *Science* 1989;246:799–803.
27. DeChiara TM, Efstratiadis A, Robertson EJ. A growth-deficiency phenotype in heterozygous mice carrying an insulin-like growth factor II gene disrupted by targeting. *Nature* 1990;345:78–80.
28. Löhler J, Timpl R, Jaenisch R. Embryonic lethal mutation in mouse collagen I gene causes rupture of blood vessels and is associated with erythropoietic and mesenchymal cell death. *Cell* 1984;38:597–607.

29. Kratochwil K, Dziadek M, Löhler J, Harbers K, Jaenisch R. Normal epithelial branching morphogenesis in the absence of collagen I. *Dev Biol* 1986;117:596–606.
30. Schnieke A, Dziadek M, Bateman J, et al. Introduction of the human pro α 1(I) collagen gene into pro α 1(I)-deficient Mov-13 mouse cells leads to formation of functional mouse-human hybrid type I collagen. *Proc Natl Acad Sci USA* 1987;84:764–768.
31. Kratochwil K, von der Mark K, Kollar EJ, et al. Retrovirus-induced insertional mutation in Mov-13 mice affects collagen I expression in tissue-specific manner. *Cell* 1989;57:807–816.
32. Bonadio J, Saunders TL, Sai E, et al. Transgenic mouse model of the mild dominant form of osteogenesis imperfecta. *Proc Natl Acad Sci USA* 1990;87:7145–7149.
33. Gridley T, Gray DA, Orr-Weaver T, et al. Molecular analysis of the Mov34 mutation: transcript disrupted by proviral integration in mice is conserved in *Drosophila Development* 1990;109:235–242.
34. Weiher H, Noda T, Gray DA, Sharpe AH, Jaenisch R. Transgenic mouse model of kidney disease: insertional inactivation of ubiquitously expressed gene leads to nephrotic syndrome. *Cell* 1990;62:425–434.
35. Herskowitz I. Functional inactivation of genes by dominant negative mutations. *Nature* 1987;329:219–220.
36. Stacey A, Bateman J, Choi T, Mascara T, Cole W, Jaenisch R. Perinatal lethal osteogenesis imperfecta in transgenic mice bearing an engineered mutant pro α 1(I) collagen gene. *Nature* 1988;332:131–136.
37. Koller BH, Smithies O. Inactivating the β 2-microglobulin locus in mouse embryonic stem cells by homologous recombination. *Proc Natl Acad Sci USA* 1989;86:8932–8935.
38. Klein J, Jretìc A, Baxevanis CN, Nagy ZA. The traditional and a new version of the mouse H-2 complex. *Nature* 1981;291:455–460.
39. Von Boehmer H. The developmental biology of T lymphocytes. *A Rev Immunol* 1988;6:309–326.
40. Marrack P, Kappler J. The T cell repertoire for antigen and MHC. *Immunol Today* 1988;9:308–315.
41. Curtis ASG, Rooney P. H-2 restriction of contact inhibition of epithelial cells. *Nature* 1979;281:222–223.
42. Bartlett PF, Edidin J. Effect of the H-2 gene complex rates of fibroblast intercellular adhesion. *J Cell Biol* 1978;77:377–388.
43. Hansen T, Stagsted J, Pederson L, Roth RA, Goldstein A, Olsson L. Inhibition of insulin receptor phosphorylation by peptides derived from major histocompatibility complex class I antigens. *Proc Natl Acad Sci USA* 1989;86:3123–3126.
44. Singh PB, Brown RE, Roser B. MHC antigens in urine as olfactory recognition cells. *Nature* 1987;327:161–164.
45. Simister NE, Mostov KE. An Fc receptor structurally related to MHC class I antigens. *Nature* 1989;337:184–187.
46. Brinckerhoff CE, Mitchell TI, Karmilowicz MJ, Kluve-Beckerman B, Benson MD. Autocrine induction of collagenase by serum amyloid A-like and β 2-microglobulin-like protein. *Science* 1989;243:655–657.
47. Zijlstra M, Bix M, Simister NE, Loring JM, Raulet D, Jaenisch R. β 2-microglobulin deficient mice lack $CD4^-8^+$ cytolytic T cells. *Nature* 1990;344:742–746.
48. Koller BH, Marrack P, Kappler JW, Smithies, O. Normal development of mice deficient in β2-m, MHC Class I proteins, and $CD8^+$ T cells. *Science* 1990;248:1227–1230.

Etiology of Human Disease at the DNA Level,
edited by Jan Lindsten and Ulf Pettersson.

14

Gene Replacement Therapy: Human and Animal Model Progress

Denis Cournoyer*, Maurizio Scarpa,* Kateri A. Moore, Frederick A. Fletcher,* Grant R. MacGregor, John W. Belmont, and C. Thomas Caskey

Institute for Molecular Genetics and Howard Hughes Medical Institute, Baylor College of Medicine, Houston, TX 77030

A wide range of inherited human disorders exist that could theoretically be corrected by the introduction of new genetic information into the proper somatic cell type (1–4). For a disease to be amenable to this approach, its pathophysiology should be well understood at the molecular level and functional DNA sequences must be available for gene transfer. Most current methods of gene transfer result in the random insertion of the incoming sequence rather than in the direct modification of the corresponding cellular gene; consequently, candidate diseases should generally be single-gene disorders of recessive inheritance in which the function of the defective gene can be supplemented or replaced by a functional copy of the sequence.

As somatic gene transfer experiments in animal models have so far been most successful via the ex vivo modification of mouse hematopoietic stem cells, it appears likely that disorders of the lympho-hematopoietic system would be among the first to be amenable to this new form of therapy. Several groups including our own have centered their efforts on the development of a gene transfer approach applicable to the correction of adenosine deaminase (ADA) deficiency. ADA deficiency is a rare autosomal recessive disorder that is responsible for approximately 15% of all cases of severe combined immunodeficiency (5). The pathophysiology of the disorder involves selective toxicity of nonmetabolized adenosine and deoxyadenosine and accumulation of dATP in immature T and to a lesser extent B cells. It has been speculated that the accumulation of dATP alters the regulation of ribonu-

*Denis Cournoyer's present address is The Montreal General Hospital, Division of Hematology and Center for Host Resistance, 1650 Avenue Cedar, Montréal, Québec H3G 1A4, Canada. Maurizio Scarpa's present address is: Department of Pediatrics, University of Padova, Via Gustiniani 3, 35100 Padova, Italy. Frederick A. Fletcher's present address is: Department of Experimental Hematology, Immunex R&D Corporation, 51 University Street, Seattle, Washington 98101.

cleotide reductase, a highly regulated enzyme required for cell replication. Alterations in DNA methylation pattern *via* inhibition of S-adenosyl homocysteine hydrolase have also been implicated. Although ADA deficiency affects only a very small number of patients worldwide, it is generally regarded as an optimal model disease to develop an approach to somatic gene therapy that could be applicable to more common disorders. The reasons for this are the following: (i) ADA deficiency is uniformly fatal if left untreated, weighing the risk/benefit ratio in favor of experimental but potentially curative approaches when conventional treatments have failed; (ii) this disorder is known to be correctable *via* allogeneic bone marrow transplantation and is therefore approachable through the genetic modification of hematopoietic stem cells; (iii) allogeneic bone marrow transplantation often fails or only partially restores the immune function; (iv) the alternative treatment with a stabilized form of ADA linked to polyethylene glycol (PEG-ADA) (6) represents remarkable progress, but it is very expensive (approximately $60,000 US/year) it requires frequent intramuscular injections, and the variability in immune function which is observed during treatment raises concern about its efficacy in preventing serious infections; (iv) in contrast to many other disorders affecting the hematopoietic system (e.g., most hemoglobinopathies), it is likely that the type of unregulated or imperfectly regulated transduced expression that can be currently achieved with retroviral vectors would be beneficial to ADA-deficient patients; and finally (v) it is plausible that an *in vivo* selection in favor of corrected cells would take place in this disorder.

To date, the most promising results toward achieving gene therapy have been obtained by using replication-defective retroviral vectors. These vectors, usually derivatives of Moloney murine leukemia virus (M-MLV), are produced by replacing the genes that encode the proteins of the retrovirus (*gag, pol,* and *env*) with the genetic information to be transferred. To complement this, several "packaging" cell lines have been constructed in mouse fibroblast cells. They are made by introducing the DNA coding sequences of the viral proteins without the information necessary for packaging of the RNA viral genome. The transfer of a recombinant retroviral vector possessing an intact packaging signal into a packaging cell line results in the production of viral particles containing RNA copies of the recombinant vector. The resultant "defective" virus is able to infect target cells and integrate into the host cell genome, achieving transfer of the new genetic information, but because it lacks the viral structural genes, it cannot replicate.

Retroviral vectors offer several advantages as vehicles for gene transfer. First, they allow very efficient gene transfer in a wide range of cell types. Second, in contrast to DNA-mediated gene transfer by physicochemical methods, the integration of the provirus (integrated DNA copy of the retrovirus) generally occurs as a single copy without gross deletion or rearrangement of the surrounding host DNA. Finally, the viral regulatory elements (enhancer and promoter) conduct efficient expression of the transduced genes in most cell types, and additional regulatory sequences can be included in the vector to obtain lineage-specific expression (7–9). Despite this, retroviruses also present certain disadvantages and potential risks. Because of the constraints on the size of viral genome that can be packaged into viral particles, a maximum of 7 to 8 kb of sequence can be added to retroviral vectors. The integration of retroviruses into the host DNA appears to be dependent on active DNA replication, limiting the efficiency of gene transfer into populations of cells that are predominantly quiescent, such as hematopoietic stem cells. Recombination events between the vector and other retroviral sequences in the packaging cell line, or possibly in the host cells, can result in the production of replication competent virus leading to undesirable active infection in the recipient of the transduced cells. Finally, the integration of the provirus may disrupt other important genes (insertional mutagenesis) or activate nearby oncogenes with the potential for malignant transformation. We will discuss here how we have addressed these risks and limitations of retroviral vectors, and developed recombinant vectors and culture conditions that result in efficient gene transfer of human ADA in mouse hematopoietic stem cells and primitive human hematopoietic progenitors.

EVALUATION OF VECTORS

One major difficulty with the use of retroviral vectors for gene therapy has been our inability to predict reliably the performance of vector constructs with regard to both virus production in packaging cell lines and expression of the transduced gene(s). Moreover, the level of expression that is obtained may vary among cell types, even in the absence of well-defined tissue-specific regulatory sequence. Undoubtedly, these phenomena reflect our relatively poor understanding of gene regulation and of the retroviral life cycle. One must therefore design and test on an empirical basis a large number of constructions to identify those that might be most useful for a particular application. In this regard, the vectors that we are currently using for ADA gene transfer experiments in mouse and human hematopoietic cells were derived from more than 20 previously evaluated ADA-transducing retroviruses (10 and K. A. Moore et al., manuscript in preparation) (Fig. 1). These vectors were designed to evaluate the effect on virus production and on the level of expression of human ADA of the following features: (i) inclusion of sequences 3′ of the Ψ packaging site, the so called *gag*$^+$ region; (ii) deletions of the promoter and/or enhancer sequences in the U3 region of the 3′ long terminal repeat (LTR) in the presence of internal transcriptional units in either orientation with respect to the vector LTRs; and (iii) inclusion of regulatory sequences that could result in lineage-specific expression in lymphoid cells. It should be noted that because the 3′ LTR serves as template

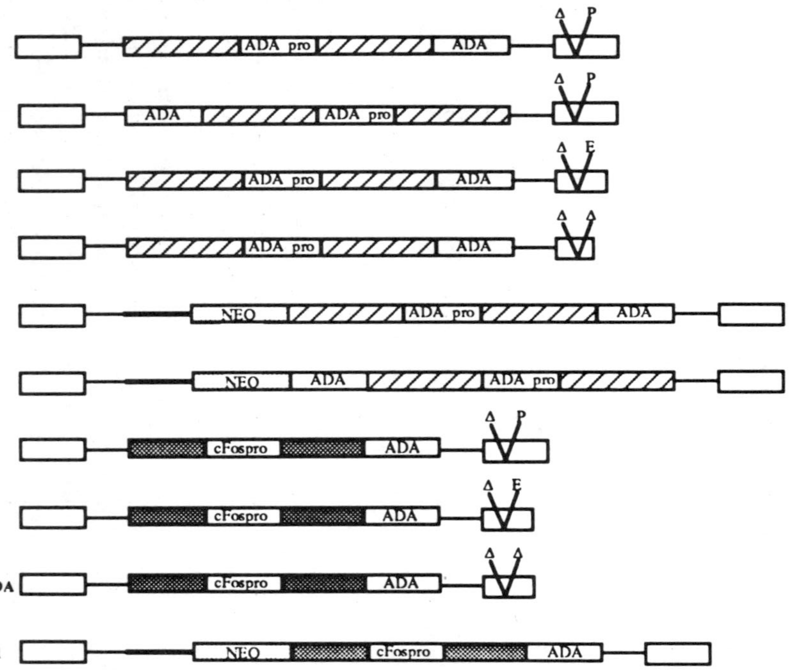

Construct	Map elements	ADA Expression		
		Transfection	Infection	Transduction
1) pADA1	ADA pro, ADA	+	-	-
2) pADA2	ADA, ADA pro	+	-	-
3) pΔPADA1	ADA pro, ADA; Δ P	+	NA	-
4) pΔPADA2	ADA, ADA pro; Δ P	+	NA	-
5) pΔEADA	ADA pro, ADA; Δ E	+	NA	-
6) pΔPΔEADA	ADA pro, ADA; Δ Δ	+	NA	-
7) pN2ADA1	NEO, ADA pro, ADA	+	+	-
8) pN2ADA2	NEO, ADA, ADA pro	+	-	-
9) pΔPFosADA	cFospro, ADA; Δ P	+	NA	-
10) pΔEFosADA	cFospro, ADA; Δ E	+	NA	-
11) pΔPΔEFosADA	cFospro, ADA; Δ Δ	+	NA	-
12) pN2FosADA1	NEO, cFospro, ADA	+	+	-

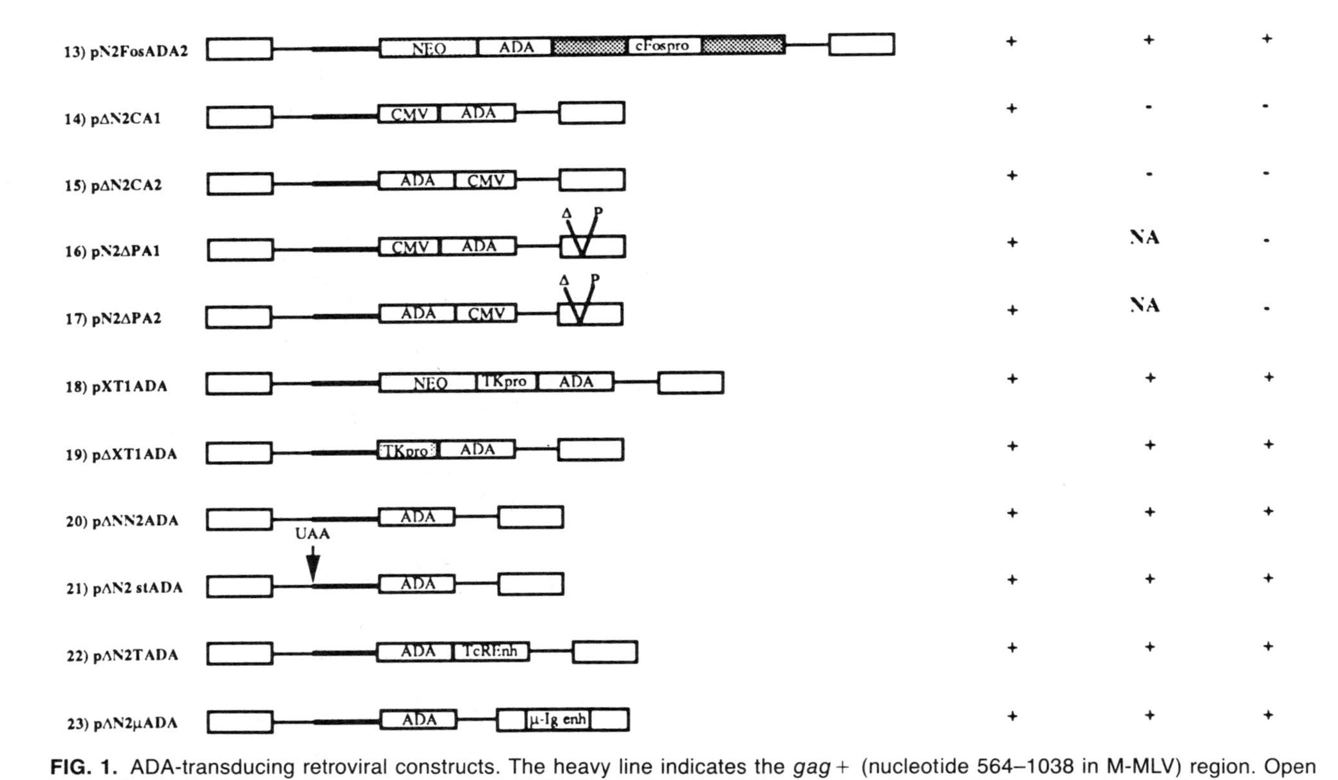

FIG. 1. ADA-transducing retroviral constructs. The heavy line indicates the *gag*+ (nucleotide 564–1038 in M-MLV) region. Open boxes denote the Moloney long terminal repeat (LTR). Plasmid vector DNAs were transfected into the packaging cell lines and virus produced from these cells was used to infect the packaging cell line of the opposite tropism. Those vectors that produced titers of virus sufficient to be detected by isoelectric focusing on unselected populations were further developed. Clonal isolates from infected populations of virus-producing cells were tested for transduction of virus to the target cell line Rat 208F (10).

for both LTRs during the process of reverse transcription, modifications in that region of the cloned vector are duplicated in the provirus that integrates into the target cell genome.

The *gag*$^{+}$ sequence has previously been shown to increase the efficiency of viral RNA packaging (11). In our experiments, only constructions that included this sequence demonstrated high titers of virus production and efficient gene transfer of human ADA. The vectors bearing alterations in the 3′ U3 region that produce self-inactivating LTRs were tested in an attempt to avoid negative regulatory effects caused by specific recognition of the LTR enhancer sequences in primitive stem cells. However, deletions of the 3′ LTR enhancer and/or promoter, even when internal promoters were included, resulted in the inefficient expression of the transduced gene. Even in the presence of an intact LTR, the inclusion of an internal promoter did not improve the level of expression in fibroblasts or in mouse hematopoietic cells. Note that, in the case of one construction with an intact LTR and an internal promoter that did perform well (pXT1ADA), more detailed studies demonstrated that the transcription was initiated predominantly from the viral regulatory sequences rather than from this internal promoter (herpes simplex thymidine kinase promoter). Finally, two vectors were evaluated for lineage-specific expression: the pΔN2TADA vector included sequences from the human T-cell receptor enhancer positioned 3′ of the hADA-cDNA (12), while in the pΔN2μADA vector the enhancer sequence of the 3′ LTR was replaced by the μ-chain immunoglobulin enhancer (13). Despite the deletion of the powerful retroviral enhancer, the pΔN2μADA construct has maintained high titers of viral production, indicating that the μ-chain immunoglobulin enhancer can substitute for this important function (K. A. Moore et al., manuscript in preparation). The resulting viruses have also proved to infect various cell types efficiently. However, comparison of expression with pΔNN2ADA in fibroblasts and B-cell lines showed similar levels with both vectors. Moreover, tissue-specific expression was not discernible *in vivo* with either pΔN2μADA or pΔN2TADA (ibid.). This result may be due to the inhibition of lymphoid-specific protein/DNA interactions in the structural context of a retroviral vector or to a dominant effect of the Moloney promoter still present in these constructs.

Thus, the most effective ADA-transducing vector identified was a relatively simple construct that includes the *gag*$^{+}$ sequence and in which the ADA cDNA is transcribed from the 5′ LTR (see Fig. 1). Other groups have reached similar conclusions regarding the relative efficiency of retroviral *vs.* heterologous regulatory sequences in the context of a retroviral construction (14). Although to date our efforts to obtain enhanced and regulated expression in lymphoid cells have been unsuccessful, we are pursuing further studies in that direction since tissue-specific expression in these cells may potentially be more effective in correcting the abnormal metabolism of ADA deficiency. One additional modification that we have recently introduced is the conversion of the natural start codon of the *gag*$^+$ sequence to a stop codon (pN2stADA vector). This decreases the risk of producing replication-competent virus should there be recombination events between the vector and other retroviral sequences. This construction did not generate replication-competent virus despite maintenance for several months in an ecotropic packaging cell line that is otherwise prone to production of helper-virus in the presence of pXM5 (N2) derivatives (15). It was also shown to have no effect on the expression of the transduced human ADA.

ADA GENE TRANSFER INTO THE HEMATOPOIETIC SYSTEM OF THE MOUSE

Evaluation of Long-Term In Vivo Expression

Our gene transfer experiments in mice have been conducted principally with the retroviral vector pΔNN2ADA (now referred to as pΔN2ADA) (see Fig. 1). These experiments were first undertaken with the vector packaged in Ψ2 cells (16). Expression of human ADA (hADA) for up to 3 months was obtained in lethally irradiated animals reconstituted with bone marrow cells that had been cocultivated with virus-producing cells (10). However, replication-competent retroviruses were present in the virus-producing cells and in the plasma of these animals. Although there was evidence that the long-term in vivo expression of hADA did not result from an ongoing propagation of the vector related to the presence of this helper-virus, additional experiments were conducted with the same vector produced in the packaging cell line GP + E-86 (17). In this cell line, the structural genes for the production of retrovirus have been split between two plasmids, increasing to three the minimal number of recombination events required to form an intact viral genome. The animals and the virus-producing cell lines in these experiments have remained free of ecotropic or amphotropic replication-competent virus (18). In addition, no other untoward effects have been observed clinically or at gross pathological examination during long-term evaluation of over 300 mice transplanted with bone marrow cells infected with replication-defective retroviruses.

In order to improve the analysis of hADA expression in mice following gene transfer, a polyclonal rabbit antibody was raised against the 15 carboxy terminal amino acids of hADA, 11 of which are absent in the mouse ADA protein (18). This antibody has also been used with success to identify cell clones producing high titers of ADA-transducing retroviral vectors.

In the initial studies with the pΔN2ADA vector introduced into GP + E-86 cells, 37 mice were transplanted with bone marrow infected with viral stocks under varied experimental conditions. The parameters evaluated were the effects of: (i) 5-fluorouracil (5-FU) pretreatment of the donor animals; and

(ii) the use of exogenous growth factors derived from conditioned media during the in vitro cocultivation period. Both of the above strategies were based on the assumption that promoting the growth of hematopoietic stem cells (HSC) would facilitate their infection with a retroviral vector. 5-FU is an antimetabolite drug that kills rapidly dividing cells and indirectly increases the fraction of the HSC that will enter the cell cycle to replenish the hematopoietic system. The conditioned media used in these experiments were derived from two cell lines: the murine myelomonocytic cell line WEHI-3B(D-) producing interkeukin-3 (IL-3) (19), and the human bladder carcinoma cell line 5637 producing numerous growth factors including IL-1α, IL-6, and granulocyte colony-stimulating factor (G-CSF) (20). IL-6, G-CSF, and, indirectly, IL-1α have been shown to act synergistically with IL-3 to promote the proliferation of mouse HSC and primitive human hematopoietic progenitors (21–23).

In these studies and in all subsequent experiments with the pΔN2ADA vector, all the mice expressed hADA in the peripheral blood for at least 9 weeks. Approximately one third of the animals expressed the human enzyme in their blood for more than 6 months. The mature hematopoietic cells that are found 6 months after transplantation are believed to be derived from primitive stem cells. Figure 2 shows an example of a Western blot demonstrating the presence of hADA in blood lysates obtained from mice 1 month after their transplantation with a genetically altered marrow. Taken together, these results suggest very efficient infection of committed erythroid progenitors and less efficient infection of the stem cells.

Mice were sacrificed after reconstitution with the donor marrow, starting 3 months after transplantation. Hematopoietic tissues were analyzed for expression of hADA and the presence of integrated provirus. Twenty five of the 37 mice (68%) expressed hADA in at least one of these tissues. Although hADA was detected in myeloid, erythroid, and lymphoid lineages, expression was variable both within and between individual mice. The polymerase chain reaction (PCR) was used to detect the presence of vector sequences. Primers were synthesized to amplify specifically the hADA cDNA encoded by the provirus and not the mouse ADA locus. Using this very sensitive technique, provirus was detected in at least two of the tissues analyzed in 81% of the 37 mice (Fig. 3). The representation of proviral sequences as determined by Southern blot analysis averaged 0.1 to 0.2 copies per cell in 22 of the 37 mice. The Southern blot analysis showed a good correlation between the presence of the provirus and the expression of hADA, as detectable by Western blot analysis. However, the Western blot proved to be more sensitive in some instances as expression of hADA could be detected in tissue samples that were negative by Southern for the presence of the provirus. This indicated that a representation of the vector sequences in less than 10% of the cells (the theoretical level of sensitivity of Southern analysis) can result in detectable levels of protein expression.

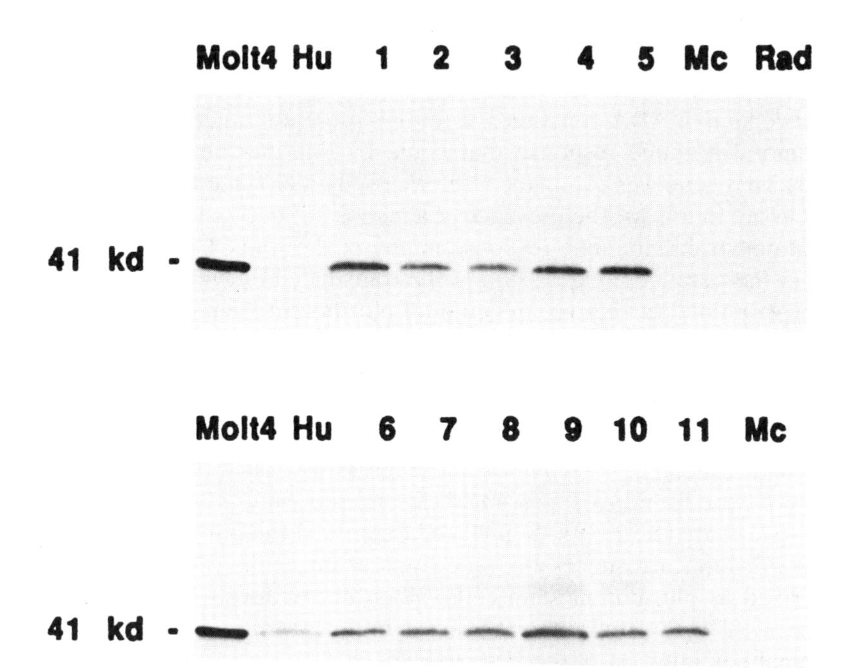

FIG. 2. Western blot of peripheral blood lysates. Mice were bled 4 weeks after transplantation with bone marrow infected by pΔN2ADA virus. Lysates (1 mg) were electrophoresed and immunoblotted with an anti-human ADA antibody as described (18). Human ADA is a 41-kd protein. Molt4 is a human leukemic T-cell line; Hu is a human blood lysate, 1–11 are lysates from transplanted mice; Mc is a nontransplant control mouse and Rad is an irradiated control mouse.

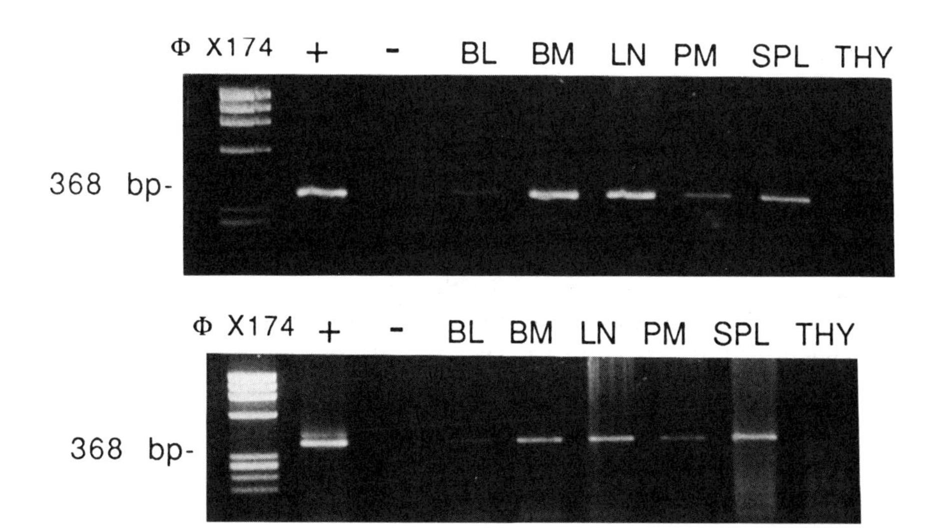

FIG. 3. PCR analysis for detection of the ADA provirus in tissues from two transplanted mice. Mice were sacrificed after reconstitution with pΔN2ADA-infected bone marrow. Genomic DNA was prepared from their hematolymphoid tissues and discrete proviral sequences were amplified as described (18). This reaction amplifies a fragment of 368 bp in presence of the provirus sequence. φ X174. *Hae*III-digested φ X174 DNA; BL, blood; BM, bone marrow; LN, lymph node; PM, peritoneal macrophages; SPL, spleen; Thy, thymus; +, ΔN2ADA cell line; −, nontransplanted mouse tissue.

Since human ADA constitutes a foreign protein in mice, the possibility of a humoral immune response that could be limiting the expression of the transferred gene was evaluated (22). We used an enzyme-linked immunosorbent assay to test for the presence of a mouse antibody to hADA. There was no demonstrable immune response in any of the transplanted mice, regardless of their status of expression of the transduced enzyme.

We examined further the infection of pluripotent stem cells by testing for their ability to repopulate secondary recipient animals and express hADA in the hematopoietic tissues and in the secondary spleen-hematopoietic colonies (CFU-S) (22). The results of these assays showed a transient and limited ability of the bone marrow cells from primary bone marrow recipients to repopulate secondary recipients and express hADA: 2 of 12 secondary recipients were found to express hADA, and this only in their bone marrow. Likewise, only 18/72 CFU-S analyzed expressed the human enzyme. These results do indicate successful gene transfer into cells that are functionally described as pluripotent stem cells, albeit at low levels. It is possible that the growth factor environment present in the conditioned media promotes terminal commitment of the HSC to a specific lineage rather than their self-renewal, thereby favoring the infection of cells that will eventually be lost from the stem cell pool.

Effects of the Leukemia Inhibitory Factor

The results of the above studies prompted us to conduct further experiments with this vector to delineate optimal in vitro cocultivation conditions that would promote cycling of the pluripotent HSC without their commitment to differentiation, thus increasing the infection efficiency of HSC and the long-term repopulating ability of the transduced marrow. We chose to study the effects of a novel cytokine, the *leukemia inhibitory factor* (LIF). LIF was first identified by its ability to induce differentiation of the murine myeloid leukemia cell line M1 (24), a property that it shares with the hematopoietic growth factors IL-6 and G-CSF (25). In addition to this differentiative effect, LIF is known to support the proliferation of the murine IL-3-dependent cell line, DA-1a (26). The apparent functional similarity of LIF with known hematopoietic growth factors, and the recent identification of unique LIF receptors on normal murine monocytes and macrophages (27), suggest a role for LIF in normal hematopoiesis. Functional identity of LIF with the murine embryonic stem (ES) cell *differentiation inhibitory factor* (DIA), which maintains their totipotentiality in culture, has also been demonstrated (28). This suggests that LIF might have a similar effect on the pluripotent HSC. We have examined the effects of recombinant LIF on: (i) the recovery and retroviral vector infection efficiency of cultured, myeloid-restricted hematopoietic stem cells (colony-forming units in the spleen at day 13, or CFU-S_{13}); and (ii) the long-term (≥6 months) in vivo expression of vector-encoded hADA in the differentiated progeny of pluripotent HSC.

The frequency of CFU-S_{13} was compared in groups of murine bone marrow cells cultured in the presence or absence of murine LIF (29). Since the recombinant LIF was derived from transfected COS cell supernatant, untransfected COS cell supernatant was included in separate control cultures. COS cell supernatants do not contain endogenous LIF activity. The recovery of CFU-S_{13} in presence of LIF or control supernatant was compared to the recovery from a control culture without any growth factor (no GF) (Fig. 4). Supplementation of cultures with COS cell supernatant did not yield a significant increase in CFU-S_{13} recovery. Cultures supplemented with 0.1U/mL and 1 U/mL LIF also yielded CFU-S_{13} recoveries similar to the

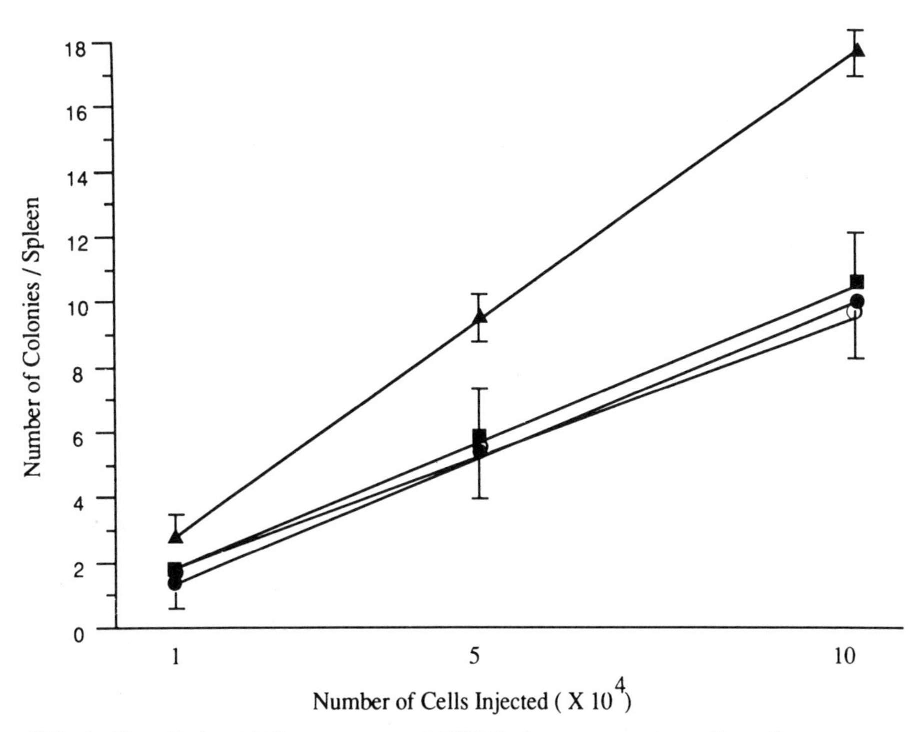

FIG. 4. The effects of LIF on recovery of CFU-S_{13} from suspension culture. Bone marrow was harvested from normal 10–12 week old FVB/N female mice and plated onto irradiated (2,500 R) monolayers of retrovirus-vector-producing fibroblasts. Irradiated (1,000 R) mice were injected IV with the indicated number of bone marrow cells cocultured as described for 72 hours in the presence of no added growth factor (closed circles)[number of spleens (n) = 30], 0.1 U/mL LIF (open circles)(n = 9), 1 U/mL LIF (closed squares)(n = 13), or 10 U/mL LIF (closed triangles)(n = 23). Each point represents the average number of macroscopic spleen colonies observed per spleen on day 13 (±SEM). (From Fletcher et al., ref. 30. Reprinted with permission from Plenum Press.)

control. A LIF concentration of 10 U/mL, however, resulted in an increase in CFU-S_{13} recovery from 9.7 per 10^5 cells to 17.8 per 10^5 cells (1.8 fold increase, $p < 0.001$).

Analysis of infection efficiency of CFU-S_{13} and determination of clonal relationships between the resulting spleen colonies provides additional evidence for the effects of LIF on these populations (29). We employed the PCR to identify infected colonies rapidly. The sensitivity of the PCR was intentionally reduced to allow discrimination between bona fide infection events of the colony-forming cell (at least a single copy of the provirus in every cell) and the presence of small numbers of provirus-bearing cells in an uninfected colony. Infection efficiencies were calculated for each experimental group (Fig. 5). The no GF control resulted in an infection efficiency of 15% (24/158), and the low volume COS supernatant control yielded similar results with 17% (18/103) infection efficiency. LIF at concentrations of 0.1 U/mL and 1 U/mL improved infection efficiency to 35% (14/40) and 73% (37/51) respectively, while a concentration of 10 U/mL increased infection efficiency to 91% (116/127) of the colonies analyzed.

Finally, quantitative analysis of hADA levels in hematopoietic tissues

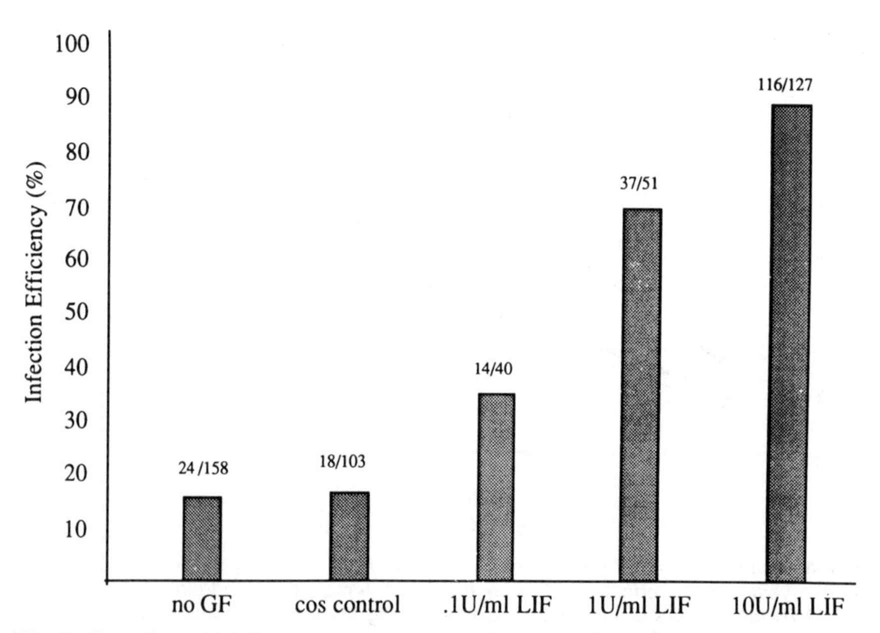

FIG. 5. The effects of LIF on *in vitro* retroviral-vector infection efficiency of CFU-S_{13}. Genomic DNA (500 ng) from individual spleen colonies was analyzed by PCR amplification, as described, for the presence of integrated provirus (29). Infection efficiency for each experimental group is reported graphically and numerically, as the number of infected colonies divided by the total number of colonies analyzed. GF, growth factor. (From Fletcher et al., ref. 30. Reprinted with permission from Plenum Press.)

from transplanted animals demonstrates that in vitro LIF stimulation of bone marrow during coculture improves long-term expression of the vector-encoded ADA in the differentiated progeny of infected stem cells (30). Expression of hADA was examined in hematopoietic tissues of recipient animals 6 months posttransplantation. Protein extracts were prepared from the spleen, thymus, bone marrow, blood, splenic B cells, and peritoneal macrophages. Western blots of these extracts were developed with the anti-hADA antibody, revealing increased levels of expression in animals receiving LIF-stimulated marrow, compared to the controls (no GF infections) (30). All tissues from the LIF animals contained higher levels of hADA than tissues derived from the control animals (Fig. 6). Our interpretation of these results is that LIF stimulation results in an increased infection efficiency of hematopoietic stem cells without affecting their long-term repopulating ability. Analysis of proviral representation in the tissues of long-term repopulated animals is proceeding to confirm this conclusion. These experiments demonstrate that LIF directly or indirectly enhances retroviral infection ef-

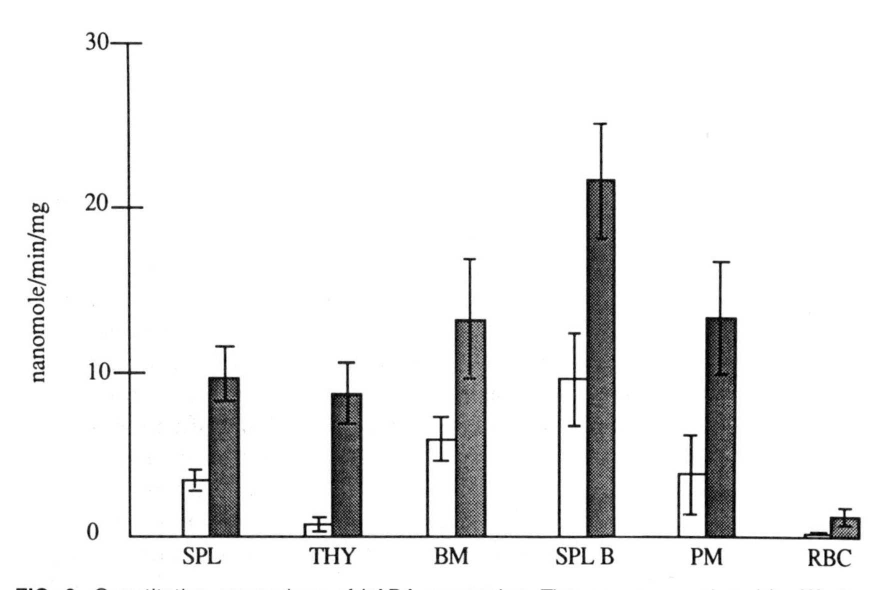

FIG. 6. Quantitative comparison of hADA expression. Tissues were analyzed by Western blotting and hADA detected with a monospecific anti-hADA antibody. The blots were scanned with a digital image analyzer and normalized to a positive control of known activity present on each blot. The level of hADA was extrapolated from a standard curve developed from purified recombinant hADA. Open bars represent tissues obtained from mice transplanted with bone marrow co-cultivated in the absence of added growth factors [number of animals (n) = 7]. Hatched bars represent tissues repopulated by LIF (10 U/mL) treated marrow (n = 10). SPL, spleen; THY, thymus; BM, bone marrow; SPL B, splenic B cells; PM, peritoneal macrophages; RBC, red blood cells. (From Fletcher et al., ref. 30. Reprinted with permission from Plenum Press.)

ficiency of hematopoietic stem cells, and might be used to improve existing gene transfer protocols.

ADA GENE TRANSFER INTO HUMAN HEMATOPOIETIC CELLS

In parallel with the experiments conducted in mice, we have evaluated the retrovirus-mediated gene transfer of human ADA in human hematopoietic cells. In order to minimize the risk of production of replication-competent virus, we have used the vector pΔN2stADA (see above) produced by GP + *env*AM-12 packaging cells (31). Similar to its ecotropic counterpart (GP + E-86) (17), this cell line has been made by separating the structural genes of M-MLV between two independent plasmids. A sensitive marker gene rescue assay for ecotropic and amphotropic retrovirus (10) and a PCR detecting the presence of transferred Moloney-like sequence in the genomic DNA of infected target cells (48) were used to monitor the potential presence of helper-virus. There was no detectable replication-competent virus production from either the virus-producing cells or from infected human cells maintained in long-term culture.

Gene Transfer into Committed Human Hematopoietic Progenitors

Aliquots of bone marrow harvest were obtained with informed consent from bone marrow transplantation donors, and light density human bone marrow cells (density $\leqslant$ 1.077) were cocultivated with irradiated virus-producing cells for 48–60 hours. A fraction of the bone marrow cells were then transferred to semisolid cultures for the growth of clonogenic hematopoietic progenitors while the remaining cells were seeded onto an irradiated human bone marrow-derived stromal layer for myeloid long-term culture. The presence of integrated viral sequence into individual hematopoietic colonies (50 to a few hundred cells) was detected by PCR amplification of crude nucleic acid extracts with vector-specific primers. Preliminary experiments demonstrated that the inclusion of recombinant human IL-3 (10 U/mL) and IL-6 (200 U/mL) during the period of cocultivation resulted in a several-fold increase in the efficiency of gene transfer into clonogenic hematopoietic progenitors (Fig. 7). These growth factors had been selected for their reported synergistic effect on the proliferation of primitive human hematopoietic progenitors (see above). The clonogenic cells that are present at the time of infection constitute committed progenitors with poor self-renewal capability and therefore do not represent our ultimate target. However, the experience with retrovirus-mediated gene transfer in the hematopoietic system of mouse suggested that conditions of infection that do not achieve high efficiency

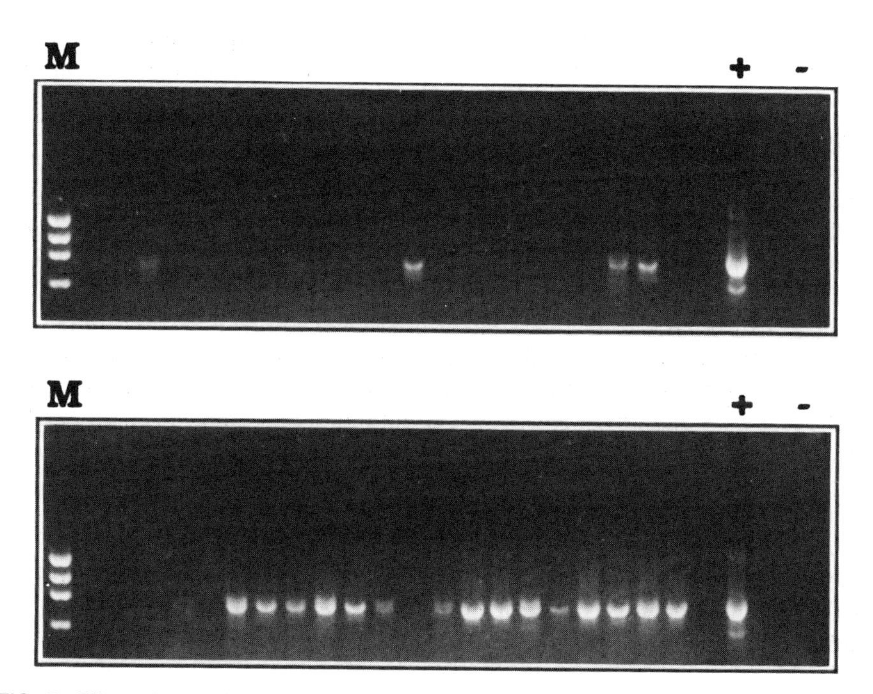

FIG. 7. Effect of recombinant human interleukin 3 (IL3) and (IL6) on retrovirus-mediated gene transfer of ADA into clonogenic human hematopoietic progenitors. Hematopoietic colonies were picked individually after 14–16 days of culture and analyzed by PCR for the presence of integrated provirus. This reaction amplifies a fragment of 787 bp in the presence of provirus sequence. Twenty colonies were analyzed following infection of human bone marrow without (*top*) or with (*bottom*) IL3 10 U/mL and IL6 200 U/mL. M, *Hae*III-digested X174 DNA; +, positive control (270 ng of genomic DNA from virus-producing cell line); –, no DNA.

gene transfer into committed progenitors were unlikely to transduce stem cells with success. In a series of seven human bone marrow infections, the gene transfer efficiency was found to be 70–100% in erythroid colonies (CFU-E) and 40–90% in granulocyte/macrophage colonies (CFU-GM) (Cournoyer et al., manuscript in preparation).

Expression of the Transduced ADA

The expression of hADA in the progeny of mouse HSC infected with the pΔNN2ADA vector gave hope that this vector (and the closely related pN2stADA) might express the transduced enzyme at high levels in human hematopoietic cells. However, this needed to be addressed directly since the level of expression from retroviral vectors can vary between cell types and between species. We have previously demonstrated expression of the transduced ADA in the supernatant cells of long-term bone marrow culture for up to 2 weeks following infection with virus-containing supernatant of bone marrow cells obtained from two ADA-deficient patients (4). However, because of the lower efficiency of gene transfer achieved with supernatant infection, it had not been possible to determine the exact subpopulation of cells in which the gene was still expressed. More recently, we have used a microradioassay (32) to compare the level of ADA activity in pooled, unselected normal CFU-E that had been infected or not with the pN2stADA vector (Cournoyer et al., manuscript in preparation). The ratio of ADA activity to the activity of another enzyme of the purine metabolism [purine nucleoside phosphorylase (PNP)] was determined to correct for the potentially unequal amounts of protein extracts obtained from the pools of hematopoietic colonies. In two separate experiments, the ADA activity was found to be increased by 67% and 79%, respectively, in normal infected vs. uninfected CFU-E. This suggests that the transduced ADA expresses at similar or slightly higher levels than each one of the two autosomal copies of the ADA gene in these cells. We are now evaluating the level of transduced ADA activity in other types of progenitors and in infected hematopoietic colonies derived from long-term bone marrow culture.

Gene Transfer into Primitive Human Hematopoietic Progenitors

One of the difficulties in demonstrating gene transfer in our ultimate target in the human (the HSC) is the lack of a perfect and definitive assay for human HSC. Currently, we employ the myeloid (Dexter) long-term bone marrow culture (LTC) system (33) to identify primitive human hematopoietic progenitors and evaluate the gene transfer efficiency in that rare population of cells. Preexisting clonogenic progenitors disappear (by death or differentiation) within 5 weeks of culture (34,35). The clonogenic cells that can be recovered at that time appear to be derived from more primitive cells, which were not in themselves clonogenic when the culture was initiated (36,37). It has been suggested that these LTC-initiating cells are functionally related to the primitive stem cells.

To our knowledge, there is only one report concerning retrovirus-mediated gene transfer into human LTC-initiating cells (38). Those experiments involved the transfer of resistance to the aminoglycoside G418 into that population of cells. One limitation that we have experienced is that present culture conditions for cocultivation appear suboptimal for the maintenance and recovery of primitive human progenitors. We are now pursuing work to correct this limitation. Nevertheless, we were recently able to show integration of the pN2stADA vector into clonogenic progenitors recovered

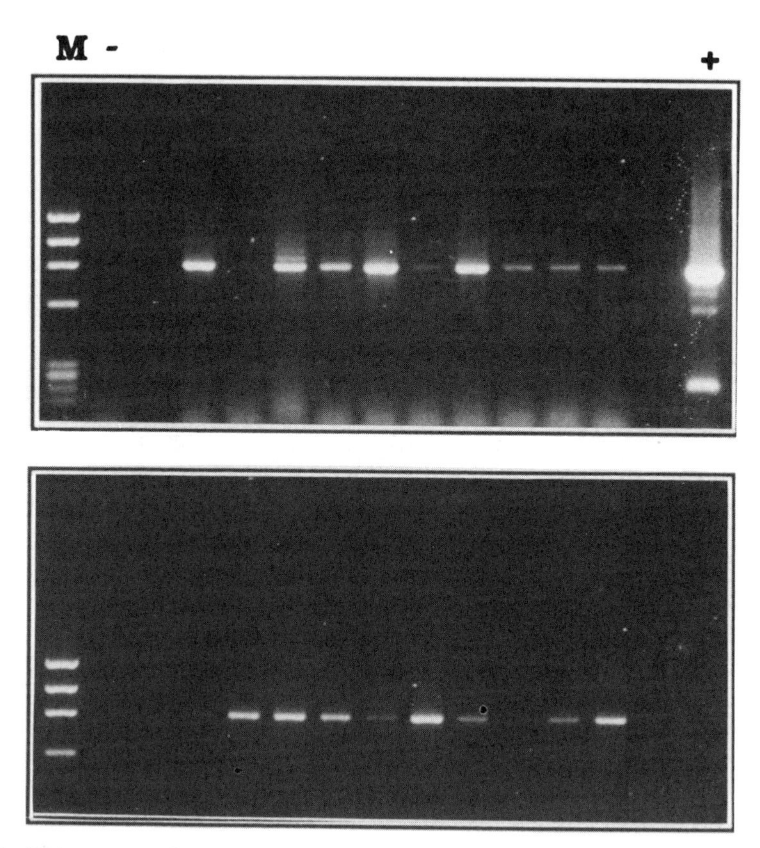

FIG. 8. ADA gene transfer into hematopoietic progenitors derived from long-term bone marrow culture 3 weeks after infection. Same analysis as Fig. 7. **Top:** Ten erythroid colonies (CFU-E) and (**bottom**) 10 granulocyte-macrophage colonies (CFU-GM) were analyzed. Same symbols as Fig. 7.

from LTC up to 9 weeks after infection (Fig. 8) (Cournoyer et al., manuscript in preparation).

We also attempted to transplant human hematopoietic cells into immunodeficient animals (39–41) to provide complementary evidence of gene transfer into primitive human hematopoietic progenitors. The system that we selected was developed by S. Kamel-Reid and J. Dick (41). It allows the engraftment of presumably primitive human hematopoietic cells into mildly irradiated triple mutant immunodeficient mice (beige, nude, and X-linked immunodeficiency), and subsequent recovery of myeloid human progenitors for extended periods. A low level of human progenitor engraftment could be detected in our initial experiment: 4% of the clonogenic progenitors recovered 4 weeks posttransplantation of uninfected human bone marrow cells

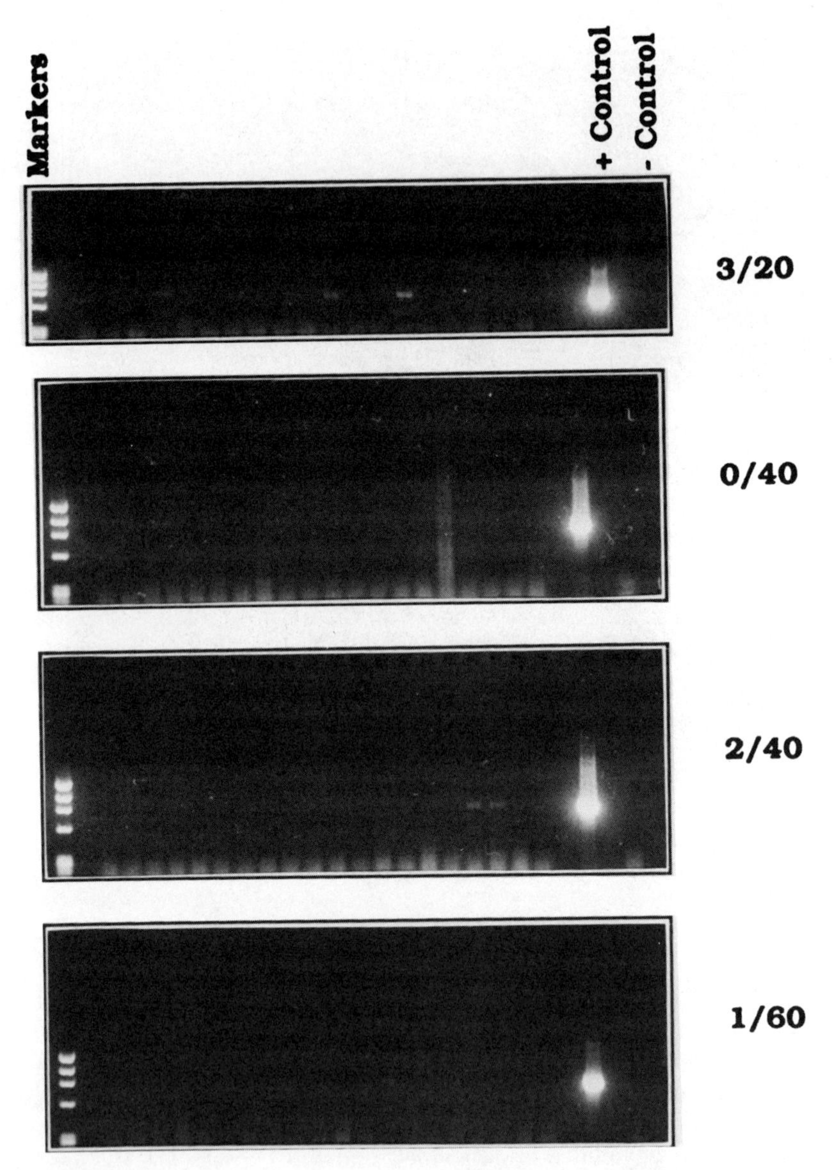

FIG. 9. Engraftment of human hematopoietic progenitors into human-immunodeficient mice. Mice were transplanted with human bone marrow cells as described (40) and were sacrificed 4 weeks later. Hematopoietic progenitors were cultivated under conditions allowing the growth of both human and mouse progenitors. Colonies were individually picked at 14–16 days and analyzed with a human-specific PCR to determine the frequency of human colonies. Each panel represents colonies analyzed from one animal. Markers, *Hae*III-digested X174 DNA; + control, human Hela cell genomic DNA 500 ng; − control, no DNA.

were found to be of human origin (Fig. 9). However, we have been unsuccessful in reproducing this result in subsequent experiments with infected or uninfected cells. Recently, it was reported that even a subclinical degree of bacterial infection of the host animals will prevent the engraftment of human hematopoietic cells, perhaps by stimulating their rejection or the expansion of the animals' own myeloid system (42). We have observed clinical signs of bacterial infection in our colony and suspect therefore that subclinical infections may have prevented the engraftment of the human cells in some of our experiments. So, although this system probably requires the maintenance of the animal colony under strictly sterile conditions that may not be widely available, it remains of promising value for the evaluation of primitive human hematopoietic cells and for other applications (42).

CONCLUSIONS

ADA deficiency is a prime candidate disease for the eventual clinical application of somatic gene therapy. Several groups (this report and 43–45) have developed efficient and safe retrovirus-mediated gene transfer systems with which to introduce vector-encoded human ADA into mouse hematopoietic stem cells. We have also obtained preliminary evidence that similar results can be reproduced in the human system (this report and D. Cournoyer et al., manuscript in preparation). The results obtained to date give hope that somatic gene therapy for ADA deficiency and other disorders of the hematopoietic system may soon become a reality.

Clinical trials involving retrovirally modified human cells are underway. Cells that had been genetically marked with a retroviral vector have recently been reinjected into cancer patients to study the role of tumor-infiltrating lymphocytes in the immunological response to cancer (46). Although this kind of experiment does not constitute a therapeutic application of the gene transfer itself, it is likely to provide valuable information for the development of gene therapy, particularly in regard to the biosafety of retroviral vectors. A formal trial of gene therapy involving retrovirus-mediated gene transfer of ADA into circulating T cells (rather than into hematopoietic stem cells) for therapeutic aims has recently been initiated by Drs. Blaese and Anderson (47). This constitutes an important landmark in the development of gene therapy. In our opinion, however, the optimal target for gene therapy in ADA deficiency remains the primitive lympho-hematopoietic stem cell with long-term repopulating ability.

ACKNOWLEDGMENTS

The authors thank the following persons for providing critical material: Drs. A. Bank and D. Markowitz from Columbia University for the packaging cell lines GP + E-86 and GP + *env*Am-12; Drs. R. H. Buckley and M. L. Markert from Duke University for bone marrow samples from ADA-deficient patients; the bone marrow transplantation team from The Methodist Hospital, Houston, Texas, for bone marrow samples from transplantation donors; Dr. D. E. Williams from Immunex R&D Corporation, Seattle, WA, for the recombinant murine growth factors; Dr. S. C. Clark from The Genetics Institute Inc., Cambridge, Massachusetts, for the human recombinant growth factors; and Dr. R. C. Mulligan, from the Massachusetts Institute of Technology and Dr. D. Givol, from the National Cancer Institute, for the plasmids containing the μ-chain immunoglobulin enhancer and T-cell receptor enhancer, respectively. We also wish to express our gratitude to our colleagues in the Caskey and Belmont laboratories who have provided technical help or advice in the conduct of those studies, and to Lillie Tanagho and Christopher Robbins for assistance in the preparation of this manuscript. This work has been supported by U.S.P.H.S. grants P01 HD21452 (C.T.C.) and R29 HD22880 (J.W.B.), Cystic Fibrosis Foundation grant MR004-9 (J.W.B.), and the Howard Hughes Medical Institute (C.T.C. and J.W.B.). D.C. was the recipient of a senior fellowship from the National Cancer Institute of Canada. M.S. was the recipient of a Cystic Fibrosis Foundation Post-doctoral Fellowship F054-9. K.A.M. was the recipient of U.S.P.H.S. Individual National Research Service Award F32 RR05034. G.R.M. was the recipient of a fellowship from the Arthritis Foundation.

REFERENCES

1. Anderson WF. Prospects for human gene therapy. *Science* 1984;226:401–409.
2. Belmont JW, Caskey CT. Developments leading to human gene therapy. In: Kucherlapati J, ed. *Gene Transfer,* New York: Plenum Publishing Corporation, 1986;411–441.
3. Friedmann T. Progress toward human gene therapy. *Science* 1989;244:1275–1281.
4. Cournoyer D, Scarpa M, Jones SN, Moore KA, Belmont JW, Caskey CT. Gene therapy: A new approach for the treatment of genetic disorders. *J Clin Pharm Ther* 1990; 47:1–11.
5. Kredich NM, Hershfield MS. Immunodeficiency diseases caused by adenosine deaminase deficiency and purine nucleoside phosphorylase deficiency. In: Scriver CR, Beaudet AL, Sly WS, Valle DM, eds. *The Metabolic Basis of Inherited Disease,* Sixth Edition, New York: McGraw-Hill, 1989;1045–1075.
6. Levy Y, Hershfield MS, Fernandez-Mejia C, et al. Adenosine deaminase deficiency with late onset of recurrent infections: Response to treatment with polyethylene glycol-modified adenosine deaminase (PEG-ADA). *J Pediatr* 1988;113:312–317.
7. Dzierzak EA, Papayonnopoulou T, Mulligan RC. Lineage-specific expression of a human β-globin gene in murine bone marrow transplant recipients reconstituted with retrovirus-transduced stem cells. *Nature* 1988;331:35–41.
8. Bender MA, Gelinas RE, Miller AD. A majority of mice show long-term expression of a human β-globin gene after retrovirus transfer into hematopoietic stem cells. *Mol Cell Biol* 1989;9:1426–1434.
9. Bodine DM, Karlsson S, Papayannopoulou, Nienhuis AW. Introduction and expression of human beta globin genes into primitive murine hematopoietic progenitor cells by retrovirus mediated gene transfer. *Prog Clin Biol Res* 1989;319:589–599.
10. Belmont JW, MacGregor GR, Wager-Smith K, et al. Expression of human adenosine deaminase in murine hematopoietic cells. *Mol Cell Biol* 1988;8:5116–5125.

11. Armentano D, Yu S-F, Kantoff PW, et al. Effect of internal viral sequences on the utility of retroviral vectors. *J Virol* 1987;61:1647–1650.
12. Luria S, Gross G, Horowitz M, Givol D. Promoter and enhancer elements in the rearranged α-chain gene of the human T cell receptor. *EMBO J* 1987;6:3307–3312.
13. Sen R, Baltimore D. Multiple nuclear factors interact with the immunoglobulin enhancer sequences. *Cell* 1986;46:705–16.
14. Hock RA, Miller AD, Osborne WR. Expression of human adenosine deaminase from various strong promoters after gene transfer into human hematopoietic cell lines. *Blood* 1989;74:876–881.
15. Keller G, Paige C, Gilboa E, Wagner EF. Expression of a foreign gene in myeloid and lymphoid cells derived from multipotent haematopoietic precursors. *Nature* 1985;318: 149–154.
16. Mann R, Mulligan RC, Baltimore D. Construction of a retrovirus packaging mutant and its use to produce helper-free defective retrovirus. *Cell* 1983;33:153–159.
17. Markowitz D, Goff S, Bank A. A safe packaging line for gene transfer: Separating viral genes on two different plasmids. *J Virol* 1988;62:1120–1124.
18. Moore KA, Fletcher FA, Villalon DK, Utter AE, Belmont JW. Human adenosine deaminase expression in mice. *Blood* 1990;75:2085–2092.
19. Bazill GW, Haynes M, Garland J, Dexter TM. Characterization and partial purification of a hemopoietic cell growth factor in WEHI-3 cell conditioned medium. *Biochem J* 1983;210:747–759.
20. McNiece IK, Kriegler AB, Quesenberry PJ. Studies on the myeloid synergistic factor from 5637: Comparison with interleukin-1 alpha. *Blood* 1989;73:919–923.
21. Ikebuchi K, Wong GG, Clark SC, et al. Interleukin 6 enhancement of interleukin 3-dependent proliferation of multipotential hemopoietic progenitors. *Proc Natl Acad Sci USA* 1987;84:9035–9039.
22. Leary AG, Ikebuchi K, Hirai Y, et al. Synergism between interleukin-6 and interleukin-3 in supporting proliferation of human hematopoietic stem cells: Comparison with interleukin-1. *Blood* 1988;71:1759–1763.
23. Ikebuchi K, Ihle JN, Hirai Y, et al. Synergistic factors for stem cell proliferation: Further studies of the target stem cells and the mechanism of stimulation by interleukin-1, interleukin-6, and granulocyte colony-stimulating factor. *Blood* 1988;72:2007–2014.
24. Hilton DJ, Nicola NA, Gough NM, Metcalf D. Resolution and purification of three distinct factors produced by Krebs ascites cells which have differentiation-inducing activity on murine myeloid leukemic cell lines. *J Biol Chem* 1988;263:9238–9443.
25. Metcalf D. Actions and interactions of G-CSF, LIF, and IL-6 on normal and leukemic murine cells. *Leukemia* 1989;3:349–355.
26. Godard A, Gascan H, Naulet J, et al. Biochemical characterization and purification of HILDA, a human lymphokine active on eosinophils and bone marrow cells. *Blood* 1988;71:1618–1623.
27. Hilton DJ, Nicola NA, Metcalf D. Specific binding of murine leukemia inhibitory factor to normal and leukemic monocytic cells. *Proc Natl Acad Sci USA* 1988;85:5971–5975.
28. Smith AG, Health JK, Donaldson DD, et al. Inhibition of pluripotential embryonic stem cell differentiation by purified polypeptides. *Nature* 1988;336:688–690.
29. Fletcher FA, Williams DE, Maliszewski C, et al. Murine leukemia inhibitory factor (LIF) enhances retroviral-vector infection efficiency of hematopoietic progenitors. *Blood* 1990;76:1098–1103.
30. Fletcher FA, Moore KA, Williams DE, et al. Effects of leukemia inhibitory factor (LIF) on gene transfer efficiency into murine hematolymphoid progenitors. In: Cooper M, Gupta S, Paul W, Rothenberg E, eds. *Proceedings of the Third International Conference on Lymphocyte Activation and Immune Regulation*. New York: Plenum Press, 1990.
31. Markowitz D, Goff S, Bank A. Construction and use of a safe and efficient amphotropic packaging cell line. *Virology* 1988;167:400–406.
32. Aitken DA, Kleijer WJ, Niermeijer MF, Herbschleb-Voogt E, Galjaard H. Prenatal detection of a probable heterozygote for ADA deficiency and severe combined immunodeficiency disease using a microradioassay. *Clin Genet* 1980;17:293–298.
33. Dexter TM, Spooncer E, Simmons P, Allen TD. Long-term marrow culture: An overview of techniques and experience. In: Wright D, Greenberg JS, eds. *Long-Term Bone Marrow Culture*. New York: Alan R. Liss, 1984;57–96.

34. Andrews RG, Takahashi M, Segal GM, et al. The L4F3 antigen is expressed by unipotent and multiponent colony-forming cells but not by their precursors. *Blood* 1986;68:1030–1035.
35. Eaves AC, Cashman JD, Gaboury LA, Kalousek DK, Eaves CJ. Unregulated proliferation of primitive chronic myeloid leukemia progenitors in the presence of normal marrow adherent cells. *Proc Natl Acad Sci USA* 1986;83:5306–5310.
36. Sutherland HJ, Eaves CJ, Eaves AC. Characterization and partial purification of human marrow cells capable of initiating long-term hemopoiesis in vitro. *Blood* 1989;74: 1563–1570.
37. Sutherland HJ, Lansdorp PM, Henkelman DH, Eaves AC, Eaves CJ. Functional characterization of individual human hematopoietic stem cells cultured at limiting dilution on supportive marrow stromal layers. *Proc Natl Acad Sci USA* 1990;87:3584–3588.
38. Hughes PFD, Eaves CJ, Hogge DE, Humphries RK. High-efficiency gene transfer to human hematopoietic cells maintained in long-term marrow culture. *Blood* 1989;74: 1915–1922.
39. Mosier DE, Gulizia RJ, Baird SM, Wilson DB. Transfer of a functional human immune system to mice with severe combined immunodeficiency. *Nature* 1988;335:256–259.
40. McCune JM, Namikawa R, Kaneshima H, et al. The SCID-hu mouse: Murine model for the analysis of human hematolymphoid differentiation and function. *Science* 1988;241: 1632–1583.
41. Kamel-Reid S, Dick JE. Engraftment of immune-deficient mice with human hematopoietic stem cells. *Science* 1989;242:1706–1709.
42. Kamel-Reid S, Letart M, Sirard C, et al. A model of human acute lymphoblastic leukemia in immune-deficient SCID mice. *Science* 1989;246:1597–1600.
43. Lim B, Apperley JF, Orkin SH, Williams DA. Long term expression of human adenosine deaminase in mice transplanted with retrovirus-infected hematopoietic stem cells. *Proc Natl Acad Sci USA* 1989;86:8892–8896.
44. Wilson JM, Danos O, Grossman M, Raulet DH, Mulligan RC. Expression of human adenosine deaminase in mice reconstituted with retrovirus-transduced hematopoietic stem cells. *Proc Natl Acad Sci USA* 1990;87:439–443.
45. Kaleko M, Garcia JV, Osborne WRA, Miller AD. Expression of human adenosine deaminase in mice after transplantation of genetically modified bone marrow. *Blood* 1990;75:1733–1741.
46. Kasid A, Morecki S, Aebersold P, et al. Human gene transfer: Characterization of human tumor-infiltrating lymphocytes as vehicles for retroviral-mediated gene transfer in man. *Proc Natl Acad Sci USA* 1990;87:473–477.
47. Culliton BJ. Gene therapy clears first hurdle (news and comments). *Science* 1990; 247:1287.
48. Scarpa M, Cournoyer D, Muzny DM, et al. Characterization of recombinant helper retroviruses from Moloney-based vectors in ecotropic and amphotropic packaging cell lines. *Virology,* in press.

Etiology of Human Disease at the DNA Level,
edited by Jan Lindsten and Ulf Pettersson.

15

Altering Genes in Animals and Humans

Oliver Smithies

Department of Pathology, University of North Carolina at Chapel Hill, Chapel Hill, North Carolina 27599-7525

Rudolf Jaenisch has already discussed some of the fascinating and surprising results that follow from the use of gene targeting to inactivate the β_2-microglobulin gene in the mouse germline (see Chapter 13). Mario Capecchi will be presenting the equally fascinating results of his experiments leading to inactivation of other genes in the mouse germline (see Chapter 16). The work in my laboratory on targeting genes in the mouse germline overlaps to some degree that of both these authors, and so I shall not describe these germline targeting results in detail. Instead, I want to review changes that have taken place over the last several years in my views concerning the potential usefulness of gene targeting in relation to human gene replacement, with special emphasis on globin genes.

My entry into the field took place a little over 8 years ago, on April 22, 1982, when I stopped the work I was doing on DNA replication and began to work on targeting a specific gene. The title of this section of the book, *Gene Replacement,* is very like the title of the relevant page in my notebook; I called it "Gene Placement," and my aim was "to place corrective DNA *in the right place*." I noted that to accomplish this I would need an "assay for developing techniques." The motivation of this proposed work is clear from the choice of the human β-globin gene cluster as a target. I was hoping that gene targeting in bone marrow stem cells might one day be valuable for the treatment of patients with hemoglobinopathies such as sickle cell anemia or thalassemia. Indeed, at that time I often commented on the sadness of having available *in the test tube* unlimited quantities of cloned β-globin gene DNA from normal individuals without having any way of using this normal DNA to correct the defective genes that were out there *in people*.

The rationale I proposed to use to direct normal DNA to the correct place is the same as that used by Hinnen et al. (1) in their yeast targeting experiments: the incoming DNA contains a region having the same nucleotide sequence as a region in the target gene, thus allowing homologous recombination between them.

Two other features that were part of this initial scheme have continued to be important in many subsequent targeting experiments, both by myself and by others. I proposed to incorporate a selectable helper gene into the incoming DNA [in my 1982 scheme it was to have been a thymidine kinase (TK) gene]. The idea, of course, was that cells not taking up the targeting DNA somewhere into their genomes where the helper gene could be expressed would be killed by using selective media. Considerable enrichment for DNA uptake should therefore be possible. A bacterial *supF* gene was also to be included in the incoming DNA, for use in identifying cells in which targeting had been accomplished. Specifically, a novel but predictable DNA fragment, diagnostic of targeting, would be generated if and only if the *supF* gene was placed next to the complete β-globin gene by the targeting event. In my first experiments this diagnostic fragment was to be detected by a complex and cumbersome assay, which depended on rescuing the diagnostic fragment in amber-containing bacteriophages (*supF* suppresses amber mutations).

Figure 1 shows the targeting event as we actually accomplished it 3 years

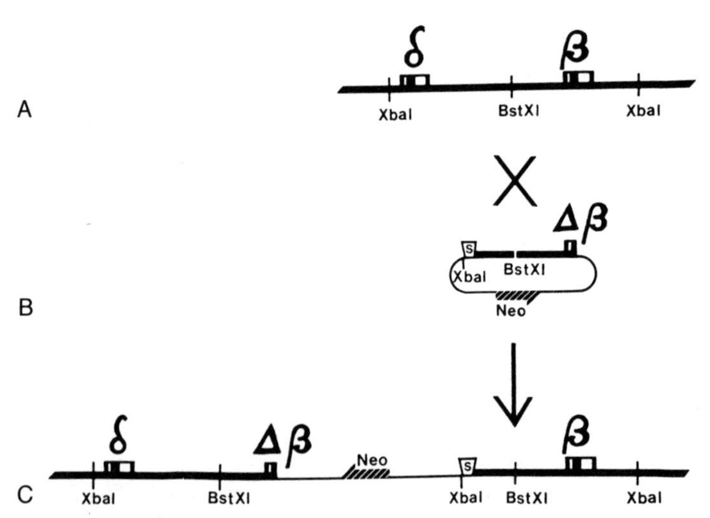

FIG. 1. Insertion of DNA into the human chromosomal β-globin locus by homologous recombination (gene targeting) **A:** Map of the unmodified human β-globin locus. Exons and introns are indicated, respectively, by solid and open bars. The three restriction sites shown are important for analyzing the results. **B:** The plasmid introduced into the cell in order to modify the target locus by homologous recombination. Note the presence of human sequences (*heavy line*) identical to sequences in the target locus, and bacterial sequences (*light line*), including a selectable helper gene, *Neo,* and a bacterial *supF* gene, S. S serves both as a marker for the incoming DNA and as a means of cloning the DNA fragment diagnostic of targeting (see text). The β-globin gene in the plasmid is *incomplete* (Δβ). **C:** Map of the locus after modification by targeting. Note that S is now next to a *complete* β-globin gene so that a novel fragment, defined by the right pair of *Xba*I sites, has been generated by the targeting. See Smithies et al. (2) for more details.

later (2). The successful experiment was carried out in mouse erythroleukemia (MEL) cells that expressed a human adult β-globin gene carried on a human chromosome 11 that had already been introduced into this particular MEL cell line.

Several things were learned from this original targeting experiment. First, gene targeting of a native chromosomal gene was possible. However, the frequency was very low (only one targeted cell was isolated out of 10^7 treated cells).

Second, a selectable helper gene included in the targeting DNA could be used for enrichment, provided that the target locus was able to support the expression of the helper gene. The helper gene used by us (an SV-40-driven *neo* gene) did not appear to work when the target locus was not expressed. This principle has subsequently been used extensively, particularly in conjunction with Dr. Capecchi's helper gene pMC1Neo (3), which has a promoter and enhancer allowing its expression in chromosomal sites in embryonic stem cells that do not support expression of SV-40-driven genes.

Third, recombinants could be identified by detecting the joining of a marker specific for the incoming DNA (*supF* in this case) with a marker specific for the target locus (the second β-globin intron in this case). This principle, in conjunction with the polymerase chain reaction (PCR), has subsequently been used extensively to detect targeted recombinants.

Fourthly, the gene targeting, which was carried out by introducing the exogenous DNA into the cells by electroporation, was achieved without the introduction of foreign sequences into other places in the genome (the corrected cells proved to have the incoming DNA only at the predicted location). As judged by restriction enzyme analysis, the targeting was absolutely precise. Subsequently, we (4) and others (5) have seen some targeting events accompanied by unexpected nearby deletions. Nevertheless, the majority of targetings continue to be precise.

In relation to my original goal of using gene targeting for gene therapy, these several results were partly encouraging and partly discouraging. The general precision of the process, and the absence of any detectable secondary alterations in other parts of the genome, were on the positive site of the balance sheet. The low frequency of targeting was negative. The ability to use a selectable helper gene was somewhat neutral; it might prove to be helpful, but at present we have no way of knowing whether or not a helper gene can be successfully used for targeting globin genes in bone marrow stem cells. The principle that correctly targeted cells can be identified by means of a diagnostic DNA fragment is likely to continue to be valuable, for it means that recombinants at any locus can be detected whether or not the targeted change leads to a detectable phenotype. In fact, as I have already indicated, many current targeting schemes depend on this principle to identify targeted cells, but the diagnostic fragment is no longer detected by the horrendous phage assay. It is now detected by PCR, in the manner demonstrated by Hyung-Suk Kim and myself (6) and later used to good effect by others.

One thing was made painfully obvious by these experiments using the phage assay—if the efficiency of the overall procedure was ever to be increased to the degree needed to make it useful for therapy, an easier system was needed in which to carry out development work. Most beneficial would be a selective system in which *only* targeted cells would survive, so that frequencies could be measured by simple colony counting. To this end, Ron Gregg and I (7) chose the *hprt* (hypoxanthine phosphoribosyl transferase) gene.

The first experiments we attempted with the *hprt* gene were gene inactivations, which will cause a cell to become 6-thioguanine (6-TG)-resistant. Since *spontaneous* mutations leading to lack of *hprt* function (and so to the acquisition of resistance to 6-TG) can also occur, we used an SV-40-driven *neo* gene as a selectable helper gene. The simultaneous loss of *hprt* function and gain of *neo* function was expected to allow easy isolation of targeted recombinants by their ability to survive in the presence of both 6-TG and G418. Despite over 2 years of work, we failed to detect targeting in the cells we used (mainly fibroblasts). This was a particularly low point in my continuously evolving estimate of the potential usefulness of targeting for gene therapy!

Two unrelated changes allowed us to proceed. First, we decided that, if all these targeting experiments were going to continue to be so difficult, we had better aim at a worthwhile *product,* not at just measuring frequencies. We therefore decided to switch our efforts from gene targeting in fibroblasts and MEL cells to embryonic stem (ES) cells of the type described by Evans and Kaufman (8). In this way, even if it took 2 years to do an experiment, we could hope to get a useful return on our investment of time—an animal having the altered gene. Second, we chose to *correct* a mutant *hprt* gene already isolated in Edinburgh by Hooper et al. (9) in ES cells. This particular mutation was especially valuable because it was caused by a deletion, and therefore gave no background of revertant *hprt*-positive cells, and because it was in an ES cell line that Hooper et al. (9) had already proved was able to contribute to the germline.

In fact, the experiment, which is illustrated in Fig. 2, did not take 2 years; Tom Doetschman in my laboratory, using a targetting plasmid designed and constructed by Nobuyo Maeda, had it working within a few months (10). Indeed a repeat of the experiment by Thompson et al. (5), our Scottish collaborators, led to the first demonstration that a targeted gene could be transmitted through the germline, although an unanticipated deletion occurred during or after the targeting event in the ES cell clone they used. Bev Koller in my laboratory demonstrated that the *hprt* gene we had corrected (10) was transmitted to subsequent generations exactly as planned (11). Some of the

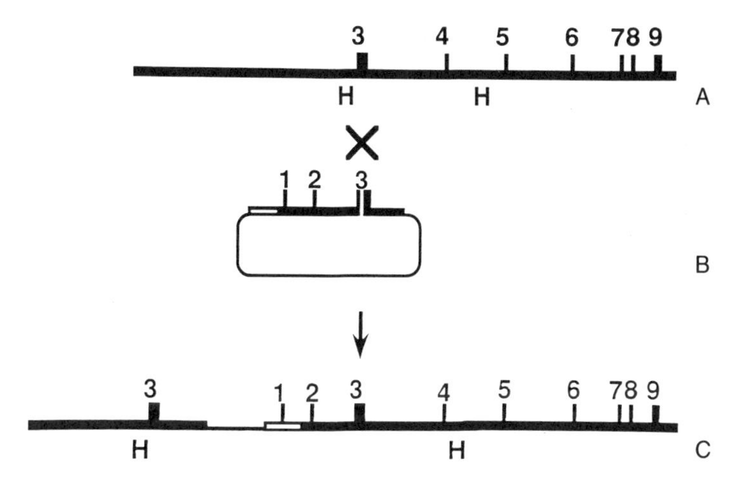

FIG. 2. Correction of a defective *hprt* gene by targeting. **A:** Map of the starting deletion-containing chromosome in which exons 3 through 9 of the *hprt* gene are present but exons 1 and 2 are missing. Two *Hind*III sites are shown by the letters H. **B:** targeting plasmid that contains sequences homologous to the target together with the missing two exons. **C:** Map of the corrected *hprt* gene, which after targeting now has all nine exons in an arrangement allowing the gene to function. Note that the two *Hind*III sites are now more widely separated. See Doetschman et al. (10) for more details.

resulting germline-competent chimeras are illustrated in Fig. 3 (see color plate).

These demonstrations that gene targeting can be used to correct a defective gene in a predictable and permanent way are obviously on the positive side of our gene therapy balance sheet. On the negative side the frequency of targeting remained low, although it had improved somewhat, perhaps because ES cells target more readily than the fibroblasts that Ron Gregg had been using (we were obtaining about one targeting event per million electroporated ES cells). Recently my student Laura Reid has made a targeting construct that corrects the deletion-containing *hprt* gene at a frequency ten times better yet. But this frequency is still only one event per hundred thousand treated cells.

Mario Capecchi and his associates have been very successful in using elegant selection schemes for making it easier to *isolate* correctly targeted cells (see Chapter 16). In particular, the positive-negative selection scheme of Mansour et al. (12) allows a helper gene driven by a powerful promoter to be used for the isolation of targeted cells without being swamped by the vast numbers of cells that are not targeted but can express the selectable helper gene. However, no really dramatic changes in the frequency of targeting, *as measured by targeting events per treated cell,* have yet been reported, except for an unconfirmed report by Zimmer and Gruss (13) that frequencies can be greatly increased when targeting DNA is introduced into the nuclei of individual cells by microinjection.

Despite these somewhat negative statements, *the exciting positive fact is that altering genes in the mouse germline by targeting in ES cells is now becoming relatively routine*. In my laboratory, we have corrected an *hprt* gene in the mouse germline, as I have already described (10,11). We have also inactivated one, the β_2-*microglobulin* gene, in the manner illustrated in Fig. 4 and described by Bev Koller and myself (14). Like Dr. Jaenisch, we have studied mice homozygous for the inactivation of this gene, and have observed the same fascinating phenotype (15): normal development, gross deficiency of major histocompatibility complex (MHC) class I proteins and of $CD4^-CD8^+$ T cells (the cytotoxic T cells). Yet the animals are healthy and fertile!

I know of published accounts of two other different genes, c-*abl* and *IGF-II*, that have now been inactivated by gene targeting in ES cells and returned to the germline. Nine more have reached the germline but not yet the press, and at least the same number of genes have been targeted, but their germline transmission is still under investigation (see Table 1).

All but two of the more than two score genes that have been targeted in

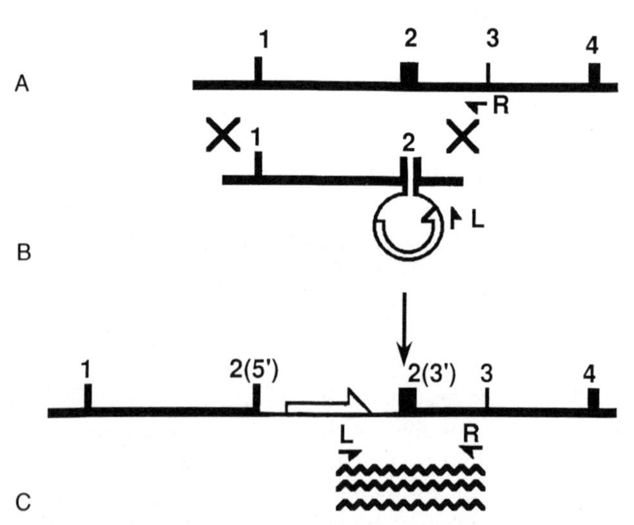

FIG. 4. Inactivation of a β_2-microglobulin gene by targeting. **A:** Map of the normal β_2-microglobulin gene showing its four exons. **B:** DNA used to modify the gene by targeting. The *open arrow* represents a selectable helper gene, Neo. **C:** Map of the modified locus in which exon 2 of the β_2-microglobulin gene has now been disrupted. The wavy lines represent a diagnostic fragment of DNA that can be amplified by PCR with the two primers (*small arrows*, L and R) only after targeting is successful. See Koller and Smithies (14) for more details.

FIG. 3. Four fertile chimeric male mice derived from embryonic stem cells containing the *hprt* gene corrected as illustrated in Fig. 2. The middle two mice transmitted the ES cell genome containing the corrected gene to some of their offspring. See Koller et al. (11) for more details.

TABLE 1. *Gene targeting in embryonic stem (ES) cells and subsequent transmission in animals*

Gene	Expression in ES cells[a]	Targeting procedure[b]	Trans-mission[c]	Reference
c-*abl*	E	I, Ω, 5′-neo	G	Schwartzberg et al. (22)
adipocyte P2	U	I, Ω, neo, tk−	nr	Johnson et al. (23)
adipsin	U	I, Ω, neo, tk−	nr	Johnson et al. (23)
en-2	VL	I, Ω, neo3′−, PCR	C	Joyner et al. (24)
c-*fos*	L	I, Ω, neo, tk−	nr	Johnson et al. (23)
*hox*1.1	nr	I, Ω, M, PCR	C	Zimmer & Gruss (13)
hprt	E	I, O, neo, hprt−	nr	Thomas & Capecchi (3)
hprt	E	I, Ω, neo, hprt−	nr	Thomas & Capecchi (3)
hprt	E	I, Ω, hprt−	nr	Doetschman et al. (4)
hprt	E	C, O, hprt+	G	Doetschman et al. (10) Koller et al. (11) Thompson et al. (5)
IGF-II	U	I, Ω, neo, tk−	G	DeChiara et al. (25)
int-2	VL	I, Ω, neo, tk−	nr	Mansour et al. (12)
β_2*m*	?N	I, Ω, neo3′−, PCR	G	Zijlstra et al. (26,27)
β_2*m*	?N	I, Ω, neo, PCR	G	Koller et al. (14,15)
N-*myc*	E	I, Ω, 5′-neo	C	Charron et al. (28)
*RPII*215	E	S, Ω, ama^r	nr	Steeg et al. (29)

[a]Expression levels indicated by the investigators are: E, expressed; L, low expression; VL, very low expression; ?N, questionably not expressed; U, undetectable expression; nr, not reported.

[b]Targeting procedures: I, inactivation; C, correction; S, substitution; O, O-type vector; Ω, Ω-type vector; neo, *neo* helper gene; 5′-neo or neo 3′, incomplete *neo* helper gene; hprt+ or hprt-, selection for or against hprt function; ama^r, selection for resistance to amanitin; M, microinjection used; PCR, PCR used to detect recombinants.

[c]Transmission: G, germline transmission; C, chimeras obtained; nr, not reported.

ES cells have been *inactivated*. Two (*hprt* and *RPII215*) have been *changed* by targeting, but not inactivated. There are two reasons for the preponderance of inactivations. First, at present it is much easier to inactivate a gene by targeting then it is to change it in a small way. Secondly, null mutations have traditionally been the most rewarding for elucidating gene function. Most gene therapy usages, in contrast, will not require gene inactivation. Some will need to correct partial or complete absence of gene function (as in the thalassemias); others will need to correct misfunction (as in sickle cell anemia). This will require the use of targeting procedures to change a gene in ways more subtle than simple inactivation. The necessary techniques for making such changes are being developed.

At this point, several questions arise: "Is it likely that gene targeting will *ever* make a contribution to gene therapy (other than the indirect one of supplying animal models of human genetic disorders)?" "What additional problems will need to be overcome before it can make such a contribution?"

The best answer that I can give to the first question is that it depends on what progress is made in other modes of therapy. Take, for example, thalassemia. Relatively long-term cure of the disease is achievable by means of bone marrow transplantation, as I expect we shall hear from Dr. O'Reilly. Prior to grafting, the patients are usually treated with high doses of myelosuppressive drugs, and long-term graft survival depends greatly on how close an immunological match is possible between donor and recipient. This in turn influences the extent to which long-term immunosuppressive treatments are required. The benefits are great, but the risks are not trivial. Yet the rate at which immunologists and cell biologists are discovering cytokines, and uncovering their negative as well as their positive influences on the immune response, makes me optimistic that graft acceptance will be more universally possible within the next decade or two than it is now. I'll retire happily from the business of gene therapy for bone-marrow-related diseases if this happens! In the interim, I intend to continue working towards finding ways in which gene targeting might contribute to gene therapy.

In fact, I am now more optimistic than I was. I have seen with my own eyes how as few as 10 targeted cells are needed to generate a living chimeric mouse able to transmit the ES cell genome to his future offspring. I know, from the work of Dr. Weissman and his colleagues (17), that 30 purified bone marrow stem cells can reconstitute the hemopoietic system of a lethally irradiated mouse. I also know from the work of Boggs et al. (18) that in mice with genetically handicapped stem cells, such as W/W^V, *one* functioning stem cell is statistically sufficient to provide a long, although possibly finite, period of cure (75% of animals were still cured after 2 years). A human child might need only 10^4 targeted bone marrow stem cells. Conceivably they could all be derived from a single targeting event! All of this is certainly neither proven nor possible at present. But I am able to define several independent problems whose solution would put us much closer to making it possible.

The first problem is the need to have a supply of purified human bone marrow stem cells. Current successes in purifying mouse bone marrow stem cells suggest that we can be optimistic that human stem cells will be purified soon.

The second problem is that we need to know how to get stem cells to proliferate clonally in vitro. This desirable goal has so far resisted long-term investments of effort by many outstanding investigators, but new approaches to its solution are on the horizon. For example, one of the best studied mutations in mouse that affect bone marrow stem cell proliferation, dominant white spotting (W), reviewed by Russell (19), has recently been shown by Chabot et al. (20) and by Geissler et al. (21) to be caused by a deficiency in a membrane-spanning growth factor receptor (c-*kit*). Isolation of the corresponding growth factor might well give us a valuable reagent for growing bone marrow stem cells in tissue culture.

A third problem is our lack of knowledge concerning the feasibility of efficiently carrying out gene targeting in bone marrow stem cells. Will they be like Ron Gregg's fibroblasts, or Tom Doetschman's ES cells? The magnitude of this problem will be difficult to assess until the first two are solved.

A fourth problem concerns providing stem cells corrected by gene targeting with an opportunity to populate the bone marrow of a treated individual. Currently "space" is made available by prior treatment with cytotoxic drugs. I would hope that gentler methods will be available before long—such as monoclonal antibodies able to kill endogenous stem cells specifically. Or perhaps it will be possible to target into the corrected stem cells a gene that will give them sufficient proliferative advantage to make prior space-making treatment unnecessary.

Over the last 8 years we have seen progress from a complete lack of ability to target *any* genes to our current position of being able to alter virtually any gene in the mouse germline via treatment of embryonic stem cells—a truly exciting and enjoyable time to have been doing science. In the next 8 years we may see comparable progress in our ability to alter genes in stem cells derived from the human bone marrow. While I feel confident in my prediction that this ability will *eventually* be acquired, I am less confident in my assessment of the time scale. I could be wrong—in either direction!

ACKNOWLEDGMENTS

The work of my laboratory is supported by grants HL37001 and GM20069 from the National Institutes of Health.

REFERENCES

1. Hinnen A, Hicks JB, Fink GR. Transformation of yeast. *Proc Natl Acad Sci USA* 1978;75:1929–1933.
2. Smithies O, Gregg RG, Boggs SS, Koralewski MA, Kucherlapati RS. Insertion of DNA sequences into the human chromosomal β-globin locus by homologous recombination. *Nature* 1985;317:230–234.
3. Thomas KR, Capecchi MR. Site-directed mutagenesis by gene targeting in mouse embryo-derived stem cells. *Cell* 1987;51:503–512.
4. Doetschman T, Maeda N, Smithies O. Targeted mutation of the HPRT gene in mouse embryonic stem cells. *Proc Natl Acad Sci* 1988;85:8583–8587.
5. Thompson S, Clarke AR, Pow AM, Hooper ML, Melton DW. Germ line transmission and expression of a corrected HPRT gene produced by gene targeting in embryonic stem cells. *Cell* 1989;56:313–321.
6. Kim H-S, Smithies O. Recombinant fragment assay for gene targetting based on the polymerase chain reaction. *Nucleic Acids Res* 1988;16:8887–8903.
7. Gregg RG, Smithies O. Targeted modification of human chromosomal genes. *Cold Spring Harbor Symp. Quant. Biol.* 1986;LI:1093–1099.
8. Evans MJ, Kaufman MH. Establishment in culture of pluripotential cells from mouse embryos. *Nature* 1981;292:154–156.
9. Hooper M, Hardy K, Handyside A, Hunter S, Monk M. HPRT-deficient (Lesch-Nyhan) mouse embryos derived from germline colonization by cultured cells. *Nature* 1987; 326:292–295.

10. Doetschman T, Gregg RG, Maeda N, et al. Targetted correction of a mutant HPRT gene in mouse embryonic stem cells. *Nature* 1987;330:576–578.
11. Koller BH, Hagemann LJ, Doetschman T, et al. Germ-line transmission of a planned alteration made in a hypoxanthine phosphoribosyltransferase gene by homologous recombination in embryonic stem cells. *Proc Natl Acad Sci* 1989;86:8927–8931.
12. Mansour SL, Thomas KR, Capecchi MR. Disruption of the proto-oncogene int-2 in mouse embryo-derived stem cells: a general strategy for targeting mutations to nonselectable genes. *Nature* 1988;336:348–352.
13. Zimmer A, Gruss P. Production of chimaeric mice containing embryonic stem (ES) cells carrying a homeobox Hox1.1 allele mutated by homologous recombination. *Nature* 1989;338:150–153.
14. Koller BH, Smithies O. Inactivating the β2-microglobulin locus in mouse embryonic stem cells by homologous recombination. *Proc Natl Acad Sci* 1989;86:8932–8935.
15. Koller BH, Marrack P, Kappler JW, Smithies O. Normal development of mice deficient in β_2M, MHC class I proteins and CD8$^+$ T cells. *Science* 1990;248:1227–1230.
17. Spangrude GJ, Heimfeld S, Weissman IL. Purification and characterization of mouse hematopoietic stem cells. *Science* 1988;241:58–62.
18. Boggs DR, Boggs SS, Saxe DP, Gress LA, Canfield DR. Hematopoietic stem cells with high proliferative potential. *J Clin Invest* 1982;70:242–253.
19. Russell ES. Hereditary anemias of the mouse: a review for geneticists. *Adv Genet* 1979;20:357–459.
20. Chabot B, Stephenson DA, Chapman VM, Besmer P, Bernstein A. The proto-oncogene c-*kit* encoding a transmembrane tyrosine kinase receptor maps to the mouse W locus. *Nature* 1988;335:88–89.
21. Geissler EN, Ryan MA, Housman D. The dominant-white spotting (W) locus of the mouse encodes the c-*kit* protooncogene. *Cell* 1988;55:185–192.
22. Schwartzberg PL, Goff SP, Robertson EJ. Germ-line transmission of a c-abl mutation produced by targetted gene disruption in ES cells. *Science* 1989;246:799–803.
23. Johnson RS, Sheng M, Greenberg ME, et al. Targetting of nonexpressed genes in embryonic stem cells via homologous recombination. *Science* 1989;245:1234–1236.
24. Joyner AL, Skarnes WC, Rossant J. Production of a mutation in mouse En-2 gene by homologous recombination in embryonic stem cells. *Nature* 1989;338:153–156.
25. DeChiara TM, Efstratiadis A, Robertson EJ. A growth-deficiency phenotype in heterozygous mice carrying an insulin-like growth factor II gene disrupted by targeting. *Nature* 1990;345:78–80.
26. Zijlstra M, Li E, Sajjadi F, Subramani S, Jaenisch R. Germ-line transmission of a disrupted β_2-microglobulin gene produced by homologous recombination in embryonic stem cells. *Nature* 1989;342:435–438.
27. Zijlstra M, Bix M, Simister NE, Loring JM, Raulet DH, Jaenisch R. β_2-microglobulin deficient mice lack CD4-8+ cytolytic T cells. *Nature* 1990;334:742–746.
28. Charron J, Malynn BA, Robertson EJ, Goff SP, Alt FW. High-frequency disruption of the N-myc gene in embryonic stem and pre-B cell lines by homologous recombination. *Mol Cell Biol* 1990;10:1799–1804.
29. Steeg CM, Ellis J, Bernstein A. Introduction of specific point mutations into RNA polymerase II by gene targeting in mouse embryonic stem cells: evidence for a DNA mismatch repair mechanism. *Proc Natl Acad Sci USA* 1990;87:4680–4684.

Etiology of Human Disease at the DNA Level,
edited by Jan Lindsten and Ulf Pettersson.

16

Creating Mice with Specific Mutations by Gene Targeting

Mario R. Capecchi

Howard Hughes Medical Institute, Department of Human Genetics and Biology, University of Utah, Salt Lake City, Utah 84112

It is no longer necessary to rely on serendipity to reveal mutations of the mouse. Mice of virtually any desired genotype can now be created using gene targeting in mouse embryo-derived stem (ES) cells (for a recent review see Capecchi, 1989a,b). First, the desired mutation is introduced by standard recombinant DNA techniques into a cloned genomic fragment of the chosen locus. Second, this DNA, the targeting vector, is introduced into a pluripotent stem cell line (ES cells). Homologous recombination between the newly introduced DNA and the cognate, chromosomal DNA sequence transfers the mutation, created in a test tube, to the genome of the recipient cell. The ES cells harboring the mutation are microinjected into mouse blastocysts and then surgically transferred into foster mothers in order to generate germ-line chimeras. Finally, interbreeding of heterozygous siblings is used to generate mice homozygous for the desired mutation.

In this approach to mouse genetics, the investigator chooses both which gene to mutate and how to mutate it. The choice of gene(s) allows the investigator to focus on specific biological problems. By creating different types of mutant alleles a comprehensive functional analysis of the chosen gene is feasible. Furthermore, complex interactions of genes underlying a chosen biological phenomenon may be dissected through the analysis of epistatic relationships among these genes. For the first time, the potential exists for subjecting the mouse to an intensive genetic analysis.

HYPOXANTHINE PHOSPHORIBOSYL TRANSFERASE

Although mammalian cells mediate recombination between homologous DNA sequences, they demonstrate an even greater propensity for mediating nonhomologous recombination. Thus the problem is to identify rare homologous recombination events in a vast arena of scattered nonhomologous recombination events. An ideal system to attack this problem is the hypoxanthine phosphoribosyl transferase (*hprt*) gene because targeted disruptions of this gene are directly selectable. Since it resides on the X-chromosome, only one mutant copy is needed to yield the recessive *hprt*$^-$ phenotype in male ES cells. Further, *hprt*$^-$ cells can be directly selected by growth in 6-thioguanine, which kills *hprt*$^+$ cells.

Targeted disruption of the *hprt* gene established the feasibility of gene targeting in ES cells and also defined some of the parameters that control the efficiency of this process (Thomas and Capecchi, 1987). It was shown (Fig. 1) that replacement and insertion targeting vectors were equally efficient in disrupting the endogenous *hprt* gene. This result was not anticipated because it was assumed that a single crossover was required to introduce an insertion vector into the homologous target whereas a replacement vector would require two crossovers. It now appears likely that insertion vectors also use two crossovers to recombine into the target locus. The outcome of this result is that the investigator can employ either sequence insertion or sequence replacement vectors with equal efficiency. Depending on the nature of the modification of the endogenous locus that is desired, each class of vector has its own intrinsic advantages. Both classes of vectors showed the same exponential dependency of the targeting frequency on the extent of homology between the exogenous and endogenous DNA sequences (Capecchi, 1989b). This strong dependency on the extent of homology appears to be unique to higher eucaryotes, since *E. coli* and yeast exhibit a linear dependency of recombination frequency on homology. Surprisingly, the amount of nonhomologous DNA being transferred from the targeting vector to the endogenous locus does not influence the gene targeting frequency (Mansour et al., 1990). For these experiments the targeting vectors each contained a 9.1-kb fragment of the *hprt* gene that was disrupted by heterologous DNA inserts of 8 bp, 1 kb, 3.4 kb, 4.3 kb and 12 kb. Homologous recombination between the targeting vectors and the endogenous gene would disrupt *hprt* coding sequences rendering the cells *hprt*$^-$. Using these vectors we found the frequency of targeted disruption of *hprt* to be the same irrespective of the length of the heterologous DNA insert transferred to the endogenous gene (see Table 1). The significance of this finding is that a much wider spectrum of designed genomic alterations is now feasible. Thus it would be reasonable to attempt substitution of one controlling region of a gene with another or to replace an entire coding sequence with another. Also, with respect to homologous recombination, introducing large insertions and generating large deletions are nearly equivalent operations. Therefore, these results suggest that generation of large genomic deletions should also be possible. Overlapping sets of deletions could be used to locate genes within mapping intervals.

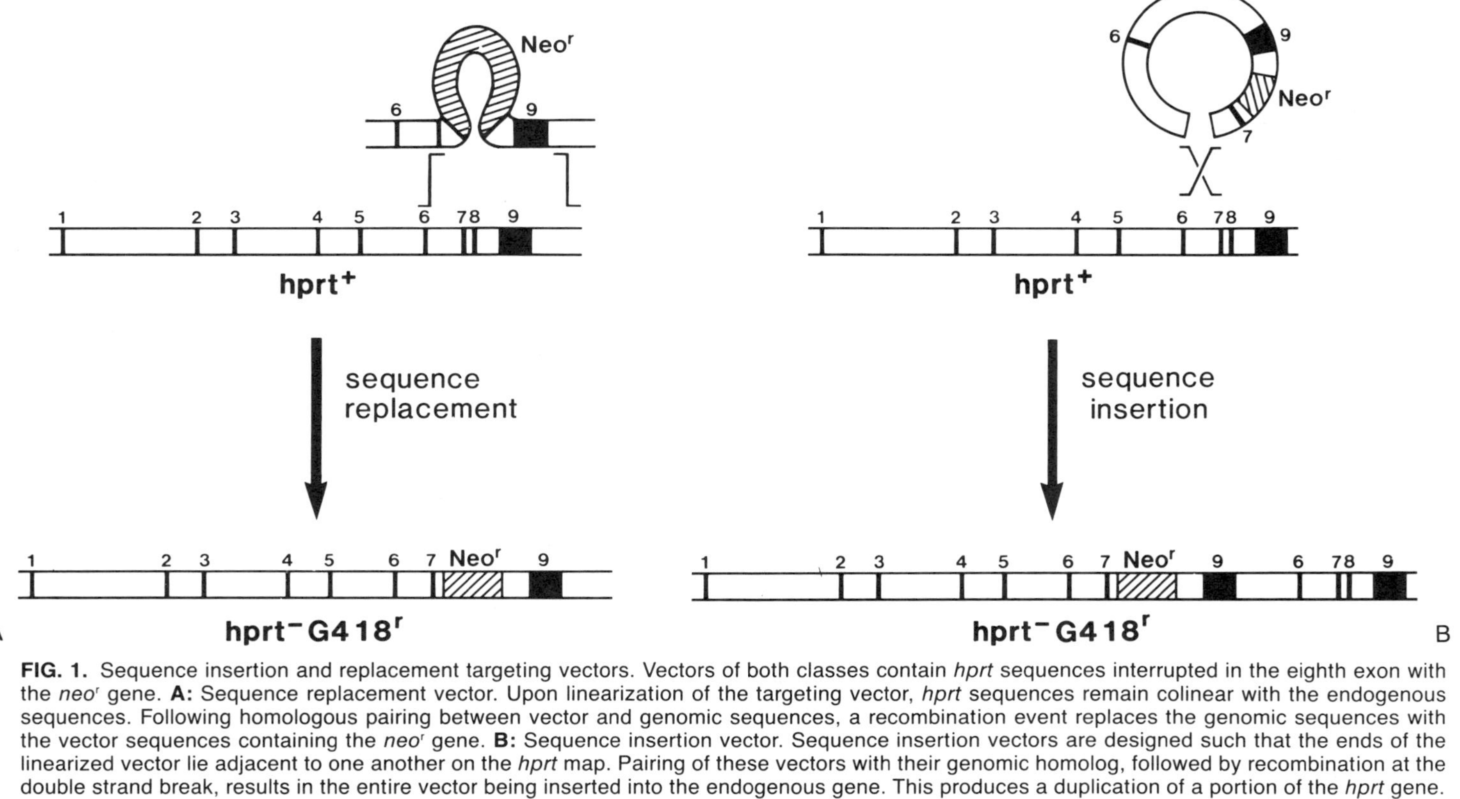

FIG. 1. Sequence insertion and replacement targeting vectors. Vectors of both classes contain *hprt* sequences interrupted in the eighth exon with the *neo*^{r} gene. **A:** Sequence replacement vector. Upon linearization of the targeting vector, *hprt* sequences remain colinear with the endogenous sequences. Following homologous pairing between vector and genomic sequences, a recombination event replaces the genomic sequences with the vector sequences containing the *neo*^{r} gene. **B:** Sequence insertion vector. Sequence insertion vectors are designed such that the ends of the linearized vector lie adjacent to one another on the *hprt* map. Pairing of these vectors with their genomic homolog, followed by recombination at the double strand break, results in the entire vector being inserted into the endogenous gene. This produces a duplication of a portion of the *hprt* gene.

TABLE 1. *Gene targeting into the* hprt *loci*[a]

Vector	Cells surviving electroporation	6-TG^r colonies	Targeted *hprt*$^-$ colonies/ 6TG^r colonies tested
pHPT^{+8}	10^6	15	6/12
pHPT^{+1k}	10^6	15	6/9
pHPT$^{+3.4k}$	10^6	15	9/15
pHPT3$^{4.3k}$	1.4×10^6	12	6/12
pHPT3^{+12k}	2×10^6	28	15/28

[a]Aliquots of cells treated with *hprt* vectors were grown in either nonselective medium to assess the total number of cells surviving electroporation (approximately 50%) or in 6-thioguanine (6-TG)-containing medium to select for *hprt*$^-$ cells. Targeting events were identified by Southern transfer analysis. pHPT^{+8}, pHPT^{+1k}, pHPT$^{+3.4k}$ contain heterologous inserts of 8 bp, 1 kb and 3.4 kb respectively in the 8th exon of *hprt,* whereas pHPT3$^{+4.3k}$ and pHPT3^{+12k} contain heterologous inserts of 4.3 kb and 12 kb in the 3rd exon of *hprt.* For these experiments only 6-TG was used as a selective agent. Therefore, *hprt*$^-$ cells are derived both from targeted disruption of the gene as well as from spontaneous mutations in the gene. We can distinguish between spontaneous and targeted *hprt*$^-$ cells by Southern blot analysis.

NONSELECTABLE GENES

Unfortunately, unlike *hprt,* inactivation or modification of the vast majority of genes do not result in a selectable cellular phenotype. Most genes are not expressed in ES cells, are autosomal and even inactivation of both copies of the gene still would not result in a selectable phenotype. Identification of the rare ES cell containing the desired targeted modification in most genes must be accomplished by screening and/or indirect enrichment procedures.

For this purpose we developed an enrichment procedure, designated as positive–negative selection (PNS), that is independent of the function and expression pattern of the target gene in the recipient cell (Mansour et al., 1988). The advantages of enrichment procedures relative to pure screening procedures are the potential for greater sensitivity and less work.

PNS uses a positive selection for cells that have incorporated the targeting vector anywhere in the ES cell genome and a negative selection against cells that have randomly integrated the vector. The net effect is to enrich for cells containing the desired targeted mutation. In a PNS targeting vector the positive selectable gene [i.e., a neomycin (*neo*^{r}) or hygromycin resistance gene] is flanked by DNA sequences homologous to the target gene (see Fig. 2). In the same vector a negative selectable gene, such as the herpes simplex virus thymidine kinase (HSVtk) gene, is placed at the ends of the vector. Following homologous recombination between the targeting vector and the target gene, the negative selectable gene is not transferred into the target locus because it is located distal to the region of homology between the vector and the target. On the other hand, cells in which a random integration of the targeting vector has occurred retain the negative selectable gene because random insertions of exogenous linear DNA into the genome occur primar-

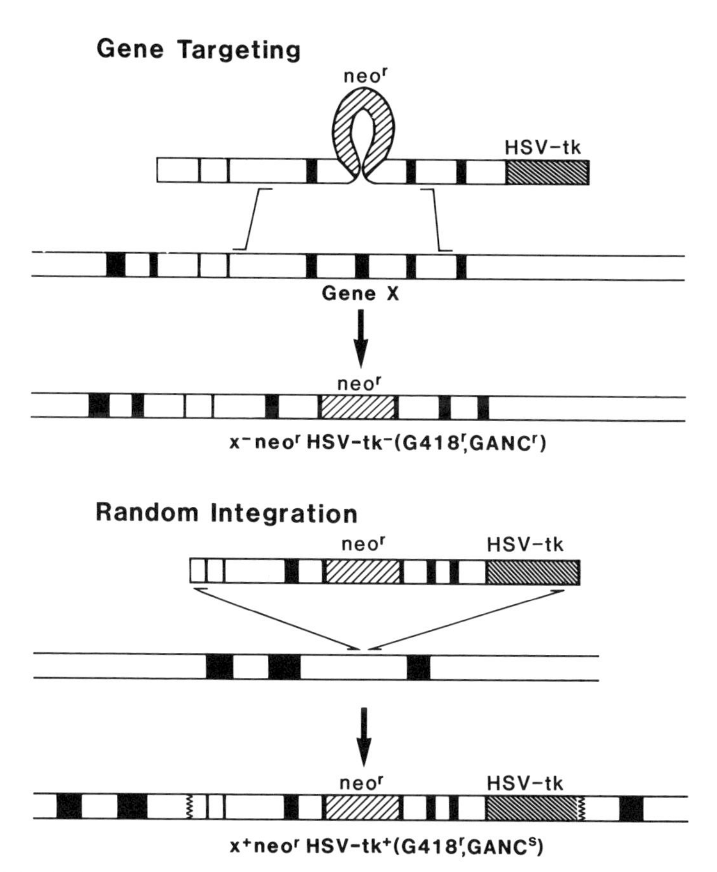

FIG. 2. The PNS procedure used to enrich for ES cells containing a targeted disruption of gene X. **A:** A gene X-replacement vector that contains an insertion of the *neo*^{r} gene in an exon of gene X and a linked HSVtk gene is shown pairing with a chromosomal copy of gene X. Homologous recombination between the targeting vector and genomic X DNA results in the disruption of one copy of gene X and the loss of HSVtk sequences. Such cells will be X$^-$ *neo*^{r} and HSVtk$^-$ and will be resistant to both G418 and GANC. **B:** Because nonhomologous insertion of exogenous DNA into the genome occurs through the ends of the linearized DNA, the HSVtk gene remains linked to the *neo*^{r} gene. Such cells will be X$^+$, *neo*^{r} and HSVtk$^+$ and therefore resistant to G418 but sensitive to GANC.

ily via their ends (see Fig. 2). Therefore, by using G418 to select for cells that contain a functional *neo*^{r} gene and by using gancyclovir (GANC) to select against cells that contain a functional HSVtk gene, we enrich for cells in which the desired homologous recombination event has occurred.

PNS can be used to isolate cell lines containing targeted disruptions in genes not expressed in ES cells. For example, Spiegelman and his colleagues (1989) have used PNS to identify ES cell lines containing targeted disruptions in the genes, adipsin and adipocyte P2, which are expressed exclusively in adipose tissue. Similarly we have used PNS to isolate ES cells containing targeted disruptions in a series of mouse homeo-box-containing genes. Most of these genes are not expressed at detectable levels in ES cells. Thus it appears that even if the neo^r gene, driven by its own strong promoter, is targeted into a transcriptionally inactive chromatin environment, sufficient neo^r product is synthesized to yield a selectable phenotype.

THE MOUSE *hox* GENES

A belief is emerging that a whole developmental program for specifying positional information in the early embryo has been conserved in *Drosophila,* mouse, and humans. The impetus for this belief is based on the finding that mice and humans contain homologs for many of the *Drosophila* genes that control early embryonic development. In humans and mice these genes are also expressed in distinct temporal and spatial patterns during embryogenesis. Further, it has been found that for the antennapedia-related homeobox-containing genes, designated as the *hox* genes, not only has the gene order in the mouse, human, and *Drosophila* been conserved, but also the correlation between the order of genes on the chromosome and anterior expression boundaries along the anterior–posterior axis of the embryo has also been conserved (Dubouele and Dolle, 1989; Graham et al., 1989). These observations suggest that the evolutionary conservation of these genes may reflect not only the mere retention of convenient DNA binding motifs but rather the inheritance of a whole transcriptional program for specifying positional information in the early embryo.

In both humans and mouse approximately 30 *hox* genes have been identified that map to four separate linkage groups (see Fig. 3; Acampora et al., 1989). DNA sequence similarity among *hox* genes in these separate linkage groups as well as the maintenance of chromosomal gene order suggests that these linkage groups arose from duplications of entire chromosomal regions (Hart et al., 1987; Graham et al., 1988; Dubouele and Dolle, 1989). No human or mouse mutations have been identified in this complex of genes. This may reflect that mutations in these genes would result in embryonic lethality. Such mutants in the mouse would not have been identified in screens for visible variation.

Through a systematic targeted disruption analysis of this complex of genes, we hope not only to associate specific phenotypes with inactivation of individual genes, but that through epistasis and molecular analysis of combinations of mutant genes, we may reveal how this network of genes functions within the embryo to designate positional value. How closely do the functions of the mouse *hox* genes reflect the functions of homologous genes in *Drosophila*? Does this complex of genes act as a hierarchical cascade of transcription factors to refine positional value in the early embryo? Do such

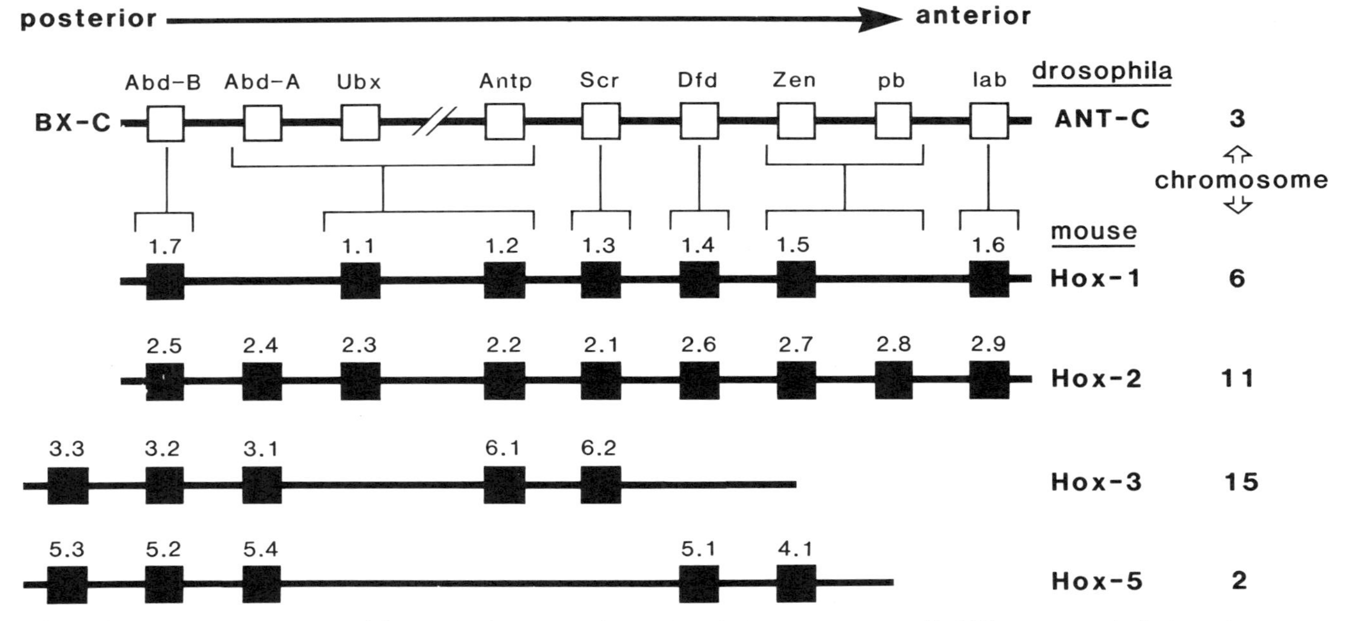

FIG. 3. Schematic representation of the murine *hox* genes. Approximately 30 mouse genes with DNA sequence similarity to the antennapedia-related homeo-box have been identified and mapped to four chromosome linkage groups. These are designated as *hox 1, 2, 3* and *5*. The DNA sequence relationship between the murine *hox* genes and *Drosophila* ultra bithorax and antennapedia genes is also indicated.

refinements, if they exist, involve the successive acquisition of the expression of combinations of these genes in groups of cells? Do corresponding members of *hox* genes on the different linkage groups perform overlapping functions? Answers to such questions may be approached through the analysis of combination of *hox* mutants generated by breeding of the appropriate mice containing single *hox* mutations.

To date, using PNS, we have generated ES cell lines containing targeted disruptions in the *hox 1.2, 1.3, 1.4, 1.5, 1.6, 1.7, 2.3* and *2.5* genes (see Fig. 3); Chisaka, Kostic, Touzuki and Capecchi, unpublished results). These cell lines have been used to generate mouse chimeras that are being evaluated for germ-line transmission of the mutant allele. Figure 4 (*see color plate*) illustrates a germ-line chimera generated from ES cells containing a targeted disruption of the *hox 1.5* gene. The ES cells were derived from 129 Sv, which is wild type at the agouti (A) locus. The recipient blastocyst was obtained from a C57Bl/6 non-agouti mouse (a). It is observed that following breeding of the chimeric mouse to a C57Bl/6 female, all of the pups are agouti, indicating that most, if not all, of the sperm produced by the chimeric mouse was derived from the ES cells bearing the *hox 1.5* mutation. As expected from Mendelian segregation, half of the offspring contained the mutant allele of *hox 1.5*. The heterozygous offspring are normal, indicating that mice heterozygous at the *hox 1.5* locus are not haplo-insufficient. Interbreeding of heterozygous siblings will permit evaluation of the phenotype resulting from inactivation of both copies of the *hox 1.5* gene.

THE *int*-RELATED GENES

The proto-oncogenes *int-1* and *int-2* are attractive candidates for genes that participate in cell–cell signaling during embryogenesis. They were first identified as genes activated in mammary tumors of mice by the nearby insertion of the mouse mammary tumor virus (Nusse and Varmus, 1982; Dickson et al., 1984). The gene products resemble growth factors and *int-2* is a member of the fibroblast growth factor family (Dickson and Peters, 1987). Their expression is highly restricted during embryogenesis. *int-1* is expressed in the 9.0 to 14.5 day old mouse embryo along the dorsal midline of the neural tube (Wilkinson et al., 1987). *int-2* expression is more complex but restricted to embryos. *int-2* message has been detected in both extra embryonic and embryonic tissues (Wilkinson et al., 1989). The diverse but highly discrete pattern of expression of *int-2* suggests multiple roles during embryogenesis, particularly in cells involved either in migration and/or induction.

int-2 appears to be restricted to mammals whereas analogs of *int-1* have been identified in both *Drosophila* (Rijsewijk et al., 1987) and *C. elegans* (H. Varmus, personal communications). The *Drosophila* homolog of *int-1* is *wingless* (*wg*), a segment polarity gene required for normal patterning within each parasegment (Baker, 1987). Mosaic analysis in *Drosophila* has shown that *wg* mutations are not cell-autonomous, which is consistent with the gene product being secreted by one set of cells and affecting the fate of neighboring cells (Morata and Lawrence, 1977; Baker, 1987). Using PNS we have generated null mutations in ES cells of both *int-1* (Thomas and Capecchi, unpublished results) and *int-2* (Mansour et al., 1988). These cell lines have been used to generate germ-line chimeras harboring either *int-1* and *int-2* mutations (Thomas and Capecchi; Mansour and Capecchi respectively, unpublished results). Colonies of these mice are presently being generated to permit analysis of homozygotes generated by interbreeding of heterozygous siblings.

INTRODUCTION OF A *lacZ* REPORTER GENE INTO SPECIFIC CHROMOSOMAL LOCI BY HOMOLOGOUS RECOMBINATION

We have introduced the *lac*Z reporter gene into the *int-2, hox 1.5, hox 1.6* and *hox 2.3* genes in ES cells (Mansour et al., 1990; and unpublished results). Homologous recombination between the *lac*Z targeting vector and the cognate chromosomal sequence results in generating an in-frame fusion between the endogenous gene product and β-galactosidase. Since the reported gene is placed among all the *cis*-acting control elements that normally mediate expression of the endogenous gene, it should accurately reflect the expression pattern of the target gene. Indeed as predicted, when the ES cells containing the *int-2-lac*Z fusion gene are differentiated in vitro to endoderm cells, β-galactosidase activity is specifically induced. Similarly, treatment of ES cells containing the *hox 1.6-lac*Z fusion with retinoic acid induces *lac*Z expression.

The ES cells containing these *lac*Z fusions are being used to generate chimeric mice. In mice heterozygous for the *lac*Z fusion, β-galactosidase activity should provide a convenient tool for following the expression of the target gene during development. This approach may be particularly instructive for the analysis of the interactions among genes, such as the *hox* genes, that participate in a developmental network. For example, by analyzing *lac*Z expression from a given *lac*Z fusion gene in the presence of different *hox* gene mutations, it may be possible to establish which genes in this network directly or indirectly interact.

The *lac*Z fusion also creates a null mutation in the corresponding gene. Mice homozygous for the *lac*Z fusion will not only exhibit the phenotype associated with inactivating both copies of the gene, but the very cells responsible for the phenotype will be tagged with β-galactosidase activity. This may facilitate interpretation of the phenotype at the cellular level.

TOWARDS A CF MOUSE

With the recent cloning and characterization of the human cystic fibrosis (CF) gene (Rommens et al., 1989; Riordan et al., 1989), we are now in a position to direct our efforts towards generating a CF mouse. Screening a mouse genomic library with human CF probes has yielded a highly conserved mouse homolog of the human gene (Thomas and Cepecchi, unpublished results). These DNA sequence are being used to make CF targeting vectors, which in turn will be used to generate CF mutations in mouse ES cells. We are attempting to create mouse mutations corresponding to human CF alleles responsible for both mild and severe forms of cystic fibrosis.

SUMMARY

I have discussed the technology for generating mice of virtually any desired genotype. We have been using this technology primarily to define the function of genes in early mouse development. Specifically we are concentrating our efforts on defining the function of the mouse *hox* genes and the *int* related genes. The former, through an undefined transcriptional network, may establish positional value along the anterior posterior axis of the early embryo. The latter, a set of proto-oncogenes participating in cell–cell signaling, may mediate cell inductive and/or migration decisions.

It is clear, however, that this technology can be applied to genetically dissect any mammalian biological problem and thereby impact disciplines other than developmental biology, including cancer, immunology, neurobiology, and human medicine. The most immediate application of this technology to human medicine will be the generation of mouse models for human genetic diseases. Such models should facilitate the analysis of the pathology of the disease as well as provide a system for the exploration of new therapeutic protocols including gene therapy.

REFERENCES

Acampora D, D'Esposito M, Farella A, et al. (1989). The human *hox* gene family. *Nucleic Acids Res.* 17, 10385.

Baker NE (1987). Molecular cloning of sequences from *wingless,* a segment polarity gene in *Drosophila:* the spatial distribution of a transcript in embryos. *EMBO J* 6, 1765.

Capecchi MR (1989a). The new mouse genetics: altering the genome by gene targeting. *Trends Genet* 5, 70.

Capecchi MR (1989b). Altering the genome by homologous recombination. *Science* 244, 1288.

Dickson C and Peters G (1987). Potential oncogene product related to growth factors. *Nature* 326, 833.

Dickson C, Smith R, Brooks S, and Peters G (1984). Tumorigenesis by mouse mammary tumor virus: proviral activation of a cellular gene in the common integration region *int-2*. *Cell* 37, 529.

Dubouele D, Dolle P (1989). The structural and functional organization of the murine *hox* gene family resembles that of *Drosophila* homeotic genes. *EMBO J* 8, 1497.

Graham A, Papalopulu N, Krumlauf R (1989). The murine and *Drosophila* homeobox gene complex has common features of organization and expression. *Cell* 57, 367.

Graham A, Papalopulu N, Lorimer J, et al. Characterization of a murine homeo-box gene, *hox 2.6* related to the *Drosophila* deformed gene. *Genes Dev* 2, 1424.

Hart CP, Fainsod A, and Ruddle FH (1987). Sequence analysis of the murine *hox 2.2, 2.3* and *2.4* homeoboxes; evolutionary and structural comparisons. *Genomics* 1, 182.

Johnson RS, Sheng M, Greenberg ME, et al. (1989). Targeting of nonexpressed genes in embryonic stem cells via homologous recombination, *Science* 245, 1234.

Mansour SL, Deng C, Thomas KR, and Capecchi MR (1990). Introduction of a *lac*Z reporter gene into the mouse *int-2* locus by homologous recombination, *Proc Natl Acad Sci USA,* in press.

Mansour SL, Thomas KR, and Capecchi MR (1988). Disruption of the proto oncogene *int-2* in mouse embryo-derived stem cells: a general strategy for targeting mutations to nonselectable genes, *Nature* 336, 348.

Morata G and Lawrence PA (1977). The development of *wingless,* a homeotic mutation of *Drosophila, Dev Biol* 56, 227.

Nusse R and Varmus H (1982). Many tumors induced by the mouse mammary tumor virus contain a provirus integrated in the same region of the host genome. *Cell* 31, 99.

Rijsewijk F, Schuermann M, Wagemaar E, et al. (1987). The *Drosophila* homolog of the mouse mammary oncogene *int-1* is identical to the segment polarity gene *wingless*. *Cell* 50, 649.

Riordan JR, Rommens JM, Kerem B, et al. (1989). Identification of the cystic fibrosis gene: cloning and characterization of complimentary DNA. *Science* 245, 1066.

Rommens JM, Iannizzi MC, Kerem B, et al. (1989). Identification of the cystic fibrosis gene: chromosome walking and jumping. *Science* 245, 1059.

Spiegelman et al., 1989.

Thomas KR and Capecchi MR (1987). Site-directed mutagenesis by gene targeting in mouse embryo-derived stem cells. *Cell* 51, 503.

Wilkinson DG, Bailes and McMahon AP (1987). Expression of the proto-oncogene *int-1* is restricted to specific neural cells in the developing mouse embryo. *Cell* 50, 79.

Wilkinson DG, Bhatt S, and McMahon AP (1989). Expression patterns of the FGF-related proto-oncogene *int-2* suggests multiple roles in fetal development. *Development* 105, 131.

Diagnosis and Therapy

Etiology of Human Disease at the DNA Level,
edited by Jan Lindsten and Ulf Pettersson.

17

Diagnosis of Genetic Disease in Preimplantation Embryos

Marilyn Monk

Medical Research Council Mammalian Development Unit, University College London, London NW1 2HE, United Kingdom

Over 1 percent of babies born are afflicted with a genetic or developmental abnormality. If it is known that a couple is at risk of having a baby with a known genetic disease, the fetus may be tested during pregnancy by amniocentesis or chorionic villus sampling and the pregnancy terminated if the fetus proves to be affected. Such couples may suffer the considerable trauma of repeated abortions in their attempt to have a normal, healthy baby. In this paper I shall present evidence to show that we can now diagnose the defective embryos before implantation. By replacing only the embryos without the defect in the mother, anxiety during pregnancy could be alleviated and abortion avoided.

We have investigated two possible approaches to preimplantation diagnosis of genetic disease—sensitive biochemical analysis to detect the altered or absent gene product and DNA analysis to detect the altered gene sequence characteristic of a particular mutation. In each case the diagnostic procedure may be carried out on a single cell sampled from the 8-cell embryo, on a few cells sampled from the trophectoderm layer of the blastocyst, or on the first polar body of the unfertilised egg. This latter approach may allow the identification of unfertilised eggs that carry the defective gene and those without the defect. By fertilising in vitro only those eggs without the defect and replacing these in the mother, the possible ethical difficulties of working with embryos would be avoided.

BIOPSY PROCEDURES

Interest in the possibility of preimplantation diagnosis has come about due to access to very early human embryos provided by the in vitro fertilisation (IVF) procedures for the alleviation of infertility. Eggs may be removed from the woman's ovaries and fertilised outside the body and two or three of the resulting 4- to 8-cell embryos may be replaced in the woman's uterus. For the purpose of diagnosis of genetic disease, direct analysis of embryo-coded enzyme or embryonic DNA must be done on a sample taken during the cleavage stages or at the blastocyst stage of development. Whatever the procedure, it must not impair the development of the embryo after replacement in the mother.

Procedures for biopsy of preimplantation embryos and removal of the first polar body were developed in mouse model systems (Fig. 1). In the mouse 8-cell embryo (Fig. 1A) the zona pellucida is removed, cell contacts loosened in medium devoid of calcium, and one cell removed by gentle pipetting. Such embryos with one cell removed show good development to term after their transfer to a foster mother (1,2). For the human embryo, a more conservative approach is taken and the removal of one cell is done through a hole made in the zona using micromanipulators (3).

Biopsy of a few trophectoderm cells of the mouse blastocyst may be performed using micromanipulation techniques to make a hole in the zona pellucida opposite the inner cell mass (Fig. 1B). This results in herniation of 5 to 10 cells, which may be removed with a micropipette. Trophectoderm bi-

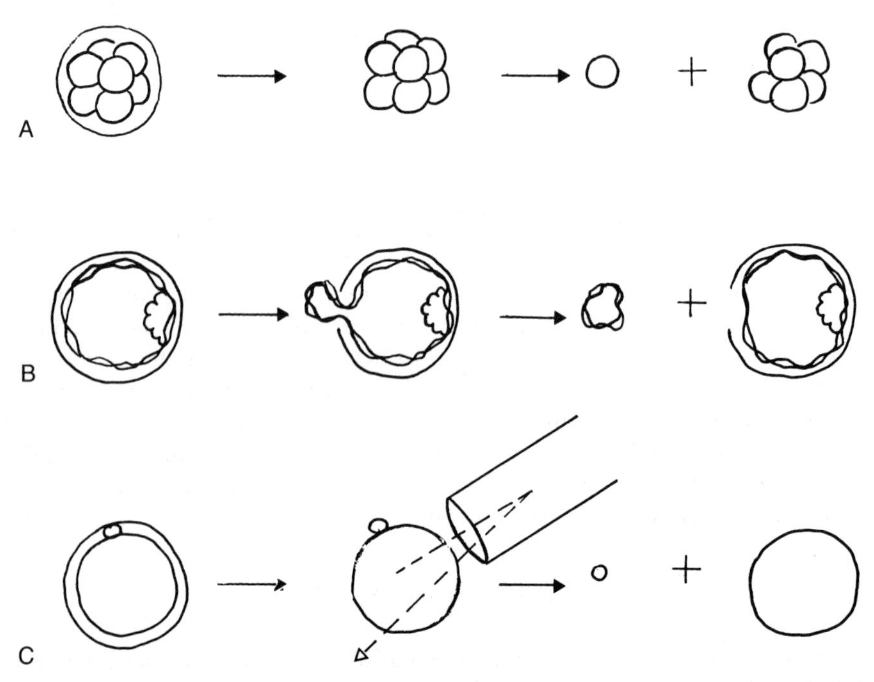

FIG. 1. Biopsy procedures for removal of a single cell from an 8-cell embryo (**A**) and a few cells from the blastocyst (**B**), and for separation of the first polar body from the unfertilised oocyte (**C**).

opsy and subsequent development of the operated embryo has been successful in the rabbit (4,5), Marmoset monkey (6) and mouse (7). In the Marmoset monkey, normal offspring were produced after the replacement of blastocysts from which up to 40% of the cells had been removed.

In the human, following IVF, embryo transfer to the mother is done at the 4- to 8-cell stage, so trophectoderm biopsy is not applicable. However, blastocysts for diagnosis by trophectoderm biopsy might be obtained nonsurgically by uterine lavage (8). This approach would have the advantages of avoiding IVF and possibly of using naturally ovulated eggs fertilised in vivo. It would also have the ethical advantage that the biopsy consists of extra-embryonic cells only. However, the efficiency of uterine lavage and the risks of ectopic pregnancy in the human have not been adequately evaluated as yet. Consequently, couples seeking preimplantation diagnosis at the present time would need to go through the procedures of IVF.

In our most recent work (9), we have removed the first polar body from fertilised human eggs (Fig. 1C). The first polar body is a tiny cell lying next to the egg under the zona pellucida. It is one of the products of the first asymmetric meiotic division and, as such, its genetic content gives information on the genetic content of the egg itself (see below). In these initial experiments, the zona pellucida was removed and the first polar body dislodged by gentle pipetting with a fine pipette of diameter slightly less than the diameter of the egg.

BIOCHEMICAL ANALYSIS OF THE BIOPSIED CELL

The Mouse Model

For preimplantation diagnosis, we need to be able to biopsy a single cell without compromising development. We also need to increase the sensitivity of the microassay for the enzyme or DNA defect under test to the single cell level. To develop these procedures we use mouse model systems. Mutant mice, which serve as model systems for the study of a number of genetic diseases, have arisen by random mutation and have been identified by genetic and biochemical screening. Using modern techniques, it is also feasible to create such mouse models. A mutation in a particular gene can be specifically induced and verified in culture in embryonal stem cells and then the mutant cells returned to a mouse preimplantation embryo, where they will participate in normal development. The application of targeted homologous genetic recombination to specific genes in embryonal stem cells theoretically makes possible the creation of mutant mice for any known gene (see Chapters 15 and 16).

In our initial experiments, we used the hypoxanthine phosphoribosyl transferase (EC 2.4.2.8) (HPRT)-deficient mouse (10). This is one of the first examples of a genetically engineered animal created for the study of genetic disease. In this case, HPRT-deficient embryonal stem cells could be directly selected in culture by treatment of the cells with 6-thioguanine. The identification of chimaeras harbouring these mutant stem cells, and thence the derivation of an HPRT-deficient mouse line, was facilitated by the use of a coat colour marker identifying the presence of mouse cells of embryonal stem cell origin, as well as by assay for HPRT activity in clonal populations of cells within the animals (10).

The enzyme HPRT is coded for by a gene on the X chromosome. In the human, HPRT deficiency causes Lesch-Nyhan disease and death by about the age of puberty (11). In the mouse, HPRT deficiency is not associated with any obvious deleterious effects. However, as a model system for the development of procedures for preimplantation diagnosis of enzyme deficiency, the HPRT-deficient mouse was ideal. Techniques were already available to measure this enzyme in single embryos (12,13) and for manipulation and biopsy of early mouse embryos (see ref. 14). In a cross between a normal male and a carrier female (heterozygous for the HPRT-positive and HPRT-negative alleles), half of the male offspring will be HPRT-deficient. The aim is to identify these affected male embryos before implantation by sampling and biochemical analysis of one or a few cells of the embryo.

Diagnostic Biochemical Analysis

We use a double microassay which simultaneously measures the activity of two enzymes in the purine salvage pathway: HPRT, which converts hypoxanthine to inosine monophosphate (IMP), and adenine phosphoribosyl transferase (EC 2.4.2.7) (APRT), which converts adenine to adenosine monophosphate (AMP) (see Fig. 2, top). The results may be expressed as the ratio of the X-linked HPRT to the autosome-linked APRT activities, which eliminates sampling error and reflects accurately the ratio of active X chromosomes to autosomes. [see Monk (12) for a full description of the assay procedure.] The assay is sufficiently sensitive to detect the presence or absence of HPRT activity in the single sampled blastomere and, at the same time, indicate that all samples, whether HPRT-positive or HPRT-deficient, have normal APRT activity. A typical analysis of single blastomeres taken from 8-cell embryos of a heterozygous carrier mother is shown in Fig. 3.

The biochemical diagnosis can be done overnight so that there is no need to freeze the embryos. The embryos, which are maintained in culture pending the results of the enzyme assay, are then transferred in groups as diagnosed to pseudopregnant recipient foster mothers to confirm the diagnoses. Of 16 fetuses resulting from transferred embryos diagnosed at the 8-cell stage and 7 resulting from diagnosed blastocysts, the diagnoses were correct in all cases (1,7). Diagnosis by direct biochemical microassay for HPRT deficiency would appear to be most effective from the results with the mouse model system.

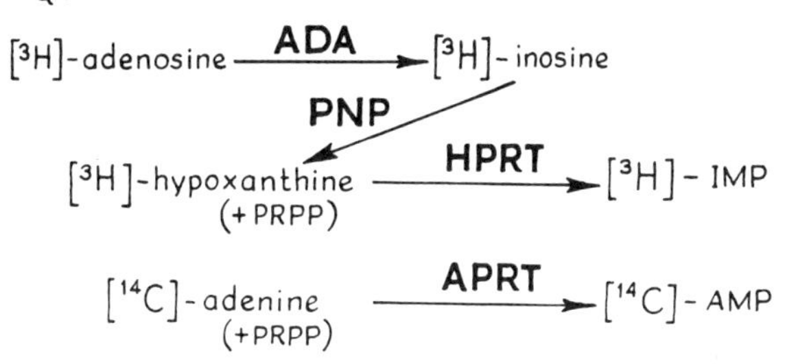

FIG. 2. Biochemical pathways showing enzymes HPRT and APRT (*top*) and ADA. PNP, HPRT, and APRT (*bottom*). (From Monk, ref. 31.)

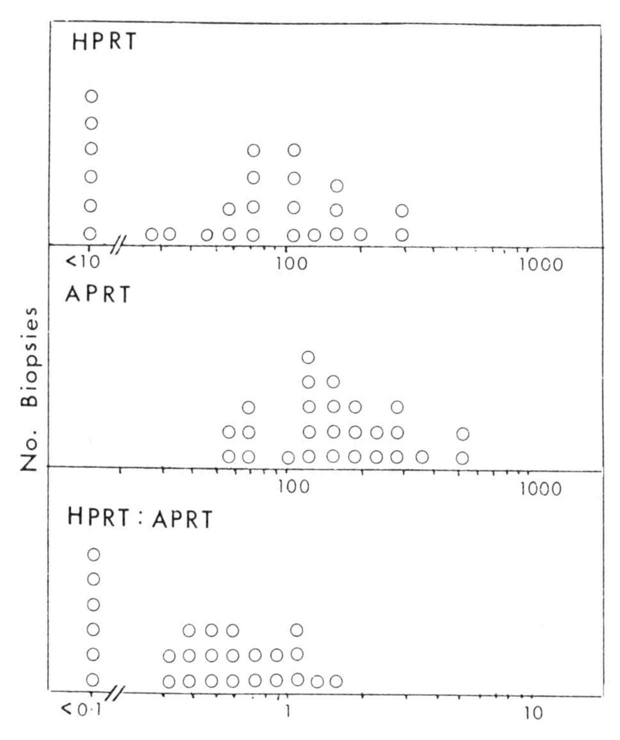

FIG. 3. HPRT and APRT activities and HPRT:APRT ratios in single blastomeres, each isolated from an 8-cell mouse embryo from a cross between a heterozygous carrier female ($HPRT^{+}/HPRT^{-}$) and a normal male ($HPRT^{+}/Y$). The HPRT-deficient blastomeres from the HPRT-deficient male embryos are clearly seen. (Data from Monk et al., ref. 1.)

Biochemical microassay of this type was developed for several other enzymes that are implicated in human genetic disease. It was found that two other enzymes in the purine salvage pathways could be assayed at the same time as HPRT and APRT by providing the substrate of adenosine deaminase, or ADA (EC 3.5.4.4), adenosine, instead of hypoxanthine in the reaction mix (see Fig. 2, bottom). In this way, a deficiency for any one of four enzymes, ADA, purine nucleoside phosphorylase or PNP (EC 2.4.2.1), HPRT, and APRT, could be detected in a single cell taken from an 8-cell mouse embryo (15). A deficiency of ADA in the human results in severe combined immunodeficiency disease (SCID), and PNP deficiency is also associated with immunological as well as neurological symptoms.

Validity of Biochemical Diagnosis

Diagnosis by direct biochemical assay of the altered or absent gene product is only possible if the activity being measured is produced by transcription and translation of the embryo's own gene for that enzyme. The time of transition from maternal enzyme (inherited in the egg cytoplasm) to embryo-coded enzyme activity varies depending on the enzyme. In general, an increase in enzyme activity during preimplantation development is an indication that the embryonic gene has been activated although this condition is not sufficient evidence. [For example, in early preimplantation development of the mouse there is an increase in HPRT activity from the 1-cell to the 2-cell stage which is attributable to translation of inherited maternal mRNA for that enzyme (16).] Further evidence for the onset of embryonic gene transcription for a particular enzyme activity is obtained by growing the preimplantation embryos in the presence of alpha-amanitin, a drug that inhibits the activity of RNA polymerase II, the enzyme responsible for transcription. In the mouse, it is clear that at least three of the enzymes measured in the quadruple microassay in Fig. 2 (bottom) are embryo-coded. Figure 4 shows that there is a marked increase in activity from the 2-cell to the morula stage which is sensitive to treatment with alpha-amanitin for HPRT (16) (Fig. 4A), for APRT (Fig. 4B, Ao and Monk, unpublished data), and for ADA (15) (Fig. 4C).

HPRT Activity in Human Preimplantation Embryos

In human preimplantation embryos the situation may well be different. It appears that the earliest gene transcription detectable by new protein synthesis in human embryos occurs at the 4-cell stage (17), one cleavage cycle later than it is found in mouse embryos (18). When a series of human preimplantation embryos were examined directly for HPRT activity there was no convincing evidence that the activity measured was due to embryonic gene

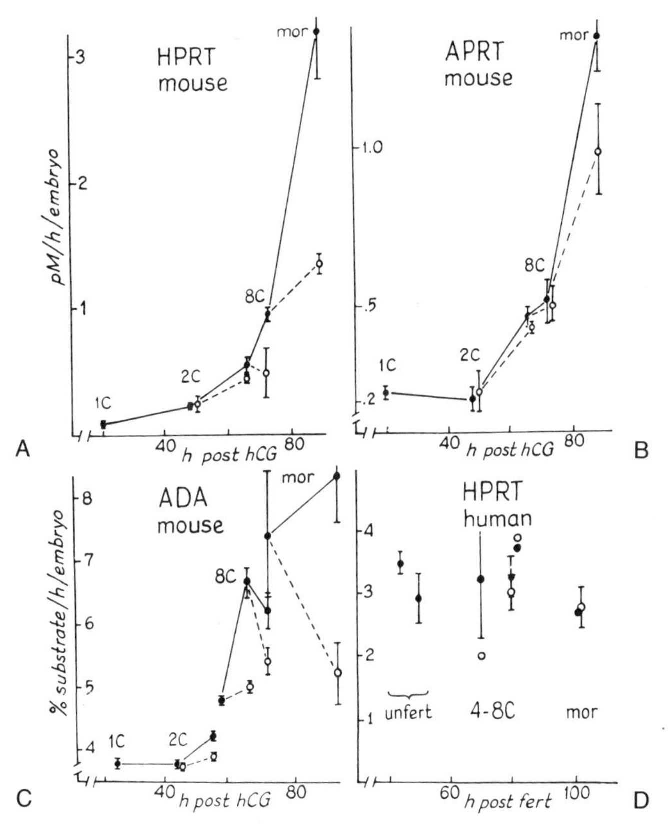

FIG. 4. Profiles of activity of enzymes (**A**) HPRT, (**B**) APRT, and (**C**) ADA during mouse preimplantation embryo development and (**D**) HPRT during human preimplantation embryo development *in vitro*. Open circles and dotted lines show values for embryos cultured in the presence of the transcription inhibitor alpha-amanitin. hCG, human chorionic gonadotropin; mor, morula. (From Monk, ref. 31.)

transcription (19). First, there was no marked increase in activity during preimplantation development (in vitro) and no significant increase in activity over and above the level inherited in the egg cytoplasm and maintained stably in unfertilised eggs during the period of culture (Fig. 4D). Second, there was no consistent effect of 24 hours' culture in alpha-amanitin on the HPRT activity level. This work emphasizes the dangers of extrapolating directly from animal models to humans and the importance of direct final analysis on human material. At this stage, the direct biochemical microassay approach could not be applied to preimplantation diagnosis of Lesch-Nyhan disease in the human.

Further work will be directed towards establishing preimplantation diagnosis based on direct biochemical microassay for other known enzyme defects. In certain cases, namely situations in which the affected protein product is known but the gene sequence is unknown, and also for genetic disease arising by frequent new mutation, direct biochemical diagnosis may be the only route.

DNA ANALYSIS OF A SINGLE CELL

The technique of polymerase chain reaction (PCR) amplification of specific DNA sequences (20) is a powerful approach to many areas of scientific endeavour including cloning, DNA sequencing, genetic mapping, detection of rare mutant or viral sequences, mRNA quantitation, evolutionary studies, forensic studies, paternity testing, detection of carrier state, and prenatal diagnosis of genetic disease. The technique is so powerful that it is possible to detect a single-copy DNA sequence in a single cell (21,22). The procedure employs two synthetic oligonucleotide primers, each about 20 nucleotides long, of opposite orientation, and homologous to sequences on opposite strands at either end of the segment of DNA to be amplified. Successive cycles at three temperatures—for denaturation of genomic DNA, annealing of the primers, and DNA synthesis—allows an exponential increase of precisely that sequence of the DNA bordered by the homologous oligonucleotide primers.

Recently, this approach has been applied to the diagnosis of sex in single blastomeres of human 6- to 10-cell embryos. The single blastomere was removed through a hole made in the zona and the sequence amplified was a repeat sequence on the Y chromosome present in about 800 to 1,500 copies per cell (3). The embryo-sexing procedure was made available to five couples at risk of having a male child afflicted with severe genetic disease. Only the diagnosed female embryos were replaced in each of these women, resulting in two pregnancies (23).

Amplification in individual unfertilised human oocytes of the single-copy sequences associated with cystic fibrosis and Duchenne's muscular dystrophy has also been reported recently (24). Moreover, there was some evidence that the amplification products could be restricted with appropriate enzymes to identify the genotype. These experiments indicate that preimplantation diagnosis of single gene defects would be feasible with a single cell biopsied from the cleavage stage embryo.

The Mouse Model

We developed procedures of amplification of a mouse beta-haemoglobin gene sequence and used the beta-thalassaemic mouse (25) as a model for preimplantation diagnosis of a defect in a single-copy gene sequence. The mutation is a spontaneously occurring deletion of about 3.3 kb including the entire beta-major haemoglobin gene sequence. Mice homozygous for the deletion survive and breed but are smaller than usual and are anaemic. Single cells are isolated from individual homozygous mutant or normal 8-cell embryos for diagnostic tests. A similar approach was taken by Lindeman et al. (26) using about five cells removed from later morula stages of development.

Diagnostic DNA Analysis

Sensitivity and specificity of the beta-haemoglobin sequence amplification is achieved with the use of nested primers (see Fig. 5, top; see also ref. 26). A 269-base pair region containing the whole of exon 3 of the mouse beta-major haemoglobin gene was at first amplified. Then an aliquot of this re-

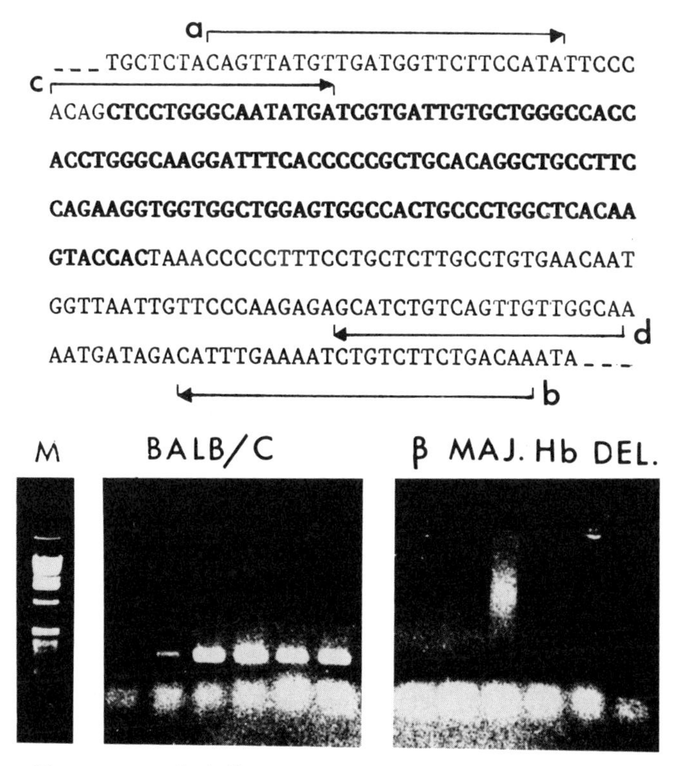

FIG. 5. Top: The sequence including exon 3 (*bold type*) of the mouse beta-haemoglobin gene and the primers (a and b; c and d) used in two sequential PCR amplifications. **Bottom:** Amplification of a 204-base pair sequence including exon 3 in single blastomeres of 8-cell embryos from normal mice (BALB/C). Blastomeres from homozygous beta-haemoglobin deletion mice (B MAJ. Hb DEL) do not give the amplification product. Water controls and wash medium controls show no contamination. M, markers. (Data from Holding and Monk, ref. 25.)

action was diluted into a new reaction mix containing a second set of primers homologous to sequences just inside the first amplified fragment. After another 30 cycles of amplification, a single strong amplified band from this single-copy gene in a single cell is detected on ethidium-bromide-stained gels. With this supreme sensitivity, contamination becomes a very difficult problem. Stringent precautions must be taken to ensure absence of contamination, the most important precautions being geographical isolation of the areas and equipment used for setting up the reactions and analysing the products.

Using this approach, 8-cell embryos from normal wild-type mice and from mice homozygous for the beta-major haemoglobin deletion could be accurately diagnosed by amplification of the DNA from just one single blastomere sampled from the embryos (Fig. 5, bottom) (27). Out of 30 blastomeres from control 8-cell embryos, 25 gave the expected amplified product whilst the remaining 5 gave a false negative response. Out of 26 "deletion" blastomeres isolated from 8-cell embryos of the deletion mutant, 25 were diagnosed correctly as negative and one blastomere gave a false positive signal.

The preliminary work in the mouse model showed that we had the sensitivity and specificity of single-copy gene detection needed to allow preimplantation diagnosis. Ideally, in a diagnostic test employing amplification of a DNA sequence identifying the presence or absence of a mutation, it would be necessary to detect two different products that unambiguously identified either the normal or the mutant DNA in the blastomere. Therefore, in our initial experiments to extend this work to the human, we chose to diagnose the mutation associated with sickle cell anaemia, where both the normal and the mutated alleles can be positively identified by restriction enzyme digest.

Beta-Haemoglobin Gene Detection in Human Preimplantation Embryos

We developed PCR amplification of a 680-base pair sequence of the human beta-haemoglobin gene spanning the site of the mutation which, in the homozygous condition, causes sickle cell anaemia. The specificity and sensitivity of amplification to the level of a single cell are again achieved by the two sequential reactions with two sets of primers (Fig. 6). Sickle cell anaemia results from a mutation in the codon in exon 1, resulting in an alteration in the sixth amino acid of the beta subunit of haemoglobin (28). The mutation destroys the *DdeI* site in exon 1, so that a restriction digest of the amplified product obtained from a single cell will positively identify the presence or absence of the sickle cell mutation in homozygous as well as heterozygous condition (29).

We applied this diagnostic test to human unfertilised oocytes (kindly provided by Dr. Peter Braude, Rosie Maternity Hospital, Cambridge, England) and the first polar bodies isolated from them (9). In the case of four fresh

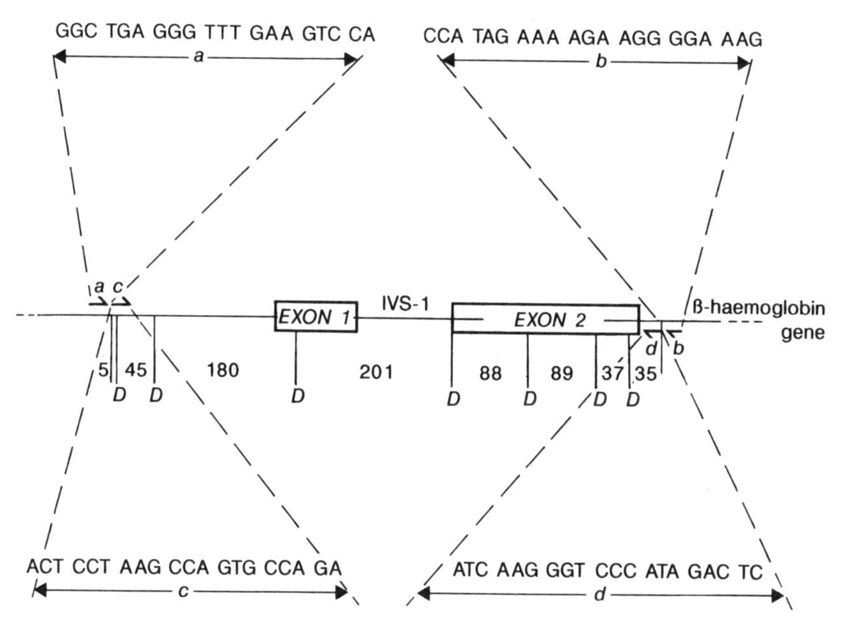

FIG. 6. The region of the human beta-haemoglobin gene amplified using two sequential PCR amplifications with the primers as shown. a and b are outer primers; c and d are nested primers. *Dde1* restriction enzyme sites (D) are shown. The site in exon 1 is destroyed by the sickle cell mutation. (From Monk and Holding, ref. 9.)

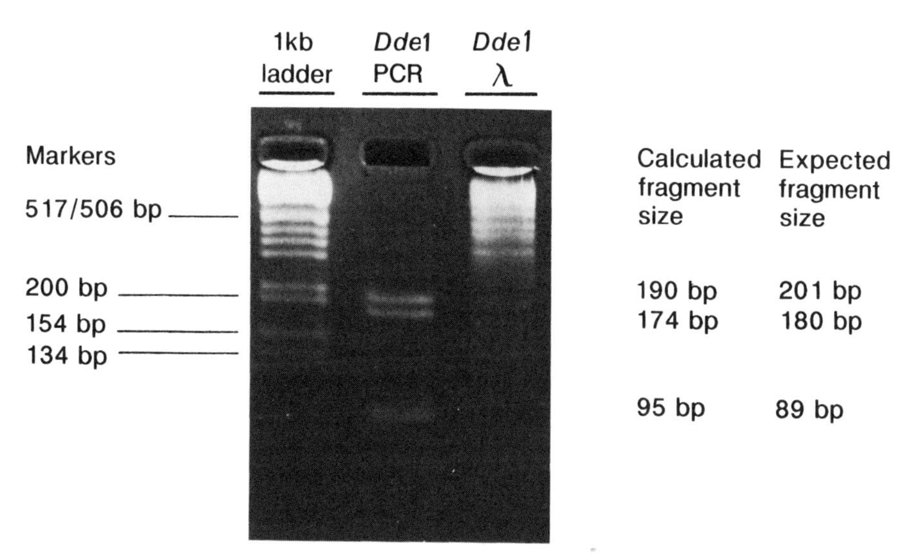

FIG. 7. A *Dde1* restriction digest of the amplified products of a single unfertilised human oocyte showing bands (201 and 180) expected from a normal (no sickle cell trait) individual. (From Monk and Holding, ref. 9.)

oocytes that had not been incubated with sperm for fertilisation, the single separated oocytes and their polar bodies all showed amplification of their beta-haemoglobin sequence. The four control samples taken from their final wash droplets were negative, confirming absence of contamination. Results from isolated polar bodies of older failed-fertilisation oocytes (incubated for 24 hours with sperm) were also highly encouraging though less reliable than fresh oocytes (9). A restriction digest of the amplified product from a single oocyte with the enzyme *DdeI* is shown in Fig. 7. Bands of the expected size for a normal individual (no sickle cell trait) are seen (201 and 180 base pairs).

This demonstration of amplification and analysis of DNA in single cells means that preimplantation diagnosis of sickle cell anaemia is possible using

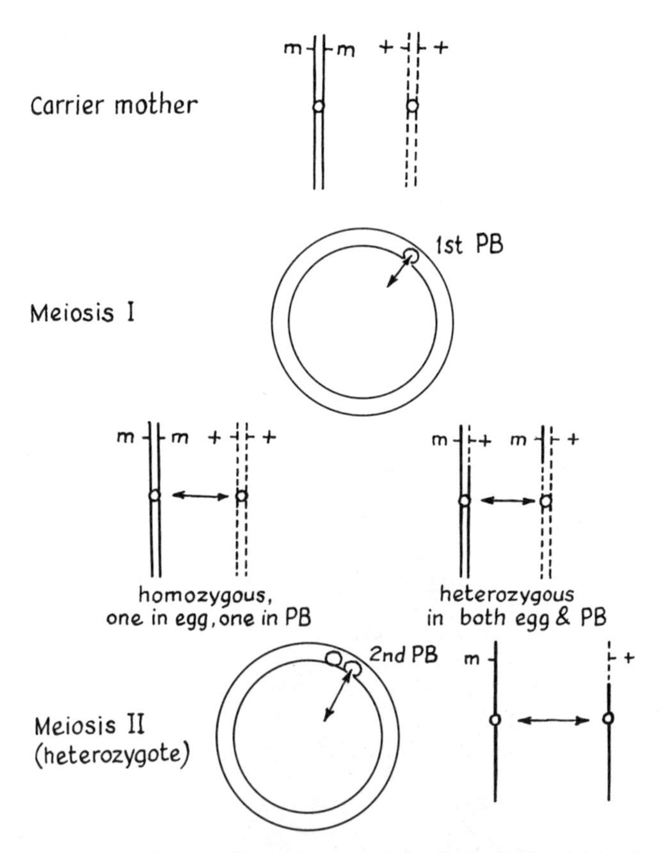

FIG. 8. Diagram showing the possible outcomes of the first meiotic division in oocyte and first polar body (PB) from a heterozygous female. Either the oocyte is normal and polar body mutant, or the oocyte is mutant and the polar body normal, or both are heterozygous if cross-over has occurred. As a procedure to prevent genetic disease, only the normal oocytes would be fertilised and replaced in the mother. m, mutation; +, normal.

a blastomere isolated from a cleavage stage embryo. In addition, the reproducible amplification of this sequence in a single first polar body demonstrates the feasibility of another approach that may circumvent some of the ethical difficulties of dealing with human embryos. The first polar body of an unfertilised oocyte contains information about its own genetic content and that of the oocyte itself (see Fig. 8). By diagnosing the genetic status of the first polar body, it is possible to determine which oocytes carry the defective gene, which carry the gene without the defect, and which are heterozygous due to a cross-over event in first meiotic prophase. Only the oocytes without the defect would be fertilised and replaced in the mother. A similar approach to genetic diagnosis by PCR analysis of the first polar body has been reported by Verlinsky et al. (30). These workers reported detection of normal and mutant gene sequences of the alpha-1-antitrypsin gene by PCR amplification of polar body DNA and hybridisation with allele-specific oligonucleotide probes.

We are currently attempting to demonstrate the efficiency with which both alleles in a single cell heterozygous for the sickle cell mutation are amplified. Previous workers have shown independent amplification of two or more minisatellite alleles from a single cell (22).

CONCLUSIONS

Using mouse model systems we have developed procedures for the diagnosis of genetic disease in a single cell of a preimplantation embryo or in the first polar body of an unfertilised egg. In the mouse it would appear that the removal of a few cells for testing will not jeopardise the future development of the embryo to term. In vitro fertilisation and subsequent development of human oocytes after removal of the first polar body has been reported by Verkinsky et al. (30). If this diagnostic test were to be used clinically, it would be necessary to remove the first polar body from beneath the zona pellucida, rather than removing the zona entirely.

In certain cases, the preferred approach to preimplantation diagnosis may be to test for the absent or altered gene product directly—for instance, in the case of new mutations or mutations for which the gene product, but not the gene sequence, is known. Diagnosis by DNA analysis using polymerase chain reaction amplification of the relevant gene sequence may be a more generally applicable approach, provided the mutant DNA base sequence is known. It has been used by us to accurately diagnose mouse embryos mutant for the beta-major haemoglobin genes by analysis of a single blastomere from the 8-cell stage, and to show that diagnosis of unfertilised human oocytes by analysis of the first polar body is a feasible proposition. Currently, the procedures are being developed further for preimplantation diagnosis in families affected by sickle cell anaemia and thalassaemia.

ACKNOWLEDGMENTS

I gratefully acknowledge the collaboration of Alan Handyside, Audrey Muggleton-Harris, David Whittingham, Peter Braude, and Martin Johnson. I am particularly grateful to Cathy Holding for her collaboration in the PCR procedures.

REFERENCES

1. Monk M, Handyside A, Hardy K, and Whittingham D. Preimplantation diagnosis of deficiency of hypoxanthine phosphoribosyl transferase in a mouse model for Lesch-Nyhan syndrome. *Lancet* 1987;ii:423–426.
2. Monk M and Handyside A. Sexing of preimplantation mouse embryos by measurement of X-linked gene dosage in a single blastomere. *J Reprod Fertil* 1988;82:365–368.
3. Handyside AH, Pattinson JK, Penketh RJA, et al. Biopsy of human preimplantation embryos and sexing by DNA amplification. *Lancet* 1989;i:347–349.
4. Edwards RG and Gardner RL. Sexing of live rabbit blastocysts. *Nature* 1967;214:576–577.
5. Gardner RL and Edwards RG. Control of the sex ratio at full term in the rabbit by transferring sexed blastocysts. *Nature* 1968;218:346–349.
6. Summers PM, Campbell JM, and Miller MW. Normal in vivo development of marmoset monkeys after trophectoderm biopsy. *Hum Reprod* 1988;3:389–393.
7. Monk M, Muggleton-Harris A, Rawlings E, and Whittingham D. Preimplantation diagnosis of HPRT-deficient male and carrier female mouse embryos by trophectoderm biopsy. *Hum Reprod* 1988;3:377–381.
8. Buster JE, Bustillo M, Rodi IA, et al. Biologic and morphologic development of human ova recovered by non-surgical uterine lavage. *Am J Obstet Gynecol* 1985;153:211–217.
9. Monk M and Holding C. Amplification of a β-haemoglobin sequence in individual human oocytes and polar bodies. *Lancet* 1990;335:985–988.
10. Hooper M, Hardy K, Handyside A, Hunter S, and Monk M. HPRT-deficient (Lesch-Nyhan) mouse embryos derived from germline colonisation by cultured cells. *Nature* 1987;326:292–295.
11. Lesch M and Nyhan WL. A familial disorder of uric acid metabolism and central nervous system function. *Am J Med* 1964;365:561–570.
12. Monk M. (1987): Biochemical microassays for X-chromosome-linked enzymes HPRT and PGK. In *Mammalian Development—A Practical Approach,* edited by M. Monk. pp. 139–161. IRL Press, Oxford.
13. Monk M and Harper M. X-chromosome activity in preimplantation mouse embryos from XX and XO mothers. *J Embryol Exp Morphol* 1978;46:53–64.
14. Monk M, editor. *Mammalian Development—A Practical Approach.* IRL Press, Oxford, 1987.
15. Benson C and Monk M. Microassay for adenosine deaminase, the enzyme lacking in some forms of immunodeficiency, in mouse preimplantation embryos. *Hum Reprod* 1988;3:1004–1009.
16. Harper M and Monk M. Evidence for translation of HPRT enzyme on maternal mRNA in early mouse embryos. *J Embryol Exp Morphol* 1983;74:15–28.
17. Braude PR, Bolton VN, and Moore S. Human gene expression first occurs between the four- and the eight-cell stages of preimplantation development. *Nature* 1988;332: 459–461.
18. Bolton VN, Oades PJ, and Johnson MH. The relationship between cleavage, DNA replication and gene expression in the mouse 2-cell embryo. *J Embryol Exp Morphol* 1984;79:139–163.
19. Braude PR, Monk M, Pickering SJ, Cant A, and Johnson MH. Measurement of HPRT activity in the human unfertilised oocyte and pre-embryo. *Prenat Diagn* 1989;9:839–850.

20. Saiki RK, Gelfand DH, Stoffel S, et al. Primer-directed enzymatic amplification of DNA with a thermstable DNA polymerase. *Science* 1988;239:487–491.
21. Li H, Gyllensten UB, Cui X, et al. Amplification and analysis of DNA sequences in single human sperm and diploid cells. *Nature* 1988;335:414–417.
22. Jeffreys AJ, Wilson V, Neumann R, and Keyte J. Amplification of human minisatellites by polymerase chain reaction: toward fingerprinting of single cells. *Nucleic Acids Res* 1988;16:10953–10971.
23. Handyside AH, Kontogianni EH, Hardy K, and Winston RM. Pregnancies from biopsied human preimplantation embryos sexed by Y-specific DNA amplification. *Nature* 1990;344:768–770.
24. Coutelle C, Williams C, Handyside A, et al. Genetic analysis of DNA from single human oocytes: a model for preimplantation diagnosis of cystic fibrosis. *Br Med J* 1989;299: 22–24.
25. Skow LC, Burkhart BA, Johnson FM, et al. A mouse model for beta-thalassaemia. *Cell* 1983;34:1043–1052.
26. Lindeman R, Lutjen J, O'Neill C, and Trent RJ. Exclusion of β-thalassaemia by biopsy and DNA amplification in mouse pre-embryos. *Prenat Diag* 1990;10:295–301.
27. Holding C and Monk M. Diagnosis of beta-thalassaemia by DNA amplification in single blastomeres from mouse preimplantation embryos. *Lancet* 1989;ii:532–535.
28. Marcotta C, Wilson J, Forget B, and Weissman S. Human beta-globin messenger RNA. *J Biol Chem* 1977;252:5040–5053.
29. Embury S, Scharf S, Saiki R, et al. Rapid prenatal diagnosis of sickle cell anaemia by a new method of DNA analysis. *N Engl J Med* 1987;316:656–661.
30. Verlinsky Y, Pergament E, Andresen P, Enriques G, and Strom C. Genetic analysis of polar body DNA: a new approach to preimplantation genetic diagnosis. *Hum Genet* 1989;Suppl 45.
31. Monk M. Preimplantation diagnosis by biochemical or DNA microassay in a single cell. In *Establishing a Successful Human Pregnancy,* edited by RG Edwards, Serono Symposium 60:183–197. Raven Press, New York, 1990.

Etiology of Human Disease at the DNA Level,
edited by Jan Lindsten and Ulf Pettersson.

18

Marrow Transplantation as a Model System for Correction of Genetic Disease

Richard J. O'Reilly

Memorial Sloan-Kettering Cancer Center, New York, New York 10021

Definition of the genes responsible for inherited diseases in humans and the isolation and cloning of their normal counterparts has focused the attention of scientists and clinicians on the possibility of correcting inherited diseases either through the administration of the normal protein produced by recombinant technology or through the introduction of genetically normal allogeneic or genetically modified autologous cells into the patient so as to provide a continuously renewable source of the gene product needed for normal function.

While administration of purified recombinant proteins has been found to be effective for several inherited disorders of plasma proteins or factors that act at the cell's surface (1,2), this approach has generally been ineffective in the treatment of genetic diseases involving enzymes active within the cell, with one notable exception (3). Even for genetic disorders of metabolism in which the biochemical abnormalities exhibited by affected cells can be corrected in vitro by introducing the deficient enzyme into the culture medium, infusions of the protein in vivo have usually been ineffective, either because the enzyme is too rapidly metabolized or is inadequately transported into affected cells or because the affected individual generates an immune response to the protein that results in its rapid elimination (4–6). Transplants of organs such as the kidney and the liver have been effective in securing long-term survival for patients with disorders such as cystinosis (7), hyperoxalurea (8), and alpha-1-antitrypsin deficiency (9), diseases inducing pathology primarily in the organs requiring replacement. Liver transplants induce a normalization of plasma levels of missing enzymes and extrahepatic manifestations of alpha-1-antitrypsin deficiency (9,10). In contrast, renal transplants have corrected only the derangements induced by uremia; extrarenal manifestations of primary metabolic defects have not been altered (7,8,11). Transplants of major organs have also necessitated indefinite treatment of the patient with immunosuppressive drugs to prevent rejection. Another approach has involved the administration of functionally normal cells such as cultured fibroblasts (12,13) or amnion cells (14,15) from normal allogeneic donors. Such transplants have produced transient biochemical improvement in certain genetic diseases of metabolism (13–15). However, the transplanted cells have a limited capacity for renewal and are also susceptible to rejection by the immunologic system of the host (13,15).

Marrow transplants circumvent several of these limitations. In this procedure, intensive myeloablative and immunosuppressive therapy is administered to destroy the patient's lymphoid and hematopoietic system. The marrow graft then repopulates the lymphohematopoietic system with cells of donor origin. Once engrafted, the progenitor cells in the marrow graft continue to generate a normal hematopoietic and immune system of donor type for the life of the transplant recipient. Thus, a continuously self-renewing population of genetically normal donor cells capable of generating deficient proteins is provided. Furthermore, since the immune system following transplantation is also of donor origin, the normal gene product does not elicit an immune response. Since marrow transplants replace the host's lymphohematopoietic system, they have been widely used to correct lethal genetic disorders of blood cell formation and function. The applicability of a marrow transplant to the correction of a genetic disorder of nonhematopoietic tissues depends on the capacity of the blood cells to produce the needed gene product and the capacity of affected cells either to incorporate this gene product appropriately into their metabolic functions, or alternatively, to mobilize substrates so as to be exposed to the normal gene product either in the plasma or within a circulating blood cell or tissue macrophage derived from a normal donor.

Transplants of marrow from histocompatible normal donors were first performed in 1968 and led to the correction of two lethal genetic disorders of immunity, severe combined immunodeficiency (SCID) (16) and Wiskott-Aldrich syndrome (17). In the following 10 years, marrow grafts were also found to correct several other lethal genetic disorders of hematopoiesis, and became recognized as a treatment of choice for aplastic anemia (18), chronic myelogenous leukemia (19,20), and acute leukemias failing an initial remission (21–23). It also became apparent that transplants administered to children or young adults in good clinical condition were associated with a markedly reduced transplant-related mortality (19,22,24–26). This finding, coupled with the development of an effective myeloablative and immunosuppressive preparative regimen for transplantation that does not include radiation (27,28) and its subsequent successful application to transplants for Wiskott-Aldrich syndrome (28) and osteopetrosis (29), has led to increased exploration of marrow transplants as a curative approach for a variety of genetic diseases of the blood system and as a treatment for certain lethal disorders of metabolism (30). In this report, I will review results of marrow transplants applied to genetic diseases and highlight important biological issues raised by this experience that are being addressed in current and planned research.

BIOLOGY OF MARROW TRANSPLANTS FOR GENETIC DISEASES

The success or failure of a marrow transplant as measured by the reestablishment of normal hematopoiesis and immunologic function in an allogeneic host, the development of durable stable chimerism, tolerance between graft and host, and sustained disease-free survival depend on many factors, principal among which are: the medical status of the prospective transplant recipient, the degree of histocompatibility existing between donor and host, the adequacy of the preparative regimen used to ablate the recipient's hematopoietic system and to eliminate the recipient's capacity to initiate an immunologic rejection of the marrow graft, the toxicity of this preparative regimen, the adequacy of the measures used to prevent and/or treat graft versus host disease (GvHD), and the success of measures used to prevent or treat infections.

The HLA antigens constitute the major determinants of histocompatibility between donor and marrow graft recipient. Disparities for these alloantigens markedly increase the risk of both marrow graft rejection and severe or lethal graft versus host disease (31–33). For these reasons, HLA-matched sibling donors have been used in most transplants applied to the treatment of genetic diseases. Despite matching for the HLA-A, -B, -C, and -D determinants, graft rejections have been observed in up to 30% of multiply transfused patients with aplastic anemia prepared for transplantation with cyclophosphamide alone (34). Furthermore, between 30 and 70% of all patients who have achieved engraftment following a transplant of HLA-matched marrow will develop acute GvHD (35–37). The minor alloantigens stimulating these T-cell-mediated reactions are still poorly understood. Mounting evidence suggests that the Y-chromosome-encoded Hy antigen may serve as a target for marrow graft rejection or GvHD in humans (38–41). Other alloantigens expressed on marrow cells have recently been described by Goulmy et al. (42); however, their role as targets for either rejection or GvHD is as yet unclear.

Durable engraftment of the marrow transplants in an allogeneic host can only be achieved if the host has been rendered incapable of mounting an immunologic response against donor-derived hematopoietic cells. Failure to ablate the host immune response places the marrow graft at risk for rejection by allospecific T cells, particularly T cells previously sensitized by transfusions (43–45). Nonspecific or autoimmune resistance systems may also contribute to graft failure as has been observed, for example, in selected patients with aplastic anemia who have failed to engraft following transplants from identical twins (46) and in certain patients with SCID who have no capacity for alloantigen-specific immune responses (47). The graft may also fail if the microenvironment is not conducive to its development, as has been observed in murine models of genetic or radiation-induced disorders of the marrow microenvironment (48–50). In addition, hematopoietic cells of the donor may compete ineffectively with host cells for space or nutrients within the host environment if the host marrow is not depleted sufficiently to prevent the regrowth of autologous marrow cells, as has been observed in a proportion of patients transplanted for thalassemia (51).

With the exception of SCID (52) and possibly reticular dysgenesis (53), the genetic diseases of the lymphohematopoietic system are not associated with immune deficiencies severe enough to permit durable engraftment of allogeneic HLA-identical marrow unless the patient is intensively treated with immunosuppressive agents prior to the marrow graft. At the present time, only cyclophosphamide and total body irradiation (TBI) have been shown to be sufficiently immunosuppressive to ensure engraftment of an allogeneic marrow transplant (54,55). For patients with disorders not producing marrow aplasia, myeloablation with agents such as TBI or myeleran must also be used to deplete the host of its hematopoietic elements in order to provide a noncompetitive microenvironment in which the donor marrow can effectively develop and expand (28,56,57). A preparative combining myeleran and cyclophosphamide at high doses has been most extensively used (28,51,57). The marrow transplant is then administered within 1–2 days of completion of this cytoreductive regimen.

The posttransplant course can be divided into three phases: a period of pancytopenia prior to engraftment (days 0–25), followed by the engraftment phase (days 26–60) during which hematopoietic reconstitution is achieved, and lastly, a period of immunologic reconstitution that may extend late into the posttransplant period. During the initial pancytopenic phase, supportive care is focused on the treatment of the acute toxic effects of the preparative regimen and the vigorous treatment of infections resulting from invasion of injured mucosal surfaces and potentiated by the leukopenic state of the host. As a result of these measures, mortality among children transplanted for genetic diseases during this period is low (2–5%) (51,58).

Engraftment, marked by the development of donor-type myeloid progenitors in the marrow and the emergence of neutrophils in the blood, is usually observed by days 12–25. However, the graft may fail to develop or may fully reconstitute hematopoiesis only to fail later in the first 3 months posttransplant. Graft failure usually reflects an immune rejection mediated by donor-reactive host T lymphocytes (43–45). It is a rare complication of unmodified HLA-matched marrow grafts administered to leukemic patients prepared with total body irradiation and cyclophosphamide, but occurs at rates ranging from 5 to 30% among similarly prepared leukemic recipients of HLA nonidentical or T-cell-depleted marrow grafts (31,38,59,60). Multiply-transfused patients with aplastic anemia who receive HLA-matched unmodified marrow grafts after preparation with cyclophosphamide alone are also at particular risk for this complication (34).

Once engraftment is established, hematopoietic reconstitution is rapid and may be complete by days 25–35. However, once engraftment has been achieved, the patient is at risk for acute GvHD. This pathologic process is initiated by donor T lymphocytes responding to host alloantigens and results in the generation of host-reactive cytolytic T cells and the conscription of monocytes and natural killer cells, which infiltrate the skin, the submucosal areas of the bowel, the liver, and the lymphoreticular system, producing secondary necrosis of cells in infiltrated tissues (61–63). While 30–70% of patients transplanted with HLA-matched unmodified marrow will develop this complication (35,36), the incidence of severe GvHD is proportional to the age of the patient and is low among children transplanted in the first decade of life (10–30%) (36,64–66). Among older individuals, however, acute GvHD contributes directly or indirectly to death in 20–40% of affected individuals (66,67). The incidence of severe GvHD and associated mortality has been reduced in recipients of HLA-matched, unmodified grafts by the prophylactic administration of methotrexate and cyclosporine (68). The risk of severe GvHD is further reduced or eliminated if the marrow is depleted of T cells (60,69,70). Approximately 30% of patients surviving an unmodified allogeneic marrow graft, irrespective of age, will develop chronic GvHD, a debilitating process pathologically distinct from acute GvHD (71). However, long-term treatment with immunosuppressive agents can ameliorate or reverse manifestations of this disorder (72). Again, the use of T-cell-depleted marrow transplants markedly reduces or eliminates the incidence of this disorder (60,69,70).

During the engraftment phase, patients are also at particular risk for interstitial pneumonias caused by cytomegalovirus (CMV), *Pneumocystis carinii,* and, less commonly, other viruses (73). Recently, however, dramatic reductions in the incidence and mortality of these complications has been achieved through the introduction of more effective prophylactic measures such as the administration of trimethoprim/sulfa prophylaxis for *Pneumocystis* before and after transplant (74), use of CMV-seronegative blood products (75) and the development of an effective treatment for CMV pneumonias (76,77).

Advances made in preparation and supportive care of marrow transplant recipients and development of more consistently effective measures for preventing or treating transplant-related complications have significantly reduced the risk of mortality following a marrow graft, particularly for patients with genetic diseases transplanted in the first decade of life. Indeed, the incidence of peritransplant mortality for children in this age range who, at the time of transplant, do not have infections or debilitating pathology in vital organs such the liver, lung, or heart, is less than 10% (51,58,78).

MARROW TRANSPLANTS APPLIED TO GENETIC DISORDERS OF HEMATOPOIESIS AND IMMUNITY

Lethal Immune Deficiencies

Over the past 20 years, marrow transplantation has become recognized as the treatment of choice for several lethal immune deficiencies, particularly the different variants of SCID (78). In a recent European survey of patients transplanted with HLA-matched marrow for SCID, 68% were surviving with reconstitution of immunity through 1986 (79). For patients with SCID diagnosed early in life who are referred for transplantation before they have acquired serious infections, the probability of a successful transplant is in excess of 90% (80).

Patients with SCID usually do not require any pretransplant cytoreduction in preparation for an HLA-matched unmodified marrow transplant, because they have little capacity to resist engraftment of HLA-matched marrow. However, in over 50% of cases, the only cells engrafted following such transplants are donor T cells and their precursors (78). Despite this limited chimerism, HLA-matched grafts usually result in a full reconstitution of both T- and B-cell function.

A series of lethal combined immunodeficiency (CID) states can be distinguished from SCID by the fact that the response of T cells to allogeneic cells is at least partially preserved, even though humoral and cell-mediated responses to microbial antigens may not be detected. In most of these disorders the etiology of the immune deficiency observed is still unknown. However, in recent years the pathogenetic bases for several syndromes have been described. These include: a congenital deficiency of the interleukin-1 (IL-1) receptor (81), congenital deficiencies of T-cell activation and IL-2 production (82–84), defects of the T-cell receptor (85), and the Bare lymphocyte syndromes, a series of genetic disorders resulting in defective transcription and ultimate expression of class I and/or class II HLA determinants on the surface of lymphoid and hematopoietic cells (86,87). In other diseases, distinctive functional abnormalities, such as aberrant antigen or mitogen-induced membrane capping (88) and cell membrane depolarization, have been described (89).

For patients with these lethal CID syndromes, in which T cells retain some capacity to respond to allogeneic cells, preparatory immunosuppression with high-dose cyclophosphamide is likely required to secure engraftment. Thus, HLA-matched transplants administered to patients with CID syndromes without preparatory cytoreduction have often failed to achieve engraftment. In contrast, patients who have been pretreated with high doses of cyclophosphamide and either cytosine arabinoside or myeleran have regularly engrafted and achieved long-term immunologic reconstitution (90).

Several lethal congenital disorders of immunity also involve other hematopoietic lineages. Correction of these disorders requires replacement of both lymphoid and other hematopoietic cells with progenitor cells derived from the normal donor. To achieve this, patients must undergo both intensive immunosuppression with cyclophosphamide and myeloablation with an agent such as myeleran, dimethylmyeleran, or TBI. The most frequently transplanted immune deficiency of this type is Wiskott-Aldrich syndrome, a sex-linked disorder characterized by deficiencies of both the T- and B-cell systems and thrombocytopenia (91). Transplants administered to such patients prepared with cyclophosphamide alone have led to engraftment of donor lymphoid cells with corrections of immune deficiencies but persistence of thrombocytopenia (17). In contrast, over 90% of patients prepared with cyclophosphamide in combination with myeleran, dimethylmyeleran, or TBI have achieved engraftment of both lymphoid and hematopoietic progenitors from the donor, full corrections of both lymphoid and platelet abnormalities, and long-term disease-free survival (30,92). Other, rare disorders of immunity that have been successfully corrected by this approach include: reticular dysgenesis, a disorder characterized by a severe CID coupled with profound leukopenia and anemia (93), cartilage-hair hypoplasia (94), and the Bare lymphocyte syndromes (95).

Congenital Disorders of Red Cell Formation and Function

Fanconi's anemia (FA) is an autosomal recessive disorder characterized by a variety of congenital malformations and anemia progressing to marrow aplasia or leukemia (96). A consistent feature of the disorder is an increased frequency of chromosomal breaks observed in metaphase preparations of peripheral blood lymphocytes upon exposure to difunctional alkylating agents, such as diepoxybutane, cyclophosphamide, or mitomycin C (97). Patients with FA rarely survive adolescence with death most often due to the complications of progressive marrow aplasia. Acute leukemia also occurs at a high frequency in these patients and has been uniformly fatal.

Bone marrow transplantation offers the potential for permanent correction of the stem cell defect of FA. Initial results with marrow transplantation for FA were poor, due, in large measure, to the severe toxicity incurred by these patients following treatment with regimens containing high doses of cyclophosphamide (98). However, in 1983, Gluckman et al. (99) described a regimen combining low-dose cyclophosphamide and thoracoabdominal radiation for preparing patients with FA for marrow transplantation. Of 19 patients treated with this regimen, 14 have survived, free of disease, 6 months to 6 years (median follow-up, 4 years) following transplantation with sustained engraftment of donor marrow elements (100). Eighteen of the 19 patients engrafted. Moderate to severe acute GvHD of grade II+ was observed in 11/18 evaluable patients despite prophylaxis with cyclosporine, possibly reflecting a limited capacity for repair of GvHD-induced lesions by cells exposed to alkylating agents. T-cell-depleted marrow grafts are currently being explored by several centers to circumvent this problem.

Congenital erythroid hypoplasia (Diamond-Blackfan syndrome) is a genetic disorder of uncertain inheritance characterized by a selective failure of erythroid progenitor development that results in a profound anemia from early infancy. Partial remissions may be achieved with steroid treatment, but are sustained in only 20–30% of cases (101). Unresponsive patients ultimately succumb to the hepatic and cardiac sequelae of transfusion-induced hemosiderosis. In 1976, August, et al. (102) reported a patient with this disease whose red cell production had been restored to normal following successful engraftment of an HLA-matched marrow transplant. Although this patient ultimately died of interstitial pneumonia, five subsequent patients transplanted for this disorder reported by Iriondo et al. (103) and Lenarsky et al. (104) have each achieved engraftment of donor hematopoietic cells, durable corrections of anemia, and long-term disease-free survival.

Thalassemia major currently represents the genetic disorder for which HLA-matched allogeneic marrow grafts are most frequently performed with over 222 patients transplanted by Luccarelli et al. (51) in Pesaro, and at least 90 patients reported so far from other centers in Europe and the United States (105–107). For patients transplanted before the age of 16 years, at least six centers have reported incidences of long-term survival and disease-free survival of 62–91% and 56–88%, respectively (105–110). Of a series of 222 such patients transplanted in Pesaro after conditioning with busulfan and cyclophosphamide, 82% survive, 75% disease-free for 1–6 years posttransplant (51). In each series, a proportion of patients (6–23%) has rejected their transplants within 3 months and experienced recovery of autologous thalassemic marrow cells. Unexpectedly, however, fewer graft rejections have been observed in the more heavily transfused patients (51). The major limitations to the success of transplants for these patients are interstitial pneumonias and GvHD, which may be less tolerable in patients with preexisting hepatic hemosiderosis (51). Largely as a result of such complications, transplants administered to older patients (>16 years of age), who have already developed significant hepatic dysfunction, have been associated with a low incidence (10–20%) of extended disease-free survival (51).

Given these results, it is clear that a transplant has the highest likelihood of inducing curative corrections of thalassemia and ensuring long-term disease-free survival if applied to a patient at some time during the first 10–15 years of life. Major controversy arises when clinicians attempt to weigh the benefits of a transplant at this age, namely a complete correction of red cell production, function, and lifespan that eliminates the need for transfusions or sustained chelation therapy, with the 10–15% risk of mortality associated with a transplant. Since recent studies suggest that with vigorous transfusion support and intensive, continuous chelation therapy with desferoxamine, approximately 67% of patients with thalassemia will survive at least 25 years (111), it is argued that transplants constitute too great a risk to a child, particularly during the first 15 years when the mortality due to transfusion-associated infections or hemosiderosis is only 2–4%. However, compliance with chelation therapy is often inconsistent, particularly among adolescents, which reduces chances for extended survival into adulthood (112). Furthermore, there is still little evidence to suggest that intensive chelation and transfusion support has significantly increased the proportion of patients surviving into the fourth decade (113). Given this situation, transplants are being increasingly invoked for the treatment of patients for whom transfusion and chelation support cannot be adequately provided, either because of limitations in the availability of medical services or patient noncompliance.

Recently, allogeneic marrow transplants have also been applied as a curative approach to the treatment of sickle cell anemia (114–116). Although experience with transplants for this disease is limited, such grafts have been consistently successful. Based on these initial promising results, Vermylen et al. (116) argue that such transplants should be considered for patients with homozygous sickle cell anemia who develop recurrent destructive vaso-occlusive events, particularly if these patients live in locales in which the accessibility of appropriate medical care is limited.

Genetic Disorders of Platelet Production and Function

Marrow transplants have been used successfully to treat congenital amegakaryocytosis, a lethal disorder producing severe sustained thrombocytopenia from birth (117). Without a transplant, such patients usually succumb to intracranial or gastrointestinal hemorrhage during the first 3 years of age. A separate disorder, the TAR syndrome, is characterized by severe thrombocytopenia, absence of the radii and other malformations. With platelet transfusions, many patients with this disorder survive the first year, after which, for unclear reasons, thrombocytopenia usually resolves (118). A proportion of patients with thrombocytopenia-absent radius (TAR) syndrome however, remain severely thrombocytopenic. Our group has recently administered an HLA-matched marrow graft to a 2 year old patient with TAR syndrome who had persistent amegakaryocytosis resulting in repeated intracranial hemorrhages. The transplant, administered after preparation of the patient with bulsulfan and cyclophosphamide, has led to full engraftment and durable correction of platelet production (119).

The autosomal recessive diseases Glanzmann's thrombasthenia types I and II constitute another series of life-threatening disorders of platelets. In the severe form (type I), the membrane glycoproteins IIb and III are absent, leading to the inability of platelets to aggregate effectively. Affected patients are at high risk of severe hemorrhage (120). Recently, Belucci et al. (121)

have corrected this disorder in a 4 year old child through a series of two HLA-matched transplants administered after myeloablation with Lomustine and procarbazine and immunosuppression with cyclophosphamide.

Disorders of Myelopoiesis and Phagocyte Function

Allogeneic marrow grafts have been applied as a corrective treatment for a series of lethal disorders of phagocyte production or function with variable results. The lethal disorder of neutrophil production, congenital agranulocytosis (Kostmann's syndrome) (122), has been corrected by allogeneic marrow grafts administered after suitable myeloablation and immunosuppression (123). In canine models, marrow grafts have also been shown to be effective in correcting cyclic neutropenia (124). Recently, however, an alternative approach has been developed for the treatment of these disorders. Studies by Bonilla et al. (125) and Welte et al. (126) indicate that production of normal numbers of functional neutrophils can be induced in the vast majority of patients with severe forms of Kostmann's agranulocytosis by the daily administration of recombinant granulocyte colony-stimulating factor (GCSF). Similarly, while treatment with GCSF does not prevent the periodic flux in neutrophil production associated with cyclic neutropenia, it can increase neutrophil counts at the nadir of each cycle (127).

Disorders of phagocyte function such as the different variants of chronic granulomatous disease (CGD) (128,129), congenital deficiency of LFA-1, the adhesion molecule on the surface of neutrophils critical to their capacity to move from the blood stream to sites of infection (130), and congenital actin deficiency (131) can also be corrected by eliminating the marrow of the diseased host and replacing it with marrow from a normal allogeneic donor. This approach has led to long-term disease-free survival for most patients transplanted for LFA-1 deficiency (130). However, in the experience accrued to date (30,79,128,129), all but one of the patients transplanted for CGD (129) or congenital actin deficiency have succumbed to preexisting or intercurrent infections early or late in the posttransplant period. Improved results would likely be achieved if such patients were transplanted when free of active infection.

In the future, other approaches not involving transplantation may also improve the prognosis of patients affected with these disorders. For example, the deficiencies of cytochrome-c characterizing neutrophils in the X-linked variant of CGD (132) can be partially corrected by activation of the cytochrome c heavy chain gene with gamma interferon (133). Preliminary results have also suggested that gamma interferon will at least partially correct deficiencies observed in vivo in this variant of CGD and thereby alter susceptibility to infection (134).

Chédiak-Higaschi syndrome, an autosomal recessive disorder characterized by pigmentary dilution of the hair and skin, deficiencies of natural killer cells, and abnormalities of microtubular assembly in the phagocytes leading to the development of characteristic giant cytoplasmic granules and impairments of bactericidal activity (135) has been repeatedly corrected by allogeneic marrow transplants (136). Engrafted patients have achieved full functional recovery of all affected lineages if prepared with busulfan and cyclophosphamide.

An alternative preparative regimen has been required to prepare patients successfully with familial erythrophagocytic lymphohistocytosis (FEL). This unique congenital myeloproliferative disorder produces pancytopenia and immunodeficiency with prominent infiltration of the spleen, liver, lymph nodes, bone marrow, and meninges with histiocytes exhibiting prominent erythrophagocytosis (137). Because the disorder is partially responsive to VP-16, a preparative cytoreductive regimen has been developed that incorporates this agent in combination with cyclophosphamide, myeleran, and cytosine arabinoside. At least three patients transplanted with this approach have achieved full and durable reversals of the disease with correction of associated lymphoid and hematopoietic abnormalities (138,139).

The juvenile, autosomal recessive form of osteopetrosis is a disorder characterized by functional deficiencies of osteoclasts that lead to an inability to resorb and reform bone (140). Encroachment of bone on the cranial cavity and its foramina produces cranial nerve palsies, blindness, deafness, and increased intracranial pressure early in life. The growth of long bones is also severely compromised. In addition, marrow spaces are gradually obliterated, ultimately eliminating intramedullary sites of hematopoiesis and inducing aplastic anemia. Allogeneic marrow transplants can correct this disorder by replacing defective host osteoclasts with normal osteoclasts derived from monocyte/macrophage pools produced by the donor marrow (141,142). Of the initial 10 patients transplanted for this disease (143), 7 have achieved long-term survival with correction of hematopoietic function. Engraftment of donor-derived osteoclasts has led to extensive resorption of sclerotic bone with subsequent normalization of bone structure and growth. A recent report of the European transplant experience with this disease has confirmed these encouraging results (80).

Marrow Transplantation for Genetic Disorders of Metabolism

Following engraftment of allogeneic marrow in a myeloablated, immunosuppressed host, donor-derived cells of the monocyte/macrophage lineage expand and migrate from the circulation, ultimately replacing populations of host-type macrophages fixed in the tissues. Cytogenetic analyses of cells in tissues biopsied from animals or from humans 2–3 months posttransplant have demonstrated that Langerhans cells in the skin (144), Kupffer cells in the liver (145), alveolar macrophages in the lung (146), osteoclasts in bone (141,142), and microglial cells in the brain (147) are of donor origin. These findings, coupled with the results of in vitro experiments demonstrating that the pathologic features of fibroblasts or glial cells derived from patients with certain mucolipidoses and mucopolysaccharidoses could be corrected by co-culture with fibroblasts, lymphocytes, or macrophages from enzymatically normal individuals (148–155), have stimulated exploration of marrow transplants as a method for introducing into affected individuals a normal, constantly renewable source of progenitor cells capable of continuously supplying tissues with enzymatically intact macrophages for correction of these diseases.

Slavin and Yatziv (156) were the first to demonstrate that a transplant of marrow could lead to normalization of levels of a lysosomal enzyme in leukocytes, liver, and plasma, when they used such grafts to correct β-glucuronidase deficiency in mice. Subsequent experiments in animal models have focused on the potential of marrow grafts to reverse or prevent the peripheral nerve and/or central nervous system pathology induced by deficiencies of enzymes necessary for mucolipid or mucopolysaccharide catabolism. Initial experiments in these models demonstrated that marrow grafts from normal donors could reverse or prevent the visceral pathology of several storage diseases, including arylsulfatase B deficiency in cats (157), type I mucopolysaccharidosis (158,159), and fucosidosis in dogs (160) and Neimann-Pick disease in mice (161). However, in the latter two models, while enzyme levels increased in the central nervous system of transplanted animals, symptoms of central nervous system degeneration progressed, and lethality was neither averted nor delayed. Similar findings were reported by Jolly (162) in mannosidotic free-martens, cows that are partially chimeric with hematopoietic cells derived from their normal fraternal twins as a result of cross circulation of blood cells during intrauterine life. These results suggested that marrow grafts would be of limited value in storage diseases involving the brain. Recently, however, it has been found that transplants administered early in life to twitcher mice (a model of galactosyl ceramidase deficiency, or Krabbe's disease in man) or to dogs afflicted with mucopolysaccharidosis type I (a model of Hurler's syndrome) can not only increase levels of the deficient enzyme in the central nervous system but also retard the development of central nervous system pathology (147). In successfully transplanted twitcher mice, significant reversals of peripheral neuropathies, with remyelinization of affected nerves, are also observed (163).

The widely disparate effects of marrow grafts in these different storage disease models may reflect any of several factors including: the timing of the transplant relative to the onset of degenerative changes in the brain, the capacity of the engrafted normal macrophages to transfer the needed enzyme to affected neural tissues, or, conversely, the capacity of diseased cells to utilize the enzyme effectively or excrete abnormal stores of substrate for catabolic degradation by neighboring, enzymatically normal donor macrophages.

Experience with transplants for the treatment of metabolic diseases in humans is still limited, but reiterates several of these issues. The transplantation programs at Westminster Hospital (164) and at the University of Minnesota (165), which have most extensively studied this approach, have each recorded sustained increments in tissue and body fluid levels of deficient enzymes following engraftment and functional development of HLA-matched marrow transplants from normal donors in patients with types I, II, III, IV, and VI mucopolysaccharidosis (164–172) and mannosidosis (173). Increments in the enzyme levels have led to reductions in tissue levels of abnormal storage products, with impressive improvements in the histological features and functions of affected organs such as the liver, spleen, corneae, heart, and skeletal system. The pathologic features of Gaucher's disease in the marrow, liver, and spleen have also been reduced following allogeneic marrow transplants (174). However, the improvements recorded have been incomplete and achieved over a much longer period of time (175). At the other end of the spectrum, little, if any improvement in the histology, concentrations of storage products, or function of affected muscles have been recorded in patients transplanted for type II glycogen storage disease (Pompe's) (176), suggesting either that normal enzyme can not be effectively transferred to these tissues or that abnormal stores of muscle glycogen cannot be mobilized or transported from the muscle cells for degradation by donor-derived macrophages.

For patients with metabolic disorders affecting the central nervous system, the effects of marrow transplants on the progression of degenerative changes induced by accumulations of storage products cannot yet be accurately assessed. However, at least two patients transplanted for metachromatic leukodystrophy have exhibited a stabilization of neurological symptoms and some improvement in developmental parameters (177,178) associated with increases in arylsulfatase A activity in the cerebrospinal fluid (178). Similar stabilizations have also been recorded in patients transplanted for Krabbe's disease, Niemann-Pick disease (165), and adrenoleukodystrophy (179), as well as for Hurler's syndrome (MPS-I) (168) and Sanfilippo's disease, type B (MPS-III) (165). In contrast, patients transplanted for Hunter's disease (MPS-II) to date have experienced progression of neurological symptoms despite a successful graft. Similarly, case reports of patients transplanted for Sanfilippo's disease type A (165,180) and Lesch-Nyhan disease (181) have failed to record any arrest in the progression of neurological symptoms. The progressive degenerative changes in the CNS associated with the latter disorders may not be altered by a marrow transplant; alternatively, different results might be achieved if a transplant were invoked earlier in the disease, before degenerative changes in CNS have become advanced.

NEW DIRECTIONS FOR MARROW TRANSPLANTATION IN THE TREATMENT OF LETHAL GENETIC DISEASES

Marrow Transplantation for Patients Lacking an HLA-Identical Sibling Donor

As marrow transplantation has become recognized as a treatment of choice for many lethal congenital and acquired disorders of hematopoiesis, the need to develop effective transplantation approaches for the 60–70% of patients who lack an HLA-matched sibling donor has escalated. Attention has been primarily focused on either the identification of adequately compatible alternative donors or the development of techniques whereby histoincompatible marrow can be successfully applied without risk of graft rejection or lethal GvHD.

Initial attempts to identify alternative donors focused on the use of HLA-haploidentical but MLC*-compatible related donors, since studies in murine models had suggested that marrow grafts administered to MHC class-II-disparate recipients were most likely to induce lethal GvHD (182). Early clinical experience with such transplants administered to patients with SCID indirectly supported this approach, since disparities for HLA-A and/or -B on one haplotype were tolerated without lethal GvHD (183,32). However, in the large series of HLA partially matched transplants administered for leukemia reported by Beatty et al. (33), no single allelic disparity (e.g., (HLA-A or -B or -D) was associated with a greater severity of GvHD. Indeed, marrow grafts administered to patients expressing single HLA allele disparities on one haplotype, while inducing severe GvHD in over 75% of cases, were associated with an incidence of long-term disease-free survival comparable to that achieved with HLA-matched marrow grafts. Recipients exhibiting two or more HLA allelic disparities on one haplotype (e.g., HLA-A and -B or HLA-B and -D), on the other hand, had a significantly increased risk of graft rejection (21% for two allele disparate grafts vs. 9% and 7% for one and zero allele disparities, respectively), and severe acute and chronic GvHD. As a consequence, the proportion of patients surviving such transplants has been low (<15%) (33).

In 1977, our group reported the first successful application of an HLA-compatible marrow graft derived from an unrelated donor to the treatment of a child afflicted with SCID (184). Thereafter, a series of case reports further documented the potential of this approach in the treatment of children with leukemia (185) and aplastic anemia (186–188). The development of a statewide registry in Iowa allowed Gingrich et al. (189) to administer unrelated marrow grafts to a series of 40 patients with heavily treated refractory forms of leukemia and aplastic anemia. Of these patients, 6 (15%) survived disease-free for more than 1 year. The incidence of severe (grade II–IV) acute GvHD was 67%; five of the six surviving patients also had chronic

*mixed lymphocyte culture.

GvHD. However, in this series, only 6/40 were ascertained to be HLA-A, -B, -D-matched with their donors. A subsequent report of the results from four centers that applied unrelated HLA-matched marrow grafts for the treatment of 37 patients with chronic myelogenous leukemia (CML) was more encouraging: 3-year survival projections for patients transplanted in chronic phase or in accelerated or blastic phase of CML were 55% and 22%, respectively (190). However, in series reported from individual centers, the incidence of both acute and chronic GvHD in recipients of unmodified marrow grafts from unrelated donors has been high, approximating that seen following transplants of marrow from one to two HLA allele-disparate related donors (188,189,191,192).

The other limitation to the use of unrelated marrow grafts is the availability of suitably matched donors. Current donor registries can provide matched donors for the 10–15% of individuals, particularly those who inherit common HLA haplotypes detected in Northern European Caucasian populations. However, donors of other ethnogeographic backgrounds are underrepresented. Even if current registries expand to include 10^6 active donors, it will still be difficult to obtain donors for individuals whose haplotypes include HLA alleles that are rare or not genetically linked (193,194). Thus, while unrelated donors may ultimately be found for an additional 30% of patients, 20–30% of patients will still lack a donor. Thus, there is a continuing need for transplantation approaches whereby consistent engraftment and functional reconstitution can be achieved in HLA-disparate recipients without severe or lethal GvHD.

The development of T-cell depletion techniques has provided one such approach to this problem. In 1980, our group demonstrated that transplants of HLA-A, -B or HLA-A, -B, -D haplotype-disparate parental marrow depleted of T cells by soybean lectin agglutination and E-rosette depletion could reconstitute hematopoietic and lymphoid function in children with SCID (195) and leukemia (196) without GvHD. Currently, of 33 patients with SCID transplanted with T-cell-depleted parental marrow, 24 survive with reconstitution of immunity 1–9 years posttransplant. The actuarial probability of extended disease-free survival (72%) associated with these haplotype-disparate grafts does not differ significantly from that achieved with unmodified, HLA-matched grafts in SCID (197,198). Early experience with such transplants in the treatment of patients with adenosine deaminase deficiency (ADA^-) SCID were discouraging because of the high incidence of graft failure in noncytoreduced patients. However, over 80% of patients with ADA^- SCID who have been prepared with cyclophosphamide and myeleran have achieved engraftment, immunologic reconstitution, and disease-free survival. Other centers and the European cooperative group have subsequently reported similar results (80,199, reviewed in 197).

Transplants of T-cell-depleted HLA haplotype-disparate marrow administered to patients for other genetic diseases or for leukemia are also associated with a low incidence of GvHD (79). However, the high incidence of graft rejection associated with such transplants has severely limited their effectiveness. In a recent European survey, of 23 children with genetic immune deficiencies other than SCID transplanted with T-cell-depleted HLA-mismatched marrow after conditioning with busulfan and cyclophosphamide, 11 failed to engraft (79). Overall, 2–4-year disease-free survival was 29% for these patients compared with a 47% disease-free survival for recipients of the HLA-matched grafts. Among leukemic patients transplanted with HLA-nonidentical T-cell-depleted marrow after cytoreduction with TBI and cyclophosphamide, the incidence of graft failures or rejection has ranged from 10 to 50% depending on the number of disparate HLA alleles unique to the donor (198,200). Recently, however, alternative approaches utilizing more intensive preparative cytoreduction (201–203), less stringent T-cell depletion (202–204), and the administration of antithymocyte globulin or T-cell-specific immunotoxins and steroids in the early posttransplant period have reduced the incidence of rejection to 0–15% without unduly increasing the risk of severe GvHD. For example, Trigg et al. (203) have reported 31 children transplanted with HLA-nonidentical marrow depleted of T cells with CT-2 monoclonal antibody and rabbit complement after preparation with cytosine arabinoside, cyclophosphamide, and TBI. In this group, 13% rejected their graft; 26% developed grade II–IV GvHD; and 54% survived 1–30 months postgrafting. Similarly, Henslee et al. (205) have reported a 50% extended disease-free survival rate in a series of patients transplanted for leukemia with monoclonal-antibody-treated, partially T-cell-depleted marrow after intensive cytoreduction and treatment posttransplant with a T-cell-specific ricin A immunotoxin. In these series, results of T-cell-depleted transplants from related, one- to two-allele disparate donors and unrelated, HLA-matched donors have yielded equivalent results. Thus, it is likely that as new approaches combining improved T-cell depletion techniques and more consistently effective cytoreductive regimens increase the incidence and quality of engraftment, the ability to extend curative transplants to the full spectrum of patients lacking a donor will be realized.

ANTENATAL CORRECTION OF GENETIC DISEASES

Several genetic disorders of blood cell production or formation, such as reticular dysgenesis, homozygous alpha thalassemia, and SCID, may dispose the affected patient to severe or life-threatening complications early in the neonatal period. Similarly, several genetic disorders of metabolism affecting neurologic development, such as metachromatic dystrophy, Niemann-Pick disease, and Tay-Sachs disease, may initiate degenerative changes of the central nervous system early in fetal development. For this reason, several investigators have begun to explore the possibility of using transplants of primitive hematopoietic cells from normal allogeneic donors for correction of genetic diseases during fetal development.

The feasibility of this approach would be expected to depend on at least five variables: the effect of a hematopoietic graft on the disease to be treated, the timing of prenatal diagnosis, the timing of the transplant in relation to the immunologic development of the fetus, the type of transplant used and its degree of compatibility with the fetus, and the space available to the graft for expansion and development.

For diseases with defined biochemical abnormalities such as alpha thalassemia or Tay-Sachs disease, diagnosis can now be performed by analysis of fetal cells derived from chorionic villous biopsies as early as 8 weeks of gestation (206,207). Other blood disorders, such as SCID and amegakaryocytosis, can be diagnosed by examining fetal blood for T-cells or platelets respectively, by 14–16 weeks gestation (208,209). Such early diagnoses would be expected to be critical to the success of the prenatal hematopoietic graft.

In the ontogeny of the human immune system, lymphocytes capable of responding to allogeneic cells are first detected at 12–14 weeks of gestation shortly after the thymus is first infiltrated with lymphoid cells (210,211). Transplants of allogeneic liver and thymus from fetuses of 8–12 weeks gestation engrafted in patients with SCID have not induced severe GvHD (212,213), suggesting that fetal lymphoid cells during this period of development have little capacity to initiate T-cell-dependent allospecific responses. However, by 14 weeks circulating lymphocytes can initiate local GvHD reactions in xenogeneic hosts (214). Furthermore, transplants of liver and thymus from 16–20-week fetuses have induced severe GvHD in patients with SCID (215, Cohen, F., personal communication). Thus, unless the fetus has a variant of SCID, it may be capable of rejecting a hematopoietic cell graft as early as 14–16 weeks' gestation. From 21 weeks' gestation on, this function is certainly fully developed, since durable engraftment of foreign leukocytes following intrauterine blood transfusions is a rare event (216).

In animal models of intrauterine transplantation, investigators have primarily used early gestational aged allogeneic fetal liver cells as a source of hematopoietic progenitors for transplantation purposes since these populations are depleted of T cells and thus unlike adult marrow have minimal potential for inducing significant GvHD. Transplants of allogeneic fetal liver cells have been administered to normal sheep and primate embryos at different stages of gestation. Durable engraftment of fully allogeneic progenitor cells has been achieved in fetuses at gestational ages approximating 14–16 weeks of human gestation (217–219). These results thus further suggest that fetuses transplanted before the full development of thymus-derived immunity may develop tolerance to the transplanted allogeneic cells.

While durable engraftment of hematopoietic cells from an allogeneic donor can be regularly achieved if the cells are administered early in fetal development, the environmental niche available for the development of these cells may be restricted. Irrespective of the dose of cells administered, the proportion of donor cells detectable in the circulation after birth has been consistently limited to 3–15% (217–219). Thus, the utility of prenatal grafts for early correction of genetic disorders may be limited to diseases in which such limited states of chimerism would be able to supply sufficient concentrations of enzyme or normal blood cells to correct the disorders of metabolism incurred. Greater degrees of chimerism may be observed in severe genetic cytopenias in which production of cells of one or more hematopoietic linkages is impaired. For example, in murine models, as few as 100 fetal liver hematopoietic cells have repopulated the red cell progenitor compartment of fetal or newborn w/w^v mice (220). Indeed, the degree of donor chimerism achieved is directly proportional to the severity of the anemia observed in the variant of w/w^v mouse transplanted, suggesting that the biological space within the hematopoietic compartment available to the donor graft influences its ultimate expansion. A similar phenomenon is observed in infants with SCID if they are transplanted without preparative cytoreduction. Chimerism is invariably limited to the lymphoid lineage; indeed, if these patients have B lymphocytes, engraftment is observed only in the T-lymphocyte lineage (198).

Because of these limitations, initial attempts to correct genetic disease by hematopoietic grafts administered during embryonal development have been focused on the treatment of lethal immune deficiencies. To date, two patients, with Bare lymphocyte syndrome and SCID, respectively, have received transplants of normal 7–10 week fetal liver cells when they were at 28 and 26 weeks of gestational age. Both children were subsequently delivered at term. Donor lymphoid cells have been detected in both patients, but constitute only 26% of the lymphoid cells. The patient with Bare lymphocyte syndrome has achieved a reconstitution of T-cell immunity; the patient with SCID has received secondary fetal liver transplants postnatally and is not evaluable (221,222). Five other patients have recorded transplants of fetal liver (one) or T-cell-depleted adult marrow (four) for thalassemia (two); metachromatic leukodystrophy (two); or Rh hemolytic disease (one) (222–224). All but one of these patients was transplanted after 23 weeks of gestation. Although the procedure was well tolerated and did not adversely affect pregnancy or the child at birth, none of these patients were engrafted.

The proportion of patients achieving engraftment and full immunologic reconstitution following T-cell-depleted marrow transplants from haplotype-disparate maternal donors is, in our own and others' experience, markedly superior to that achieved with fully allogeneic fetal liver grafts (225), possibly reflecting the advantages of HLA haplotype homology between donor and host for ultimate reconstitution of immunity. Since such transplants have a comparably low risk of inducing GvHD, such preparations may offer a more consistent and readily available source of hematopoietic progenitors for future transplants of this type.

CONCLUSIONS

Transplants of marrow from genetically normal allogeneic donors have been used to correct a variety of lethal disorders of blood formation and function and show considerable promise as a therapeutic approach for the correction of severe lethal genetic disorders of metabolism. Future research in animal models and clinical trials must focus on the development of a clearer understanding of the mechanism available for intercellular enzyme transport and mobilization of abnormal concentrations of substrates to better define and predict the utility of marrow transplants in the treatment of metabolic diseases, particularly disorders affecting the central nervous system. Experience with T-cell-depleted marrow transplants applied to the treatment of SCID has demonstrated that such grafts can reconstitute hematopoietic and immune function in HLA-disparate hosts without risk of GvHD. However, development of integrated unrelated donor pools and improved methods for securing durable engraftment of HLA-nonidentical marrow without risk of severe GvHD are needed before marrow transplants can be extended to the 60–70% of patients with genetic diseases that do not involve the immune system who lack an HLA-compatible sibling donor. Lastly, research in animal models suggests the possibility that in the future, safe and consistently effective transplantation approaches may be developed for correction of certain of the most severe genetic diseases prior to birth.

REFERENCES

1. Rasmussen LH, et al. Authentic recombinant human growth hormone. Results of a multicenter clinical trial in patients with growth hormone deficiency. *Helv Paediatr ACTA* 1989;43:443–448.
2. White GC, McMillan C, Kingdon HS, Shoemaker CB. Use of recombinant antihemophilic factor in the treatment of two patients with classic hemophilia. *N Engl J Med* 1989;320:166–170.
3. Hershfield MS, Buckley RH, Greenberg ML, et al. Treatment of adenosine deaminase deficiency with polyethylene glyco-modified adenosine deaminase. *N Engl J Med* 1987;316:589–595.
4. Green HL, Hug G, Schubert WK. Metachromatic leukocystrophy: Treatment with arylsulfatase A. *Arch Neurol* 1969;147.
5. Achord DT, Brot FE, Bell CE, Sly WS. Human beta-glucuronidase: *In vivo* clearance and *in vitro* uptake by a glycoprotein recognition system on reticuloendothelial cells. *Cell* 1978;15:269–278.
6. Boyer SH, Siggers DC, Krueger LJ. Caveat to protein replacement therapy for genetic disease. Immunologic complications of accurate molecular diagnosis. *Lancet* 1973;12: 654–658.

7. Langlois RP, O'Regan S, Pelletier M, Robitaille P. Kidney transplantation in uremic children with cystinosis. *Nephron* 1981;28:273.
8. O'Regan P, Constable AR, Joekes AM, et al. Successful renal transplantation in primary hyperoxaluria. *Postgrad Med J* 1980;56:288.
9. Hood JH, Koep LJ, Peters RL, et al. Liver transplantation for advanced liver disease with alpha-1 antitrypsin deficiency. *N Engl J Med* 1980;302:272.
10. Starzl TE, Iwatsuki S, Van Thiel DH, et al. Evolution of liver transplantation. *Hepatology* 1982;2:614.
11. Breed A, Chesney R, Friedman A, et al. Oxalosis-induced bone disease: A complication of transplantation and prolonged survival in primary hyperoxaluria. *J Bone Joint Surg* 1981;63A:310.
12. Gibbs DA, Spellach E, Tompkins R, et al. A clinical trial of fibroblast transplantation for the treatment of mucopolysaccharidosis. *J Intern Metab Dis* 1983;6:62–81.
13. Adinolfi M, McColl I, Chase D, et al. Transplantation of fetal fibroblasts and correction of enzymatic deficiencies in patients with Hunter's or Hurler's disease. *Transplantation* 1986;42:271–274.
14. Tylski-Szymanska A, Maciejko D, Jablonska-Budaj V, et al. Amniotic tissue transplantation as a trial of treatment in some lysosomal storage diseases. *J Inherit Metab Dis* 1985;8:101–104.
15. Yeager AM, Singer HS, Buck JR, et al. A therapeutic trial of amniotic epithelial cell implantation in patients with lysosomal storage diseases. *Am J Med Genet* 1985;22:347.
16. Gatti RA, Meuwiissen HJ, Allen HD, et al. Immunological reconstitution of sex-linked lymphopenic immunological deficiency. *Lancet* 1968;2:1366–1369.
17. Bach FH, Albertini RJ, Joo P, et al. Bone marrow transplantation in a patient with the Wiskott-Aldrich syndrome. *Lancet* 1968;2:1364–1366.
18. Camitta BM, Thomas ED, Nathan DC, et al. A prospective study of androgens and bone marrow transplantation for treatment of severe aplastic anemia. *Blood* 1979;504–514.
19. Thomas ED, Cliff RA, Fefer A, et al. Marrow transplantation for the treatment of chronic myelogenous leukemia. *Ann Intern Med* 1986;104:155–163.
20. Goldman JM, Apperley JF, Jones L, et al. Bone marrow transplantation for patients with chronic myeloid leukemia. *N Engl J Med* 1986;314:202–207.
21. Brochstein JR, Kernan NA, Groshen S, et al. Allogeneic bone marrow transplantation after hyperfractionated total-body irradiation and cyclophosphamide in children with acute leukemia. *N Engl J Med* 1987;317:1618–1624.
22. Dinsmore R, Kirkpatrick D, Flomenberg N, et al. Allogeneic bone marrow transplantation for patients with acute nonlymphocytic leukemia. *Blood* 1984;63:649–656.
23. Weisdorf DJ, McGlave PB, Ramsay NKC, et al. Allogeneic bone marrow transplantation for acute leukemia: Comparative outcomes for adults and children. *Br J Haematol* 1988;69:351–358.
24. Clift RA, Buckner CD, Thomas ED, et al. The treatment of acute nonlymphoblastic leukemia by allogeneic marrow transplantation. *Bone Marrow Transplantation* 1987; 2:243–258.
25. Gratwohl A, Zwaan FG, Hermans J, et al. Bone marrow transplantation for leukemia in Europe. Report from The Leukemia Working Party. *Bone Marrow Transplantation* 1986;1:177–181.
26. McGlave PB, Haake R, Miller W. Therapy of severe aplastic anemia in young adults and children with allogeneic bone marrow transplantation. *Blood* 1987;70:1325–1330.
27. Tutschka PJ, Santos GW. Bone marrow transplantation in the busulfan-treated rat. III. Relationship between myelosuppression and immunosuppression for conditioning bone marrow recipients. *Transplantation* 1977;24:52–62.
28. Kapoor N, Kirkpatrick D, Blaese RM, et al. Reconstitution of normal megakaryocytopoiesis and immunologic functions in Wiskott-Aldrich syndrome by marrow transplantation following myeloablation and immunosuppression with busulfan and cyclophosphamide. *Blood* 1981;57:692–696.
29. Sorrell M, Kapoor N, Kirkpatrick D, et al. Marrow transplantation for juvenile osteopetrosis. *Am J Med* 1981;70:1280–1287.

30. O'Reilly RJ, Brochstein J, Dinsmore R, Kirkpatrick D. Marrow transplantation for congenital disorders. *Semin Hematol* 1984;21:188–221.
31. Anasetti C, Amos D, Beatty PG, et al. Effect of HLA compatibility of engraftment of bone marrow transplants in patients with leukemia or lymphoma. *N Engl J Med* 1989;320:197–204.
32. Kenny AB, Hitzig WH. Bone marrow transplantation for severe combined immunodeficiency. *Eur J Pediatr* 1979;131:155–177.
33. Beatty PG, Clift RA, Michelson EM, et al. Marrow transplantation from related donors other than HLA-identical siblings. *N Engl J Med* 1985;313:765–771.
34. Storb R, Prentice RL, Thomas ED. Marrow transplantation for treatment of aplastic anemia: An analysis of factors associated with graft rejection. *N Engl J Med* 1977;296:61.
35. Glucksberg H, Storb R, Fefer A, et al. Clinical manifestations of GvHD in human recipients of marrow from HLA matched sibling donors. *Transplantation* 1974;18:295.
36. Storb R, Prentice RL, Thomas ED. Treatment of aplastic anemia by marrow transplantation from HLA identical siblings. *J Clin Invest* 1977;59:625–632.
37. Bross DS, Tutschka PJ, Farmer ER. Predictive factors for acute graft-versus host disease in patients transplanted with HLA-identical bone marrow. *Blood* 1984;63:1265–1270.
38. Kernan NA, Bordignon C, Heller G, et al. Graft failure after T-cell depleted human leukocyte antigen identical marrow transplants for leukemia: I. Analysis of risk factors and results of secondary transplants. *Blood* 1989;74:2227–2236.
39. Goulmy E, Termijtzlen A, Bradley BA, et al. Y-antigen killing by T-cells of woman is restricted by HLA. *Nature* 1977;226:544.
40. Voogt PJ, Goulmy WE, Fibbe WE, et al. Minor histocompatibility antigen H-Y is expressed on human hematopoietic progenitor cells. *J Clin Invest* 1988;82:906.
41. Van Els CACM, DeBuerger MM, Kempenaar J, Ponec M, Goulmy E: Susceptibility of human male keratinocytes to MHC restricted H-Y specific lysis. *J Exp Med* 1989;170:1469.
42. Goulmy E. Minor histocompatibility antigens in man and their role in transplantation. In Morris J, Tilney NL (eds): *Transplant Reviews, 1988,* Philadelphia: W.B. Saunders 1988, pp 29–53.
43. Kernan NA, Flomenberg N, Dupont B, et al. Graft rejection in recipients of T-cell depleted HLA-non-identical marrow transplants for leukemia. *Transplantation* 1987; 43:842.
44. Bunjes D, Heit W, Arnold R, et al. Evidence of the involvement of host-derived OKT8 positive T-cells in the rejection of T-depleted HLA-identical bone marrow grafts. *Transplantation* 1987;43:501.
45. Bordignon C, Keever CA, Small TN, et al. Graft failure after T-cell depleted human leukocyte antigen identical marrow transplants for leukemia: II. *In vitro* analysis of host effector mechanisms. *Blood* 1989;74:2237–2243.
46. Royal Marsden Hospital Bone Marrow Transplantation Team. Failure of syngeneic bone marrow graft without pre-conditioning in post-hepatitis marrow aplasia. *Lancet* 1977;2:742–744.
47. O'Reilly RJ, Brochstein J, Collins N, et al. Evaluation of HLA-haplotype disparate parental marrow grafts depleted of T lymphocytes by differential agglutination with a soybean lectin and E-rosette depletion for the treatment of severe combined immunodeficiency. *Vox Sang* 1986;51:81–86.
48. McCulloch EA, Siminovitch L, Till JE, et al. The cellular basis of the genetically determined hemopoietic defect in anemic mice of genotype SL/SL_D. *Blood* 1969;26:399–410.
49. Yoshida H, Hayashi SI, Kunisada T, et al. The murine mutation osteopetrosis is in the coding region of the macrophage colony stimulating factor gene. *Nature* 1990;345:442–444.
50. Knospe WH, Blum J, Crosby WH. Regeneration of locally irradiated bone marrow-I. Dose-dependent long-term changes in the rat with particular emphasis upon vascular and stromal reaction. *Blood* 1966;28:398–415.

51. Luccarelli G, Galimberti M, Polchi P, et al. Bone marrow transplantation in patients with thalassemia. *N Engl J Med* 1990;322:417–421.
52. Bortin MM, Rimm A. Severe combined immunodeficiency disease. Characteristics of the disease and results of transplantation. *JAMA* 1977;238:591–600.
53. Devaal OM, Seyhaeve V. Reticular dysgenesia. *Lancet* 1959;2:1123–1125.
54. Santos GW: Immunosuppression of clinical marrow transplantation. *Semin Hematol* 1974;11:341.
55. Thomas ED, Buckner CD, Rudolph RH, et al. Allogeneic marrow grafting for hematologic malignancy using HLA matched donor recipient sibling pairs. *Blood* 1971;38: 267.
56. Parkman R, Rappeport J, Geha R, et al. Complete correction of the Wiskott-Aldrich syndrome by allogeneic bone marrow transplantation. *N Engl J Med* 1978;298:921.
57. Hobbs JR, Hugh-Jones K, Shaw PJ, et al. Engraftment rates related to busulfan and cyclosphosphamide dosages for displacement. Bone marrow transplantation in fifty children. *Bone Marrow Transplantation* 1986;1:201–208.
58. Krivit W, Whitley CB, Chang PN, et al. Lysosomal storage diseases treated by bone marrow transplantation. In Gale RP, Champlin RE (eds): *Bone Marrow Transplantation. Current Controversies,* New York: Alan R. Liss Inc., 1989, pp 367–378.
59. Martin PJ, Hansen JA, Buckner CD, et al. Effects of *in vitro* depletion of T-cells in HLA-identical allogeneic marrow grafts. *Blood* 1985;66:664.
60. Hale G, Cobbold S, Waldmann H. T-cell depletion with Campath-1 in allogeneic bone marrow transplantation. *Transplantation* 1988;45:753–759.
61. Paller AS, Nelson A, Steffen L, Gottschalk L, Kaizer H. T-lymphocyte subsets in the lesional skin of allogeneic and autologous bone marrow transplant patients. *Arch Dermatol* 1988;124:1795–1801.
62. Kasten-Sportes C, Masset M, Varrin F, Devergie A, Gluckman E. Phenotype and function of T lymphocytes infiltrating the skin during graft-versus-host disease following allogeneic bone marrow transplantation. *Transplantation* 1989;47:621–624.
63. Ferrara JL, Guillen FJ, Dijken PJ, et al. Evidence that large granular lymphocytes of donor origin mediate acute graft-versus-host disease. *Transplantation* 1989;47: 50–54.
64. Bruss DS, Tutschka PJ, Farmer ER, et al. Predictive factors for acute graft-versus-host disease in patients transplanted with HLA-identical bone marrow. *Blood* 1984;63:1265–1270.
65. Brochstein J, Kernan NA, Groshen S, et al. Allogeneic bone marrow transplantation after hyperfractionated total body irradiation and cyclophosphamide in children with acute leukemia. *N Engl J Med* 1987;317:1618–1624.
66. Storb R, Prentice RL, Buckner CD, et al. Graft vs. host disease and survival in patients with aplastic anaemia treated by marrow grafts from HLA-identical siblings. *N Engl J Med* 1983;308:302–307.
67. Vogelsang GB, Hess AD, Santos GW. Acute graft-versus-host disease: Clinical characteristics in the cyclosporine era. *Medicine* 1988;67:163.
68. Storb R, Deeg HJ, Farewell V, et al. Marrow transplantation for severe aplastic anemia: Methotrexate alone compared with a combination of methotrexate and cyclosporine for prevention of acute graft-versus-host disease. *Blood* 1986;68:119–125.
69. O'Reilly RJ, Kernan NA, Cunningham I, et al. T-cell depleted marrow transplants for the treatment of leukemia. In Gale RP, Champlin R (eds). *Bone Marrow Transplantation: Current Controversies,* New York, Alan R. Liss Inc., 1989 pp 477–493.
70. Racadot E, Herve P, Beaujean F, et al. Prevention of graft-versus-host disease in HLA-matched bone marrow transplantation for malignant diseases: A multicentric study of 62 patients using 3 PAN-T monoclonal antibodies and rabbit complement. *J Clin Oncol* 1987;5:426–435.
71. Shulman HM, Sullivan KM, Weiden PL, et al. Chronic graft-versus-host syndrome in man: A clinico-pathological study of 20 longterm Seattle patients. *Am J Med* 1980;69:204.
72. Sullivan KM, Shulman HM, Storb R, et al. Chronic graft versus host disease in 52 patients: Adverse natural course and successful treatment with combination immunosuppression. *Blood* 1981;57:267.

73. Meyers JD, Flournoy N, Thomas ED. Risk factors for cytomegalovirus infection after human marrow transplantation. *J Infect Dis* 1986;153:478–488.
74. Winston DJ, Gale RP, Meyer DV, Young LS. Infectious complications of human bone marrow transplantation. *Medicine* 1979;58:1–31.
75. Bowden RA, Sayers M, Flournoy N, et al. Cytomegalovirus immune globulin and seronegative blood products to prevent primary cytomegalovirus infection after marrow transplantation. *N Engl J Med* 1986;314:1006–1010.
76. Emanuel D, Cunningham I, Jules-Elysea K, et al. Cytomegalovirus pneumonia after bone marrow transplantation successfully treated with the combination of ganciclovir and high dose intravenous immune globulin. *Ann Intern Med* 1988;109:777–782.
77. Reed EC, Bowden RA, Dandliker PS. Treatment of cytomegalovirus pneumonia with ganciclovir and intravenous cytomegalovirus immunoglobulin in patients with bone marrow transplants. *Ann Intern Med* 1988;109:783–787.
78. O'Reilly RJ, Keever CA, Small TN, Brochstein J. The use of HLA-non-identical T-cell depleted marrow transplants for correction of severe combined immunodeficiency disease. *Immunodefic Rev* 1989;1:273–309.
79. Fischer A, Friedrich W, Levinsky R, et al. Bone marrow transplantation for immunodeficiencies and osteopetrosis: European Survey 1968–1985. *Lancet* 1986;2:1080–1083.
80. Fischer A. Bone marrow transplantation in immunodeficiency and osteopetrosis. *Bone Marrow Transplant* 1989;4(Suppl 14):12–14.
81. Chu E, Rosenwasser LH, Dinareud CA, et al. Immunodeficiency with defective T-cell response to interleukin-1. *Proc Natl Acad Sci USA* 1984;81:4945–4949.
82. Pahwa R, Paradise C, Pahwa S, et al. Recombinant IL-2 therapy in severe combined immunodeficiency disease. *Proc Natl Acad Sci USA* 1989;86:5069–5073.
83. Weinberg K, Parkman R. Severe combined immunodeficiency due to a specific defect in the production of interleukin-2. *N Engl J Med* 1990;322:1718–1723.
84. Disanto J, Keever CA, Small TN, et al. Absence of interleukin-2 production in a severe combined immunodeficiency disease syndrome with T-cells. *J Exp Med* 1990;171:1697–1704.
85. Alarcon B, Regueiro JR, Arnaiz-Villena A, et al. Familial defect in the surface expression of the T-cell receptor-CD3 complex. *N Engl J Med* 1988;319:1203–1208.
86. Touraine JL, Betuel H, Souillet G, et al. Combined immunodeficiency disease associated with absence of cell surface HLA-A and B antigens. *J Pediatr* 1978;93:47–51.
87. Hadam MR, Dupfer R, Peter HH, et al. Congenital agammaglobulinemia associated with lack of expression of HLA-D region antigens. In Griscelli C, Vussen J (eds): *Progress in Immunodeficiency Research and Therapy*, Amsterdam: Excerpta Medica, 1984 pp 43–50.
88. Gehrz RC, McAuliffe JJ, Linner KM, Kersey JH. Defective membrane function in a patient with severe combined immunodeficiency disease. *Clin Exp Immunol* 1980;39: 344–348.
89. Gelfand EW, Oliver JM, Schuurman RK, et al. Abnormal lymphocyte capping in a patient with severe combined immunodeficiency disease. *N Engl J Med* 1980;301:1245–1249.
90. Fischer A, Griscelli C. Which conditioning regimen should be used for bone marrow transplantation in children with inherited diseases? *Exp Hematol* 1983;11(Suppl 1): 84–95.
91. Cooper MD, Chase HP, Lowman JT, et al. Immunologic defects in patients with Wiskott-Aldrich syndrome. *Birth Defects* 1968;4:378.
92. Brochstein JA, Gillio A, Ruggiero M, et al. Bone marrow transplantation (BMT) from HLA-identical or haploidentical donors for correction of Wiskott-Aldrich syndrome. *Pediatr Res* 1989;25(Suppl 1):160A.
93. Levinsky RJ, Tiedman K. Successful bone marrow transplantation for reticular dysgenesis. *Lancet* 1983;1:671–673.
94. Sorell M, Kirkpatrick D, Kapoor N, et al. Bone marrow transplant for combined immune deficiency (CID) and agranulocytosis associated with cartilage-hair hypoplasia (CHH). *Pediatr Res* 1979;13:455.
95. Touraine JL, Betuel H, Phillipe N. The Bare lymphocyte syndrome. I. Immunological studies before and after bone marrow transplantation. *BLUT* 1980;41:198–202.

96. Fanconi G. Familial constitutional panmyelocytopathy Fanconi's anaemia. Clinical aspects. *Semin Hematol* 1967;4:233–240.
97. Auerbach AD, Adler B, Chaganti RJK. Prenatal and postnatal diagnosis and carrier detection of Fanconi anemia by a cytogenetic method. *Pediatrics* 1981;67:128–135.
98. Gluckman E, Devergie A, Schaison G, et al. Bone marrow transplantation in Fanconi anemia. *Br J Haematol* 1980;45:557–564.
99. Gluckman E, Devergie A, Dutreix J. Radiosensitivity in Fanconi anemia: Application to the conditioning regimen for bone marrow transplantation. *Br J Haematol* 1983;54:431–440.
100. Gluckman E. Bone marrow transplantation for Fanconi anemia. In Shahidi NT (ed): *Aplastic Anemia and Other Bone Marrow Failure Syndromes,* New York: Springer-Verlag, 1990, pp 134–144.
101. Diamond LK, Allen DM, Magill FB. Congenital hypoplastic anemia. *Adv Pediatr* 1976;22:348.
102. August CS, King E, Githens JH, et al. Establishment of erythropoiesis following bone marrow transplantation in a patient with congenital hypoplastic anemia (Diamond-Blackfan syndrome). *Blood* 1976;48:491.
103. Iriondo A, Garijo J, Baro J, et al. Complete recovery of hemopoiesis following bone marrow transplant in a patient with unresponsive congenital hypoplastic anemia (Blackfan-Diamond syndrome). *Blood* 1984;64:348–351.
104. Lenarsky C, Weinberg K, Guinan E. et al. Bone marrow transplantation for constitutional pure red cell aplasia. *Blood* 1988;71:226–229.
105. DiBartolomeo P, DiGirolamo G, Angrilli F. Bone marrow transplantation for thalassemia in Pescara. In Buckner CD, Gale RP, Lucarelli G (eds) *Advances and Controversies in Thalassemia Therapy,* New York: Alan R. Liss, 1988 pp 193–200.
106. Hugh-Jones K, Vellodi A, Jones ST, et al. Bone marrow transplantation for thalassemia: Westminster Children's Hospital and United Kingdom experience. In Buckner CD, Gale RP, Lucarelli G (eds) *Advances and Controversies in Thalassemia Therapy,* New York: Alan R. Liss, 1988, pp 201–205.
107. Frappaz D, Gluckman E, Souillet G, et al. Bone marrow transplantation (BMT) for thalassemia major (TM). The French experience. In Buckner CD, Gale RP, Lucarelli G (eds) *Advances and Controversies in Thalassemia Therapy,* New York: Alan R. Liss, 1988, pp 207–216.
108. Barret AJ, Lucarelli G, Gale RP, et al. Bone marrow transplantation for thalassemia—A preliminary report from the International Bone Marrow Transplant Registry. In Buckner CD, Gale RP, Lucarelli G (eds) *Advances and Controversies in Thalassemia Therapy,* New York: Alan R. Liss, 1988, pp 173–180.
109. Thomas ED, Buckner CD, Sanders J. Marrow transplantation for leukemia. *Lancet* 1982;2:279.
110. Brochstein JA, Kirkpatrick D, Giardina P, et al. Bone marrow transplantation in two multiply transfused patients with thalassemia major. *Br J Haematol* 1986;63:445–456.
111. Gabutti V, Piga A, Sacchetti L, et al. Quality of life and life expectancy in thalassemic patients with complications. In Buckner CD, Gale RP, Lucarelli G (eds) *Advances and Controversies in Thalassemia Therapy,* New York: Alan R. Liss, 1988, pp 35–41.
112. Vullo C, DiPalma A. Compliance with therapy in Cooley's anaemia. In Buckner CD, Gale RP, Lucarelli G (eds) *Advances and Controversies in Thalassemia Therapy,* New York: Alan R. Liss, 1988, pp 43–49.
113. Borgna-Pignatti C, Zurlo MG, Destefano P, et al. Survival in thalassemia with conventional treatment. In Buckner CD, Gale RP, Lucarelli G (eds) *Advances and Controversies in Thalassemia Therapy,* New York: Alan R. Liss, 1988, pp 27–33.
114. Johnson FL, Look AT, Gockerman J, et al. Bone marrow transplantation in a patient with sickle cell anemia. *N Engl J Med* 1984;311:780–783.
115. Milpied N, Harousseau JL, Garand R, et al. Bone marrow transplantation for sickle cell anaemia. Letters to the Editor. *Lancet* 1988;II:328–329.
116. Vermylen C, Fernandez-Robles E, Ninane J, et al. Bone marrow transplantation in five children with sickle cell anaemia. *Lancet* I:1988;1427–1428.
117. Saunders LEF, Freedman MH. Constitutional aplastic anemia: Defective haematopoietic stem cell growth *in vitro*. *Br J Haematol* 1978;40:277–287.

118. Alter B. The bone marrow failure syndromes. In Nathan D, Oski F (eds): *Hematology of Infancy and Childhood,* Philadelphia: W.B. Saunders Co., 1987, pp 159–241.
119. Brochstein JA, Kernan NA, Laver J, et al. Bone marrow transplantation for thrombocytopenia—absent radii (TAR) syndrome. *Proc XXI Congress Int Soc Haematolo,* Sydney, Australia, 1986.
120. Belluci S, Tobelem G, Caen JP. Inherited platelet disorders. *Prog Hematol* 1983;13: 223–263.
121. Belucci S, DeVergie A, Gluckman E, et al. Complete correction of Glanzmann's thrombasthenia by allogeneic bone marrow transplantation. *Br J Haematolo* 1985;59:635–641.
122. Kostmann R. Infantile genetic agranulocytosis. *ACTA Paediatr Scand* 1975;64:362.
123. Rappeport JH, Parkman R, Newburger P, et al. Correction of infantile agranulocytosis (Kostmann's syndrome) by allogeneic bone marrow transplantation. *Am J Med* 1980;68:605–609.
124. Dale DC, Graw RG. Transplantation of allogeneic bone marrow in canine cyclic neutropenia. *Science* 1974;183:83.
125. Bonilla MA, Gillio AP, Ruggiero M, et al. Effects of recombinant human granulocyte colony-stimulating factor on neutropenia in patients with congenital agranulocytosis. *N Engl J Med* 1989;320:1574–1580.
126. Welte K, Zeidler C, Reiter A, et al. Differential effects of granulocyte-macrophage colony-stimulating factor and granulocyte colony-stimulating factor in children with severe congenital neutropenia. *Blood* 1990;75:1056–1063.
127. Hammond WP, Price TH, Souza LM, Dale DC. Treatment of cyclic neutropenia with granulocyte colony stimulating factor. *N Engl J Med* 1989;320:1306–1311.
128. Curnette JT, Babior BM. Chronic granulomatous disease. *Adv Hum Genet* 1987;16: 229–297.
129. DiBartolomeo P, DiGirolamo G, Angrilli F, et al. Successful allogeneic bone marrow transplantation (BMT) in a patient with chronic granulomatous disease (CGD). *Bone Marrow Transplantation* 1987;2 (Suppl 1):131.
130. Fischer A, Lisowska-Grospierre B, Anderson DC, Springer TA. Leukocyte adhesion deficiency: Molecular basis and functional consequences. *Immunodefici Rev* 1988;1: 39–54.
131. Camitta BM, Quesenberry PS, Parkman R, et al. Bone marrow transplantation for an infant with neutrophil dysfunction. *Exp Haematol* 1977;5:109–116.
132. Dinauer MC, Orkin SH, Brown R, et al. The glycoprotein encoded by the X-linked chronic granulomatous disease locus is a component of the neutrophil cytochrome b complex. *Nature* 1987;327:717–720.
133. Newburger PE, Ezekowitz RAB, Whitney C, et al. Induction of phagocyte cytochrome b heavy chain gene expression by interferon gamma *Proc Natl Acad Sci USA* 1988;85:5215–5219.
134. Ezekowitz RAB, Dinauer MC, Jaffe HS, et al. Partial correction of the phagocyte defect in patients with X-linked chronic granulomatous disease by subcutaneous interferon gamma. *N Engl J Med* 1988;319:46–52.
135. Chédiak M. Nouvelle anomalie leucocytaire de caractère constitutional et familial. *Rev Hematol* 1952;7:362.
136. Vierlizier JL, Lagrue A, Durandy A, et al. Reversal of natural killer defect in a patient with Chédiak-Higashi syndrome after bone marrow transplantation. *N Engl J Med* 1982,306:1055–1056.
137. Macmahon HE, Bedizel M, Ellis CA. Familial erythrophagocytic lymphohistiocytosis. *Pediatrics* 1963;32:868.
138. Fischer A, Cerf-Bensussan N, Blanche S, et al. Allogeneic bone marrow transplantation for erythrophagocytic lymphohistiocytosis. *J Pediatr* 1986;108:267–270.
139. O'Reilly RJ. Experience of the Memorial Sloan-Kettering Cancer Center Transplantation Service.
140. Brown DM, Dent PB. Pathogenesis of osteopetrosis: A comparison of human and animal spectra. *Pediat Res* 1971;5:181–191.
141. Coccia PE, Krivit W, Cervenka J, et al. Successful bone-marrow transplantation for infantile malignant osteopetrosis. *N Engl J Med* 1981;70:1280–1287.

142. Sorell M, Kapoor N, Kirkpatrick D, et al. Marrow transplantation for juvenile osteopetrosis. *Am J Med* 1981;70:1280–1287.
143. Gerritsen EJA, Van Loo IHG, Fischer A. European results of allogeneic bone marrow transplantation in juvenile osteopetrosis. *Bone Marrow Transplantation* 1990;5(Suppl 2):12.
144. Stingl S, Tamaki K, Katz SI, Katz S. Origin and function of epidermal Langerhans cells. *Immunol Rev* 1980;53:149–174.
145. Gale RP, Sparkes RS, Golde DW. Bone marrow origin of hepatic macrophages (Kupffer cells) in humans. *Science* 1978;201:937.
146. Thomas ED, Ramberg RE, Sale GE. Direct evidence for a bone marrow origin of the alveolar macrophages in man. *Science* 1976;192:1016.
147. Hoogerbrugge PM, Suzuki K, Suzuki K, et al. Donor-derived cells in the central nervous system of twitcher mice after bone marrow transplantation. *Science* 1988;239: 1035–1038.
148. Neufeld EF. Replacement of genotype specific proteins in mucopolysaccharidosis Enzyme therapy in genetic disease. In Desnick RJ, Bernlohr LRW, Krivit W (eds): *Birth Defects March of Dimes Original Series,* vol. IX, no. 2, New York: Alan R. Liss, 1973, pp 27–30.
149. Kihara H, Porter MT, Fluharty AL. Enzyme replacement in cultured fibroblasts from metachromatic leukodystrophy. (Ibid), pp 19–26.
150. Weismann VN, Rossi EE, Hirschowitz NN. Treatment of metachromatic leukodystrophy fibroblasts by enzyme replacement. *N Engl J Med* 1971;204:672–673.
151. Conzelmann E, Sandhoff K. Partial enzyme deficiencies; residual activities and the development of neurological disorders. *Dev Neurosci* 1983/1984;6:58–71.
152. Olsen I, Muir H, Smith R, Fensome A, Watt DJ. Direct enzyme transfer from lymphocytes is specific. *Nature* 1983;306:75–77.
153. Abraham LD, Muir H, Olsen I, Winchester B. Direct enzyme transfer from lymphocytes corrects a lysosomal storage. *Biochem Biophys Res Commun* 1985;129:415–417.
154. Brooks SE, Adachi M, Hoffman LM, Amsterdam D, Schneek L. Enzymatic biochemical and morphological correction of Tay-Sachs disease glial cells *in vitro*. In Calahan W, Lowden JL (eds): *Lysosomes and Lysosomal Storage Diseases,* New York: Raven Press, 1981, pp 195–203.
155. Gruber HE, Koienker R, Luchtman A, et al. Glial cells metabolically cooperate: Potential requirements for gene replacement therapy. *Proc Natl Acad Sci USA* 1985;82: 6662–6666.
156. Slavin S, Yatziv S. Correction of enzyme deficiency in mice by allogeneic bone marrow transplantation with total lymphoid irradiation. *Science* 1980;210:1150–1152.
157. Gasper PW, Thrall MA, Wenger DA, et al. Correction of feline arysulfatase B deficiency (mucopoly-saccharidosis VI) by bone marrow transplantation. *Nature* 1984; 312:467–469.
158. Shull RM, Hastings HE, Selcer RR, et al. Bone marrow transplantation in canine mucopolysaccharidosis I. *J Clin Invest* 1987;79:435–443.
159. Shull RM, Breider MA, Constantopoulos GG. Longterm neurological effects of bone marrow transplantation in a canine lysosomal storage disease. *Pediatr Res* 1988;24: 347–352.
160. Taylor RM, Stewart GJ, Farrow BRH. Enzyme replacement in nervous tissue after allogeneic bone marrow transplantation for fucosidosis in dogs. *Lancet* 1986;2:722.
161. Sakiyama T, Tsuda M, Owada M, et al. Bone marrow transplantation for Niemann-Pick mice. *Biochem Biophys Res Commun* 1983;113:605–610.
162. Jolly RD, Thompson KG, Murphy CE, et al. Enzyme replacement therapy—an experiment of nature in a chimeric mannosidosis calf. *Pediatr Res* 1976;10:219–224.
163. Yeager AM, Brennan S, Tiffany C, Moser NW, Santos GW. Prolonged survival and remyelination after hematopoietic cell transplantation in the twitcher mouse. *Science* 1984;225:1052–1054.
164. Hobbs JR, Hugh-Jones K, Chambers JD, et al. Lysosomal enzyme replacement therapy by displacement bone marrow transplantation with immunoprophylaxis. *Adv Clin Enzymol* 1986;3:184–201.

165. Krivit W, Whitley CB, Chang PN. Lysosomal storage diseases treated by bone marrow transplantation: Review of 21 patients. In: Johnson FL, Pochedly C (eds): New York: Raven Press, 1990, pp 261–287.
166. Hobbs JR, Hugh-Jones K, Barrett AJ, et al. Reversal of clinical features of Hurler's disease and biochemical improvement after treatment by bone marrow transplantation. *Lancet* 1981;2:709.
167. Whitley CB, Ramsay NKC, Kersey JH, Krivit W. Bone marrow transplantation for Hurler syndrome: Assessment of metabolic correction. *Birth Defects* 1986;22:7–24.
168. Hugh-Jones K. Psychomotor development of children with mucopolysacchidosis type I-H following bone marrow transplantation. *Birth Defects* 1986;22:25.
169. Warkentin PI, Dixon MS, Schafer I, et al. Bone marrow transplantation in Hurler syndrome: A preliminary report. *Birth Defects* 1986;22:31–39.
170. Desai S, Hobbs JR, Williamson S, et al. Morquio's disease (mucopolysaccharidosis IV) treated by bone marrow transplant. *Exp Hematol* 1983;11(Suppl):98–100.
171. Krivit W, Pierpont ME, Ayaz K, et al. Bone marrow transplantation in the Maroteaux-Lamy syndrome (mucopolysaccharidosis type VI). *N Engl J Med* 1984;311:1606.
172. Hugh-Jones K, Kendra J, James DCQ, et al. Treatment of San Filippo B disease (MPS III B) by bone marrow transplant. *Exp Hematol* 1982;10(Suppl 1):50–51.
173. Will A, Cooper A, Hatton C, et al. Bone marrow transplantation in the treatment of alpha-mannosidosis. *Arch Dis Child* 1987;62:1044–1049.
174. Rappeport JM, Ginns EL. Bone marrow transplantation in severe Gaucher's disease. *N Engl J Med* 1984;311:84–92.
175. Ringden O, Groth, C-G, Erikson A, et al. Longterm follow up of the first successful bone marrow transplantation in Gaucher's disease. *Transplantation* 1988;46:66–70.
176. Harris RE, Hannon D, Vogler C, Hug G. Bone marrow transplantation in type IIa glycogen storage disease. *Birth Defects* 1986;22:119–132.
177. Bayever E, Philippart M, Nuwer M, et al. Bone marrow transplantation for metachromatic leukodystrophy. *Lancet* 1985;2:471–473.
178. Krivit W, Shapiro E, Kennedy W, et al. Treatment of late infantile metachromatic leukodystrophy by bone marrow transplantation. *N Engl J Med* 1990;322:28–32.
179. Weinberg K, Moser A, Watkins P, et al. Bone marrow transplantation (BMT) for adrenoleukodystrophy (ALD). *Pediatr Res* 1988;23:334A.
180. Hugh-Jones K, Hobbs JR, Pot C, et al. Comparison of serial psychomotor testing of children with various mucopolysaccharidoses following bone marrow transplantation. *Bone Marrow Transplantation* 1986;1(Suppl 1):342.
181. Nyhan WL, Page T, Truber AB, et al. Bone marrow transplantation in Lesch-Nyham disease. *Birth Defects* 1986;22:41–53.
182. Klein J, Park JM. Graft-versus-host reaction across different regions of the H-2 complex of the mouse. *J Exp Med* 1973;137:1213–1225.
183. Dupont B, O'Reilly RJ, Pollack MS, Good RA. Use of genotypically different donors in bone marrow transplantation. *Transplant Proc* 1979;11:219–224.
184. O'Reilly RJ, Dupont B, Pahwa S, et al. Reconstitution in severe combined immunodeficiency by transplantation of marrow from an unrelated donor. *N Engl J Med* 1977;297:1311.
185. Hansen JA, Clift RA, Thomas ED, et al. Transplantation of marrow from an unrelated donor to a patient with acute leukemia. *N Engl J Med* 1980;303:565–567.
186. DuQuesnoy RJ, Zeevi A, Marrari M, et al. Bone marrow transplantation for severe aplastic anemia using a phenotypically HLA-identical, SB-compatible unrelated donor. *Transplantation* 1983;35:566–571.
187. Gordon-Smith EC, Fairhead SM, Chipping PM, et al. Bone marrow transplantation for severe aplastic anemia using histocompatible unrelated volunteer donors. *Br Med J* 1982;2:835–837.
188. Hows JM, Yin JL, Marsh D, et al. Histocompatible unrelated volunteer donors compared with HLA non-identical family donors in marrow transplantation for aplastic anemia and leukemia. *Blood* 1986;68:1322–1328.
189. Gingrich RD, Ginder GD, Goeken D, et al. Allogeneic marrow grafting with partially mismatched, unrelated donors. *Blood* 1988;71:1375–1381.

190. Beatty P, Ash R, Hows JM, McGlave PB. The use of unrelated bone marrow donors in the treatment of patients with chronic myelogenous leukemia: Experience of four marrow transplant centers. *Bone Marrow Transplantation* 1989;4:287–290.
191. Gluckman E, Bourdeau H, Belanger C. Bone marrow transplantation using matched unrelated transplant. Prevention of graft versus host disease by a rIL-2 monoclonal antibody. *Exp Hematol* 1990;18:683A.
192. Beatty PG, Hansen JA, Anasetti C, et al. Significance of different levels of histocompatibility in patients receiving marrow grafts from unrelated donors. *Exp Hematol* 1990;18:683A.
193. Beatty PG, Dahlberg S, Mickelson EM, et al. Probability of finding HLA-matched unrelated marrow donors. *Transplantation* 1988;45:714–718.
194. Sonnenberg FA, Eckman MH, Pauker SG. Bone marrow donor registries: The relation between registry size and probability of finding complete and partial matches. *Blood* 1989;74:2569–2578.
195. Reisner Y, Kapoor N, Kirkpatrick D, et al. Transplantation for severe combined immunodeficiency with HLA-A,B,D, Dr incompatible parental marrow fractionated by soybean agglutinin and sheep red blood cells. *Blood* 1983;61:341–348.
196. Reisner Y, Kapoor N, Kirkpatrick D, et al. Transplantation for acute leukemia with HLA-A and B non-identical parental marrow cells fractionated with soybean agglutinin and sheep red cells. *Lancet* 1981;2:327–331.
197. O'Reilly RJ, Keever CA, Small TN, Brochstein J. The use of HLA-non-identical T-cell-depleted marrow transplants for correction of severe combined immunodeficiency disease. *Immunodefic Rev* 1989;1:273–309.
198. O'Reilly RJ, Keever C, Kernan NA, et al. HLA non-identical T-cell depleted marrow transplants: A comparison of results in patients treated for leukemia and severe combined immunodeficiency disease. *Transplant Proc* 1987;19:55–60.
199. Bluetters-Sawatzki R, Friedrich W, Ebell W, et al. HLA-haploidentical bone marrow transplantation in three infants with adenosine deaminase deficiency: Stable immunological reconstitution and reversal of skeletal abnormalities. *Eur J Pediatr* 1989;149: 104–109.
200. O'Reilly RJ, Collins NH, Kernan NA, et al. Transplantation of marrow depleted of T-cells by soybean lectin agglutination and E-rosette depletion: Major histocompatibility complex-related graft resistance in leukemic transplant patients. *Transplant Proc* 1985;17:455–459.
201. Bozdech MJ, Sondel PM, Trigg ME, et al. Transplantation of HLA-haploidentical T-cell depleted marrow for leukemia: Addition of cytosine arabinoside to the pretransplant conditioning prevents rejection. *Exp Hematol* 1985;13:1201–1210.
202. Cahn JY, Herve P, Flesch M, et al. Marrow transplantation from HLA nonidentical family donors for the treatment of leukaemia: A pilot study of 15 patients using additional immunosuppression and T-cell depletion. *Br J Haematol* 1988;69:345–349.
203. Trigg ME, Gingrich R, Goeken N, et al. Low rejection rate when using unrelated or haploidentical donors for children with leukemia undergoing marrow transplantation. *Bone Marrow Transplantation* 1989;4:431–436.
204. Ash RC, Casper JT, Chitambas CR, et al. Successful allogeneic transplantation of T-cell depleted bone marrow from closely HLA-matched unrelated donors. *N Engl J Med* 1990;322:485.
205. Henslee PJ, MacDonald JS, Messino MJ. Freedom from relapse following histoincompatible marrow transplantation in patients with high risk acute lymphoblastic leukemia. *Exp Hematol* 1989;17:547.
206. Old JM, Ward RHT, et al. First trimester fetal diagnosis of hemoglobinopathies: Three cases. *Lancet* 1982;2:1413.
207. Pergament E, Ginsberg N, Verlinsky Y, et al. Pre-natal Tay-Sachs diagnosis by chorionic villi sampling. *Lancet* 1983;2:286.
208. Durandy A, Dumez Y, Guy-Grand E, et al. Prenatal diagnosis of severe combined immunodeficiency. *J Pediatr* 1982;101:995–997.
209. Alter B. Bone marrow failure syndromes. In Nathan D, Oski F (eds): *Hematology of infancy and childhood*. Philadelphia: W.B. Saunders, 1974, pp 159–241.

210. August CS, Berkel AI, Driscoll S, Merier E. Onset of lymphocyte function in the developing human fetus. *Pediatr Res Sci* 1971;5:539–547.
211. Carr MC, Stites DP, Fudenberg HH. Dissociation of responses to phytohemagglutinin and adult allogeneic lymphocytes in human foetal lymphoid tissues. *Nature* 1973;241: 279–281.
212. O'Reilly RJ, Kapoor N, Kirkpatrick D. Fetal tissue transplants for severe combined immunodeficiency—their limitations and functional potential. In: Seligmann M, Hitzig, WH (eds): *Primary immunodeficiencies*. Elsevier–North Holland, 1980, pp 419–433.
213. Touraine JL. European experience with fetal tissue transplantation in severe combined immunodeficiency (SCID). *Birth Defects* 1983;19:139–142.
214. Asantila T, Survari T, Hirvonen T, Toivanen P. Xenogeneic reactivity of human fetal lymphocytes. *J Immunol* 1973;111:984–987.
215. Kay HEM. Foetal thymus transplants in man. In: *Ontogeny of Acquired Immunity*. A CIBA Foundation Symposium. New York: Associated Scientific Press, p 249.
216. Bowman JM, Friesen RF, Bowman WD, et al. Fetal transfusion in severe Rh isoimmunization. *JAMA* 1969;207:1101.
217. Zanjani ED, Lim G, McGlave PB, et al. Adult haematopoietic cells transplanted to sheep fetuses continue to produce adult globins. *Nature* 1982;295:244–246.
218. Flake AW, Harrison MR, Adzick S, Zanjani ED. Transplantation of fetal hematopoietic stem cells in utero: The creation of hematopoietic chimeras. *Science* 1986;233:776–778.
219. Harrison MR, Slotnick RN, Crombleholme TM, et al. In utero transplantation of fetal liver haemopoietic stem cells in monkeys. *Lancet* 1989;2:1425–1427.
220. Fleischman RA, Custer RP, Mintz B. Totipotent hematopoietic stem cells: Normal self-renewal and differentiation after transplantation between mouse fetuses. *Cell* 1982; 30:351.
221. Touraine JL. New strategies in the treatment of immunological and other inherited diseases: Allogeneic stem cell transplantation. *Bone Marrow Transplant* 1989;4(Suppl 4):139–141.
222. Touraine JL, Raudrant D, Rebaud A, et al. *In vitro* stem cell transplantation in human fetuses. *Exp Hematol* 1990;18:657A.
223. Linch DC, Rodeck CH, Nicolaides K, et al. Attempted bone marrow transplantation in a 17-week fetus. *Lancet* 1986;2:1453.
224. Slavin S, Naparstek E, Ziegler M, et al. Intrauterine bone marrow transplantation for correction of genetic disorders in man. *Exp Hematol* 1990;18:658A.
225. O'Reilly RJ, Brochstein J, Kernan NA, et al. Fetal liver transplantation in man. *Report of the Human Fetal Tissue Transplantation Research Panel NIH*, vol II. Bethesda, MD: National Institutes of Health Documents, 1988; appendix D, p 230.

Etiology of Human Disease at the DNA Level,
edited by Jan Lindsten and Ulf Pettersson.

19

Gene Mapping by Fluorescent In Situ Hybridization and Digital Imaging Microscopy

*,**David C. Ward, *†Peter Lichter, **Ann Boyle,
*Antonio Baldini, *Joan Menninger, and *S. Gwyn Ballard

**Department of Human Genetics and ** Molecular Biophysics and Biochemistry,
Yale University School of Medicine, 333 Cedar Street,
New Haven, Connecticut 06510*

INTRODUCTION

The technique of in situ hybridization originated over 20 years ago when Pardue and Gall (1) and John et al. (2) first used radiolabeled ribosomal RNA as a probe to delineate rDNA genes in HeLa cells and Xenopus oocytes, respectively. Although the advent of recombinant DNA technology in the mid 1970s provided a plethora of cloned DNA fragments suitable for cytological studies, it was not until 1981 that gene sequences present in the DNA of mammalian cells at the single copy level could be visualized by in situ hybridization (3,4). During the ensuing decade, technical improvements were devised in numerous laboratories that enhanced the speed, reproducibility and spatial resolution of the original, time-consuming autoradiographic procedure. The development of methods for labeling DNA and RNA non-isotopically (reviewed in 5–6) made colorimetric, fluorometric, and chemiluminescent detection systems, with their inherently better spatial resolution, feasible. Parameters that influence hybridization efficiency and signal specificity were more precisely defined and optimized (7–11). Efficient methods to suppress hybridization signals from ubiquitous repetitive sequence elements present in the genomes of higher eukaryotes were also established (12–14). This both circumvented laborious subcloning procedures, previously used to prepare suitable, repeat-free, probes for hybridization,

†*Present Address:* Deutsches Krebsforschungszentrum, Im Neuenheimer Feld 280, D-6900 Heidelberg, Germany.

and provided the opportunity to use DNA of virtually any genetic complexity directly as the probe. Advances in digital imaging technology made it possible to record extremely weak hybridization signals, which can be subsequently enhanced by computer (reviewed in 15,16).

Efforts to construct high resolution genetic and physical maps of the genomes of man and other species have intensified significantly over the past few years. To achieve the goals of such genome initiatives, it will be necessary to establish reference landmarks along each chromosome to guide subsequent fine mapping endeavors; the higher the resolution of the map desired, the greater the number of landmark clones need to be put in place. Since in situ hybridization is the most direct method both for making an initial chromosomal assignment of a probe and for its regional localization, it is not surprising that there is considerable interest in the potential utility of in situ hybridization in establishing the physical order of DNA markers on metaphase chromosomes. Such information could be used to assist the development of more detailed genetic and physical maps via other analytical methods.

In this presentation I give a progress report on the efforts in our laboratory over the past two years to apply in situ hybridization to the physical mapping of human and murine DNA clones. The advantages and disadvantages of this approach to physical mapping are discussed and some experimental strategies to enhance its utility are outlined.

MATERIALS AND METHODS

The methods and materials used to generate the data summarized in this presentation have all been recently published and are cited in the text or figure legends.

The DNA clones mapped to date have been provided by many collaborators, whose cooperation has been instrumental to our own research endeavors. These include members of the laboratories of S. Weissman, K. Kidd, A. Horwich and B. Forget (Yale), D. Housman (MIT), G. Evans (Salk Institute), E. Hildebrand and R. Stallings (Los Alamos National Laboratory), D. Barker (Utah), S. Lux and L. Kunkel (Children's Hospital, Boston), J. Gusella (Massachusetts General Hospital), K. Klinger and G. Landis (Integrated Genetics Labs, Inc.), I. Dawid and G. Ventor (NIH), and D. Nelson and D. Ledbetter (Baylor).

RESULTS

A suppression hybridization strategy was considered to be the most efficient means of developing a rapid, high throughput method for mapping the chromosomal coordinates of multiple fragments of genomic DNA on metaphase chromosome spreads. The ability to suppress hybridization signals from the ubiquitous repetitive sequences found in higher eukaryotic DNA, such as the LINE and SINE elements, by preannealing the probe with an appropriate competitor DNA should make it possible to use probes of any genetic complexity without the necessity of subcloning to generate unique sequence subsets. The feasibility of physically ordering large numbers of clones along a given chromosome was originally tested using a set of 50 cosmid clones containing DNA inserts from human chromosome 11. These studies, a collaborative effort between our laboratory and those of Glen Evans (Salk Institute) and David Housman (Massachusetts Institute of Technology), demonstrated that it was indeed possible to produce a high resolution physical map by hybridizing multiple biotin- or digoxigenin-labeled cosmid clones to extended (prometaphase) chromosome spreads, either singly or in combination (17). Specific hybridization signals were observed on both chromatids of both chromosome 11 homologs in over 85% of the spreads after stringent posthybridization washes and incubation with fluorescein-labeled avidin or fluorophore-labeled anti-digoxigenin antibodies. The high efficiency of signal production markedly reduced the statistical analysis required to precisely define the chromosomal locus of each probe.

In these initial experiments, propidium iodide was used as a (red) fluorescent counterstain for chromosome visualization: chromosome identification was generally done by DAPI-banding. Alternatively, chromosome assignment was established by cohybridization with a differentially-labeled chromosome 11-specific alphoid repeat or an Alu-repeat clone which generated an R-banding pattern (see below). Both the hybridization signal and the propidium iodide counterstain were imaged electronically using a laser scanning confocal microscope system (BioRad–MRC 500) in a photon counting mode. An example of a metaphase spread hybridized with a cloned DNA probe and detected by digital imaging is shown in Fig. 1A. DAPI-banding fluorescence was evaluated separately using a wide-field microscope, since it could not be excited by the 480 or 514 mm emission lines of the argon ion laser. While conventional fluorescence and photomicroscopy is adequate for detecting and recording the results of many such hybridization experiments, we opted to use digital imaging microscopy for several reasons. First, the detection limit (S/N ratio) of electronic images greatly exceeds that of film. Second, images can be processed electronically, for example by averaging multiple scans or thresholding, to further maximize signal contrast. The availability of computer algorithms for measuring chromosome contour lengths and intersignal distances facilitates establishment of a probe's map coordinate. Third, digital images can be placed directly into a computer database for subsequent data retrieval or archiving. The latter point is of significant importance when data acquisition on a large scale is being contemplated.

The map coordinate of each hybridized probe was expressed in terms of its fractional length (FL) distance from a chromosomal reference point, which we arbitrarily chose as the terminus of the p arm (pter), relative to the total length of the chromosome displaying the signal. These FLpter measurements were made directly on the computer monitor using a computer-generated segmented ruler (Fig. 1B). A summary of the FLpter values obtained with over 70 cosmid clones mapped to human chromosome 11 is given in Fig. 2. While expressing mapping data in terms of FLpter values does not permit the assignment to discrete cytologic bands of the ISCN convention (and no extrapolation from FLpter values to bands should be made since bands are not normalized to fractional chromosome length), it does provide a rapid means of establishing the linear order of probes on a chromosome with a reasonably high degree of resolution. The precision of the mapping data depends on the degree of chromosome condensation, with extended prometaphase chromosomes exhibiting much tighter FLpter intervals than more condensed metaphase chromosomes. Using extended chromosome preparations and cohybridization experiments with two clones, each labeled for detection by a separate fluorophore, it has been shown that probes separated by as little as 600-800 kbp on the genome can be spatially resolved. However, unequivocal ordering of two closely spaced probes on metaphase

FIG. 1. A: Biotinylated c-myc plasmid hybridized to female mouse metaphase chromosome preparation. The probe was detected on chromosome 15 with FITC-avidin (yellow) and the chromosomes were counterstained with propidium iodide (red). Probe provided by John Sedivy. **B:** Normal human metaphase chromosomes hybridized with an ankyrin genomic DNA probe depicting the computerized measurement technique utilized to determine FLpter location of the ankyrin gene on chromosome 8. **C:** L1md hybridization karyotype. Biotinylated KS13A, a plasmid containing 1.3 kb from the middle of L1md, the long interspersed repetitive element, was hybridized to mouse chromosomes and detected with FITC-avidin (white). Chromosomes were counterstained with propidium iodide (red). Clone provided by Thomas Fanning. **D:** Karyotype of human chromosomes hybridized, in situ, with digoxigenin-labeled cosmid clone c7.24 (locus SRPR: signal recognition particle receptor, see reference 17, mapped on 11q24-q25), shown in red, and biotin labeled Alu-PCR products, in yellow. Alu-PCR products [obtained using the Alu primer #517 described in Nelson et al (25)] are rich in Alu sequences and generate an R-banding-like pattern that allows the identification of chromosomes and the regional assignment of probes in simultaneous, dual labeling experiments (26). Digoxigenin-labeled DNA was detected using a primary anti-digoxigenin antibody and a secondary antibody conjugated with Texas Red. Biotin-labeled DNA was detected using FITC-conjugated avidin. The signals from the two fluorochromes were recorded separately as gray scale images and then pseudocolored and merged electronically. **E:** Mapping of the cholecystokinin (CCK) gene to distal chromosome 9. Cohybridization of digoxigenin-labeled KS13A and biotinylated CCK plasmid to mouse metaphase chromosomes. Hybridization banding was detected with antidigoxigenin antibodies labeled with FITC (yellow) and CCK was detected with Texas Red conjugated avidin (red). Chromosomes were counterstained with DAPI (here, pseudocolored yellow). Clone provided by Jeff Friedman. **F:** Metaphase spread from which the karyotype shown in Figure D was derived. **G:** Normal human lymphocyte nucleus co-hybridized with a c-myc genomic probe labeled with digoxigenin and a biotinylated ankyrin genomic DNA clone. The ankyrin probe was detected with FITC (yellow) and the c-myc clone with Texas Red (red); nucleus is counterstained with DAPI (blue). **H:** An interphase nucleus from a patient with hereditary spherocytosis co-hybridized with the same probes as in Figure G. Note that there is only one ankyrin (yellow) signal.

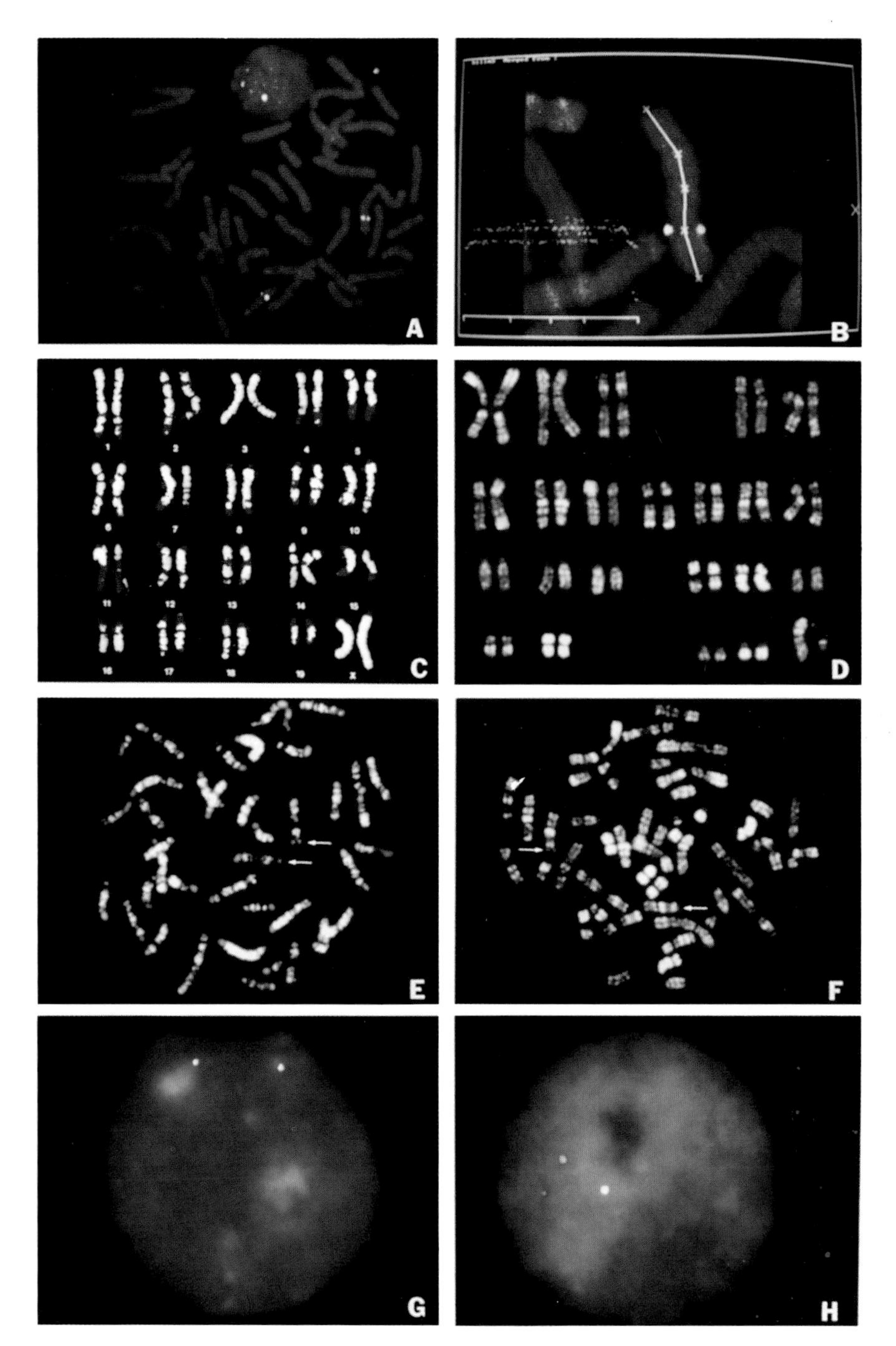

FIG. 1.

chromosomes requires that the probes be separated by approximately 1–2 million base pairs (Y. Endo and P. Lichter, unpublished results). Similar results have been reported by Lawrence et al. (18)

We have recently extended our mapping efforts to other human chromosomes as well as to the mouse genome. An overall summary of the results obtained up to the point of this symposium* is presented in Table 1. Probes have been mapped to virtually every human and murine chromosome, although the majority of clones have been localized on human chromosomes 11 and 16. The latter probes, cosmids constituting member clones of cosmid contigs being constructed by the Human Genome Research Group at Los Alamos National Laboratory, are being mapped in a blind fashion without prior knowledge as to which contig they belong. By mapping multiple clones from each contig we both confirm the continuity of the contig and establish its regional location on the chromosome. To date ~ 400 cosmid clones (37 contigs with an average size of 11) containing about 5 Mega bp of chromosome 16 DNA have been mapped. A significant number of the other mapped clones are linking clones, predominantly on chromosomes 11 and X, that are being used to construct long range restriction maps by Southern blot analysis of DNA fractionated by pulsed field gel electrophoresis. A large number of new genes have also been localized. These include a group of 40 zinc finger protein genes (clones provided by Igor Dawid and Patricia Bray, NIH), G-protein genes, hormone receptor genes, and erythrocyte membrane protein genes (19,20). To assist genetic linkage studies, we have also mapped a number of polymorphic DNA markers whose chromosomal origin had not been previously established.

When a region of a metaphase chromosome becomes saturated with probes to the point that physical order can no longer be easily deduced (e.g., see the 11q23.3 region in Figure 2), it should be possible to gain additional information on map order by extending the analysis to the interphase nucleus. Both Lawrence et al. (18,21) and Trask et al. (22) have demonstrated that the physical distance between two probe signals (in microns) is linearly related to their physical separation (in kilobasepairs), at least over the range of several hundred kbp. Thus by measuring the intrasignal distances in interphase nuclei after hybridizing multiple probes in pairwise combinations it should be possible to deduce clone order with a resolution significantly greater than that obtained using metaphase chromosomes. Preliminary studies with clones derived from the q23.3 region of chromosome 11 strongly support this contention (J. Lu, G. Evans, and D. Ward, unpublished results).

Genetic loci have for many years been described in terms of chromosomal (cytogenetic) bands, most commonly at the 550 or 800 band resolution level. Although expressing map coordinates as FLpter values offers an intrinsi-

---

*Nobel Symposium on the "Etiology of Human Disease at the DNA Level," Björkborn, Sweden, June 1990—Ed.

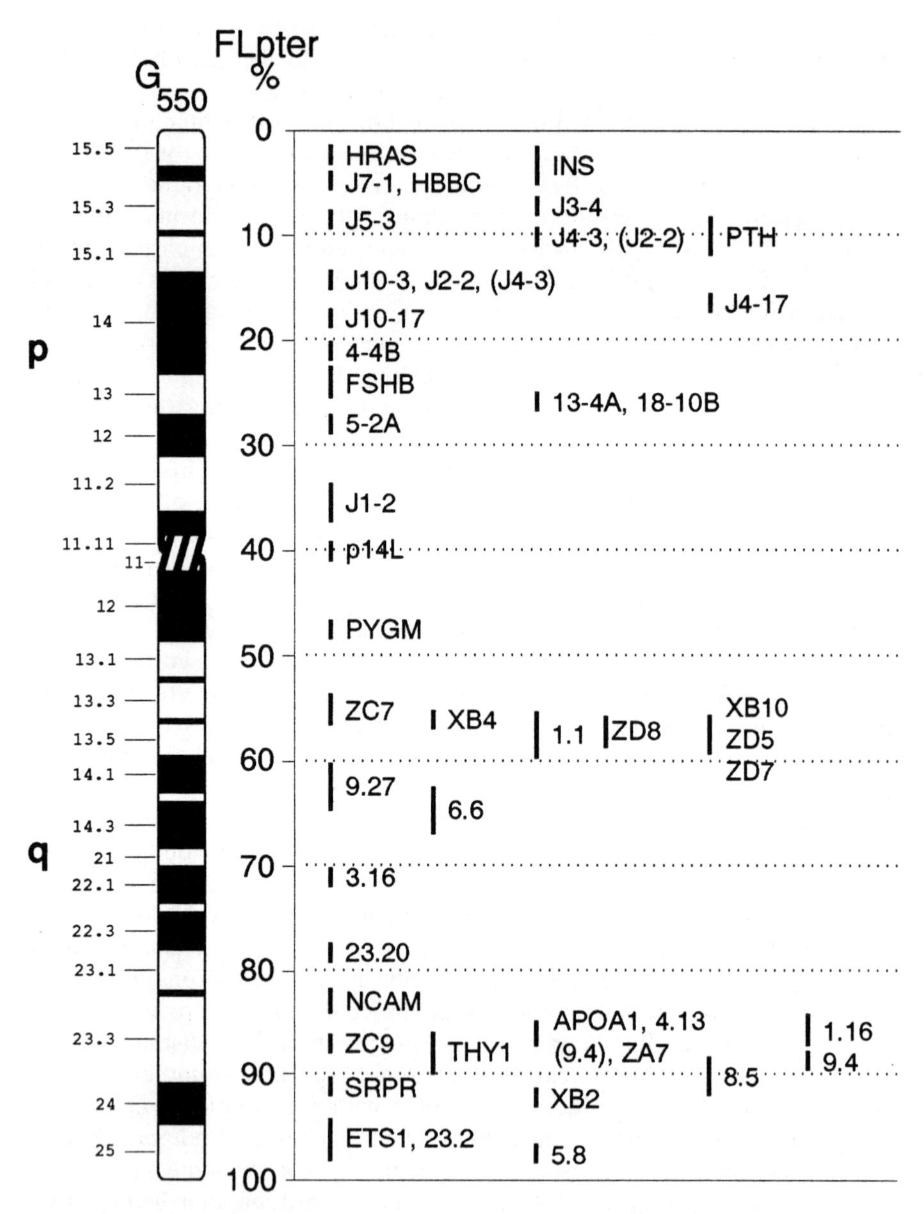

FIG. 2. Diagramatic summary of the mapping data for cosmid clones localized on human chromosome 11. All clones provided by the Evans and Housman laboratories. See reference 17 for experimental details.

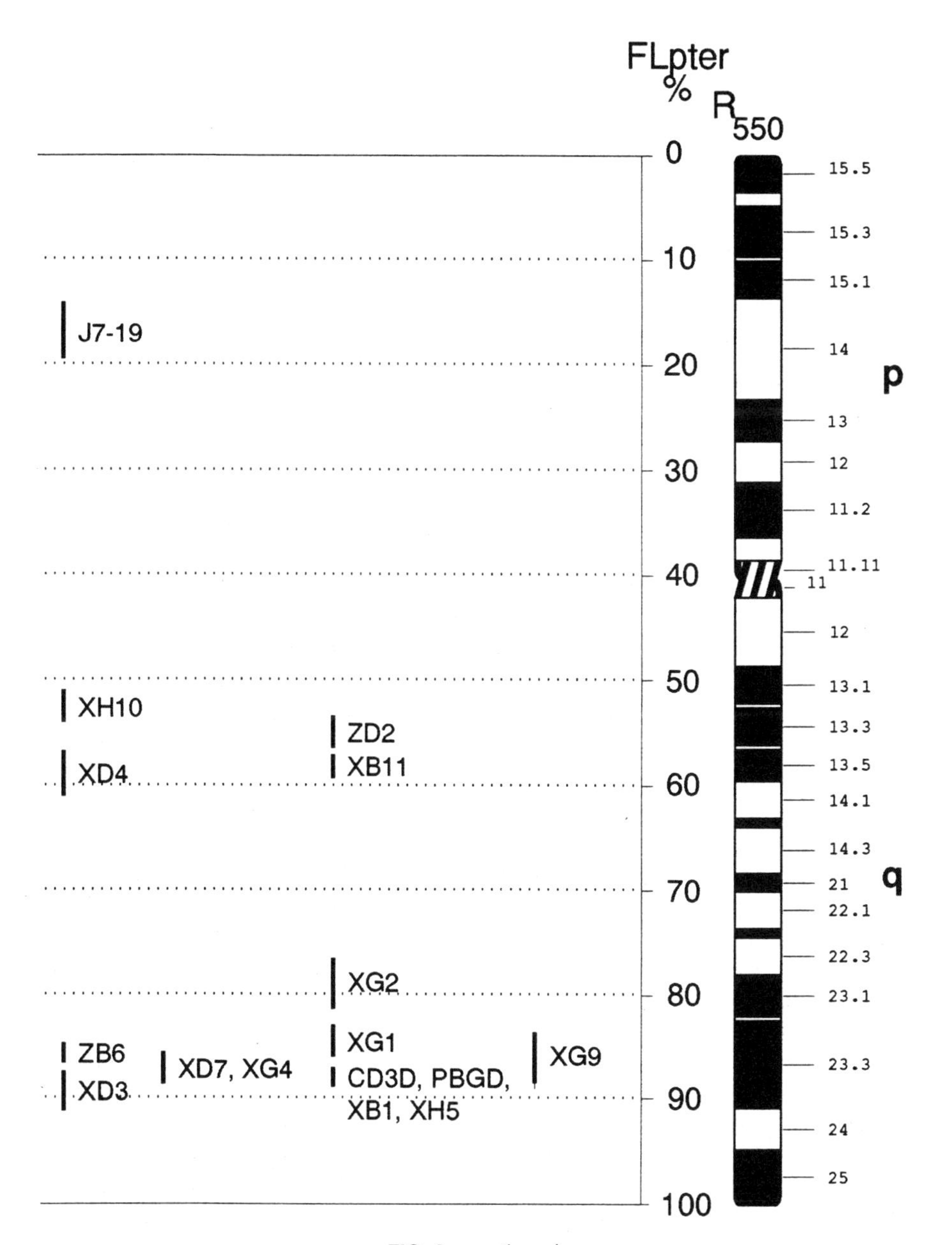

FIG. 2, *continued.*

TABLE 1. *Summary of* in situ *mapping data as of 6/1/90.*

Type of clone	Human		Mouse	
Known genes or pseudogenes	76		27	
HTF island linking clones	109		0	
Polymorphic clones	42		12	
YACs	17		3	
Cosmid contigs*	36		0	
Repetitive sequences	22		10	
Integration sites (virus V; transgenics T)	40	(V)	6	(T)
Other miscellaneous clones	25		38	
Totals without contig size :	367		96	
Totals with contig size :	727		96	
Human + Mouse total with contigs :		823		

*Average contig. sizes: Human, 11.

cally higher resolution potential, since several probes that are assigned to a single cytogenetic band (and thus unordered) can be spatially resolved (and ordered) when their coordinates are expressed as FLpter values, it is important that both mapping terminologies be interrelated. Indeed, it would seem desirable to combine cytogenetic banding conventions with geometric ordering by assigning fractional length indices to each band; the recent report of Maurer and Weinberg (23) provides a potential means to achieve this objective. We have, therefore, focused some attention on developing a simple and reproducible method of chromosome banding that is fully compatible with our fluorescent in situ hybridization protocol. Previous studies have shown that the SINE and LINE repetitive sequences are differentially distributed in the genomes of higher eukaryotes. SINE (Alu) sequences are localized predominantly in the DNA found in chromosomal regions which are lightly stained by Giemsa, i.e., the reverse (R)-bands, while the LINE sequences are found predominantly in chromosomal regions which are darkly stained by the dye, i.e., Giemsa positive (G) bands. We have shown that a clone of a murine LINE repeat (24) gives a G-banding profile when hybridized to metaphase chromosomes (Fig. 1C) which, with the exception of chromosomes 11 and X, is very similar to a conventional Giemsa stained preparation. Similarly, a clone of the murine B2 DNA family gives a characteristic R-banding pattern when hybridized to mouse metaphase chromosomes. Interestingly, neither clones of the B1 Alu family nor the human LINES gave discrete banding patterns and therefore were not suitable for hybridization karyotyping. Clones of the human Alu repeat also gave an R-banding pattern on human metaphase chromosomes; however, we observed that the amplification products generated from a single Alu DNA primer (25) by the polymerase chain reaction gave sharper and more reproducible R-bands when used as a complex hybridization probe (26; see Fig. 1D). Thus, in situations where cytogenetic band information is of paramount importance, such as mapping known genes, we carry out dual label hybridization experiments with the gene probe labeled with one reporter and the repetitive element karyotype probe with a second reporter. Examples of mapping known murine and human genes using these in situ hybridization banding techniques are shown in Figures 1E and 1F, respectively.

The PCR products generated from Alu or LINE primers also have proven extremely useful in characterizing the DNA content of hybrid cell lines (27). Human–rodent somatic cell hybrids, particularly those containing only a single human chromosome, have been very important resources for the production of chromosome-specific DNA libraries or the construction of DNA panels for the chromosomal assignment of probes by Southern or dot blot analysis. As such they play an important role in the overall genome initiative. Since these cell lines are often unstable with regard to chromosomal complement and contain chimeric chromosomes that possess both rodent and human DNA, it is important that their true chromosomal content be known with a high level of confidence. In collaboration with David and Susan Ledbetter (Baylor), we have shown that the PCR products generated from hybrid cell DNA using human specific Alu or LINE primers readily detect human chromosomal material with speed and precision when hybridized to normal human metaphase spreads. The analysis is simple to perform and can be used to screen the DNA from numerous cell lines in rapid fashion. The Alu and LINE primers also provide an alternative way of generating complex DNA probe sets by PCR for analyzing defined subchromosomal regions by in situ or molecular hybridization techniques.

One of the human genes that we recently mapped was for the erythrocyte protein ankyrin, which links β-spectrin to the anion exchange protein (band 3 or AEI). The map position of this gene was determined to be on chromosome 8 in band 8p11.2 (FLpter values, 0.28–0.31). Interestingly, a genetic locus (SPH2) for hereditary spherocytosis (HS), one of the most common inherited haemolytic anemias, had been mapped to the p11.1–11.2 region of chromosome 8 through studies of families with HS and balanced translocations involving chromosomes 8 and 12 or 8 and 3 (28,29). Although some patients with recessive HS have a mutation in the spectrin α-2 domain (Sally Marchesi, Yale, unpublished results) and a few dominant HS patients have an unstable β-spectrin which weakens spectrin–actin interactions (30), in most patients the cause of HS is unknown. In a collaborative effort with the groups of Sam Lux (Children's Hospital, Boston) and Bernard Forget (Yale) we addressed the question of whether there was an association between the HS disease and the ankyrin gene (19). The DNA from two unrelated children with HS were compared by Southern blot analysis with DNA of normal individuals. Gene dosage studies suggested that both patient samples had only about 50% of the normal level of the ankyrin gene. To confirm and extend this observation, lymphoblastoid cell cultures were established from one of the HS patients and both metaphase spreads and interphase nuclei examined by in situ hybridization with an ankyrin gene probe. As an internal control, a c-myc probe (which maps to chromosome 8q24.1) was differentially labeled and cohybridized with the ankyrin probe. As expected, normal individuals show two hybridization signals from both the ankyrin and myc genes, both on metaphase chromosomes and interphase nuclei. In contrast, the HS patient samples gave two myc signals per metaphase plate or nucleus but only a single ankyrin gene signal. Examples of the hybridization data obtained with normal and HS patient nuclei are shown in Figures 1G and 1H, respectively. These results strongly suggest that an ankyrin deficiency, caused by gene deletion or mutation, may be responsible for at least part of the HS phenotype in man.

The data illustrated in Figure 1G and 1H also highlight the relative advantage of in situ hybridization for gene (or chromosome) dosage studies. The method is considerably faster than Southern blot analysis and provides a ready means of obtaining quantitative data on a cell by cell basis. It can be applied to the detection of diseases, or carrier status for diseases, which result from partial or complete deletion of the gene in question (e.g., for dystrophin or retinoblastoma) as well as the detection of chromosome aneuploidies associated with specific neoplastic or inherited disorders (31,32). The application of such fluorescence in situ hybridization techniques is expanding rapidly in the fields of prenatal diagnosis and clinical cytogenetics.

DISCUSSION

The data summarized in this report demonstrate the utility of fluorescent in situ hybridization for creating medium resolution physical maps of human or murine chromosomes. Genomic DNA sequences cloned in plasmid, phage, cosmid or yeast artificial chromosomes have been used successfully under suppression hybridization conditions without having to remove the DNA insert from the vector. Each step of the mapping process—probe production, probe labeling, preannealing to suppress signal from interspersed repeat DNA and the in situ hybridization reaction itself—can be carried out with multiple clones at the same time. The total time required to identify the chromosomal location of a set of 15–20 clones, apart from that needed to prepare the probe DNA and metaphase spreads, is 2 to 3 days, thus large numbers of samples can be processed in a relatively short period, dramatically shorter than similar experiments conducted with isotopically labeled probes. The high efficiency of signal generation on both chromatids of both chromosome homologs markedly reduces the requirement for extensive statistical analysis to identify a hybridization locus, although fine mapping on prometaphase spreads does entail additional effort to complete the necessary chromosomal measurements.

Map coordinates can be expressed either as FLpter values, which interrelate locus positions by fractional chromosome length, or in terms of conventional chromosome banding terminology. To eliminate the need for either pre- or post-hybridization treatments to produce banding, we have identified probe sets that generate G or R banding patterns during the hybridization procedure itself. In the case of murine chromosomes, a cloned member of the LINE family yields the most reproducible karyotype as G-banded chromosomes. In contrast, with human chromosomes the Alu-PCR products generated from genomic DNA gave the sharpest and most reproducible banding, in this instance R-bands. This necessitates the use of two fluorophores, one to detect the probe locus and one to visualize the bands for karyotyping. Obviously, the development of additional fluorophore-reporter combinations could further simplify the mapping process by facilitating the analysis of multiple probes simultaneously, with each one being discerned as a different fluorescent color.

The laser scanning confocal microscope, at least as presently marketed, is not the optimal digital imaging system for simultaneous multiprobe mapping, since the 480 and 514 nm emission lines of the argon ion laser can excite only a limited number of fluorophores. In addition, present commercial confocal scanners are notorious for their low optical efficiency. By comparison, the charge coupled device (CCD) camera systems, now available from a number of manufacturers, are exquisitely sensitive to photons over a broad spectral range (400 to 1000 nm) and permit fluorophore excitation at any wavelength. Additional advantages of CCD cameras, especially the thermoelectrically cooled type, include a dynamic operational range that is linear over a five log interval, an extremely high photon-detection efficiency and, relative to the laser scanners, a more modest price. We have recently set up two digital imaging microscopes, one equipped with an intensified CCD camera and one with a cooled CCD camera, to both enhance our detection sensitivity and to better exploit the potential of multicolor imaging. Examples of data generated using the more sensitive cooled CCD system are shown in Figures 1G and 1H. It should be noted that when multiple fluorescence images of a given specimen are taken, each fluorophore signal is generally recorded using a different set of optical filters. It is imperative that each image be obtained in precise registration in order to guarantee the correct spatial representation of each after they are merged to construct the final (composite) image. Loss of registration can easily occur due to movement during the positioning of optical filters and also result from imperfections in the emission filters themselves. The image registration problem is not a trivial one and the development of methods to circumvent the problem is an area of active investigation in our lab.

Challenges presented in mapping the human genome occur at many levels. Mapping a large number of clones rapidly, efficiently, and accurately to develop a regularly spaced array of clones spanning the entire genome is an important initial goal. The generation of more detailed maps of specific chromosomal regions to facilitate the identification of genes of medical importance is a second major challenge. The data presented here indicate that fluorescent in situ hybridization has the potential to contribute significantly to such endeavors.

ACKNOWLEDGMENTS

This work was supported by grants HG-00246, HG-00272 and HG-00307 from the Center for Human Genome Research, National Institutes of Health. We acknowledge with gratitude the numerous collaborators who contributed to the work summarized here. We thank D. Greenberg for word processing.

REFERENCES

1. Gall JG and Pardue ML. *Proc Natl Acad Sci (USA)* 1969; 63:378–383.
2. John H, Birnstiel ML, and Jones KW. *Nature (London)* 1969; 223:582–587.
3. Harper ME and Saunders FG. *Chromosoma (Berlin)* 1981; 86:431–439.
4. Gerhard DS, Kawasaki ES, Bancroft FC, and Szabo P. *Proc Natl Acad Sci (USA)* 1981; 78:3755–3759.
5. Lichter P, Boyle AL, Cremer T, and Ward DC. *Gen Anal Techn Appl,* in press.
6. Raap AK, Dirks RW, Jiwa NM, Nederlof PM, and van der Ploeg. In *Modern Pathology of AIDS and Other Retroviral Infections,* P Racz, AT Haase, and JC Gluckman (Eds.). Basel: S. Karger Press 1990, pp. 17–28.
7. Lawrence JB and Singer RH. *Nucleic Acids Res.* 1985; 13:1777–1799.
8. Manuelidis L. *Focus* 1985; 7:4–8.
9. Albertson DC. *EMBO J.* 1985; 4, 2493–2498.
10. Pinkel D, Straume T, Gray JW. *Proc Natl Acad Sci (USA)* 1986;83:2934–2938.
11. Raap AK, Marijnen JGJ, Vrolijk J, and van der Ploeg M. *Cytometry* 1986; 7:235–242.
12. Landegent JE, Jansen in de Wal N, Dirks RW, Baas F, and van der Ploeg M. *Hum Genet* 1987; 77:366–370.
13. Lichter P, Cremer T, Borden J, Manuelidis L, and Ward DC. *Hum Genet* 1988; 80:224–234.
14. Pinkel D, Landegent J, Collins C, et al. *Proc Natl Acad Sci (USA)* 1988; 85:9138–9142.
15. Jovin TM and Arndt-Jovin DJ. *Ann Rev Biophys Biochem* 1989; 18:271–308.
16. Jones SJ, Taylor ML, Baarslag MW, et al. *J Microscop* 1990; 158:235–248.
17. Lichter P, Tang C-C, Call, K. et al. *Science* 1990; 247:64–69.
18. Lawrence JB, Singer RH, and McNeil JA. *Science* 1990; 249:928–932.
19. Lux SE, Tse WT, Menninger JC, et al. *Nature* 1990; 345:736–739.
20. Tse WT, Menninger JC, Yang-Feng TL et al. *Genomics,* in press.
21. Lawrence JB, Villnave CA, and Singer RH. *Cell* 1988; 52:51–61.
22. Trask B, Pinkel D, and van den Engh G. *Genomics* 1989; 5:710–717.
23. Maurer A and Weinberg J. *Cytometry,* in press.
24. Boyle AL, Ballard SG, and Ward DC. *Proc Natl Acad Sci (USA)* 1990; 87:7757–7761.
25. Nelson DL, Ledbetter SA, Corbo L, et al. *Proc Natl Acad Sci (USA)* 1989; 86:6686–6690.
26. Baldini A and Ward DC. *Genomics,* in press.
27. Lichter P, Ledbetter SA, Ledbetter DH, and Ward DC. *Proc Natl Acad Sci (USA)* 1990; 87:6634–6638.
28. Kimberling WJ, Taylor RA, Chapman RG, and Lubs HA. *Blood* 1978; 52:859–867.

29. Bass EB, Smith SW Jr, Stevenson RE, and Rosse WF. *Ann Intern Med* 1983; 99:192–193.
30. Goodman SR, Shiffer KA, Casoria LA, and Eyster ME. *Blood* 1982; 60:772–784.
31. Lichter P and Ward DC. *Nature* 1990; 345:93–95.
32. Lichter P, Jauch A, Cremer T, and Ward DC. Detection of Down Syndrome by in situ hybridization with chromosome 21 specific DNA probes. In: *Molecular Genetics of Chromosome 21 and Down Syndrome*. C Epstein and D Patterson, Eds. New York: Wiley-Liss, 1990, pp. 69–78.

Subject Index

Subject Index